Joko Widodo's Foreign Policy and Diplomatic Legacy

Indriana Kartini · Khanisa
Editors

Joko Widodo's Foreign Policy and Diplomatic Legacy

BRIN
BADAN RISET
DAN INOVASI NASIONAL

Springer

Editors
Indriana Kartini
Research Center for Politics
National Research and Innovation Agency (BRIN)
Jakarta, Indonesia

Khanisa
Research Center for Politics
National Research and Innovation Agency (BRIN)
Jakarta, Indonesia

ISBN 978-981-95-2084-8 ISBN 978-981-95-2085-5 (eBook)
https://doi.org/10.1007/978-981-95-2085-5

Jointly published with BRIN Publishing

This Springer imprint is published by the registered company Springer Nature Singapore Pte Ltd.
The registered company address is: 152 Beach Road, #21-01/04 Gateway East, Singapore 189721, Singapore

Foreword

It is an honor for me to have served my country as the first female Foreign Minister under the leadership of President Joko Widodo (Jokowi). Throughout the decade, Indonesia's diplomacy was directed toward fulfilling the national interest, advancing the people's interest, and playing an active role in promoting and maintaining peace. Priorities of Indonesia's Foreign Policy were centered on the 4+1:

- First: economic diplomacy;
- Second: protection of Indonesian citizens;
- Third: sovereignty diplomacy;
- Fourth: diplomacy for world peace and security;
- Fifth: infrastructure diplomacy.

During 2014–2024, the world was encountering rapid global change: shifting power dynamics, rising geopolitical tensions, the erosion of multilateralism, and emerging global challenges such as pandemics, climate change, and digital disruption. Against this backdrop, Indonesia's foreign policy was guided not by lofty idealism, but by what we often referred to as *down-to-earth diplomacy*—diplomasi yang membumi. It was diplomacy that delivered, diplomacy that served the people.

A defining feature of this approach was the strong emphasis on *economic diplomacy*. Foreign policy became a tool to unlock trade, investment, and technology partnerships that support national development. Indonesian missions abroad were mobilized to promote our industries, secure access to strategic markets, and support economic resilience. Economic diplomacy was no longer the domain of specialists—it became the business of every diplomat, every day.

At the same time, the protection of Indonesian citizens abroad was elevated to a core diplomatic priority. During President Jokowi's terms, the Ministry of Foreign Affairs institutionalized citizen protection not as a reactive service, but as a proactive and structured mission. From large-scale evacuations in conflict zones to consular assistance for migrant workers, we redefined diplomacy as a responsibility to those beyond our borders, but still under our care.

This volume brings together reflections and analyses from multiple perspectives, shedding light on how Indonesia's foreign policy evolved in a fast-changing world.

Whether in Indonesia's chairmanship of ASEAN, our leadership in the G20, our principled stance on Palestine, our contribution on Afghanistan, Myanmar, Women, Peace and Security issues, as well as our work to bridge divides at the UN, this was a decade when Indonesia showed that an emerging country can lead with both conviction and moderation.

Indonesia's diplomacy was increasingly anchored in service, in substance, and in sovereignty. The quiet resolve behind many initiatives—the late-night negotiations, the behind-the-scenes coordination, the balancing of national interest with global responsibility—reflects the grounded vision of Indonesia's foreign policy.

May the work of Indonesia's Foreign Policy during 2014–2024 serve as a source of knowledge and inspiration for those who continue the task of shaping Indonesia's future in the world.

Jakarta, Indonesia Retno L. P. Marsudi

Preface

President Joko Widodo, the seventh president of Indonesia, brings a practical, business-oriented perspective to his role. This pragmatic approach is evident in his foreign policy and diplomacy over the past decade. His down-to-earth diplomacy, always rooted in the interests of the Indonesian people, is most notably seen in his strong focus on economic diplomacy. The tangible results of this approach, such as increased investment and financial gain, provide reassurance about the effectiveness of his decisions.

President Joko Widodo builds relationships with neighboring countries and international partners by prioritizing national interests. In addition to maintaining good relations with old partners, he continues exploring the potential for new relationships to produce profitable cooperation. In responding to international issues—such as conflicts in various countries, including (the Palestine-Israel conflict), Indonesia's navigation between China and the United States, climate change, refugee crises, border disputes, internationalization of Papua, and the fluctuations of the global economy—President Joko Widodo ensures that Indonesia remains steadfast in upholding the values stated in the constitution, namely independence, eternal peace, and social justice. In the 10 years of his presidency, the COVID-19 pandemic tested the limitations of conventional diplomacy and the country's ability to adapt. In dealing with this, Indonesia is fighting for health diplomacy, which aims to provide certainty of vaccine access for the Indonesian people. This has implications for creating the same spirit in various other countries. During Jokowi's presidency, Indonesia has leveraged its diplomatic capabilities and strategic vision to shape global and regional policies by engaging as a middle power in both multilateralism and minilateralism.

The book, "Joko Widodo's Foreign Policy and Diplomatic Legacy," will delve into these dynamics. Its seventeen chapters will comprehensively review strategic issues in Indonesian foreign policy and diplomacy. This book does not discuss the relation of Jokowi's foreign policy and the environment or national debts as these topics are addressed in another book. Importantly, this book will not only celebrate the successes of existing foreign policy agendas but also critically examine the shortcomings. This balanced approach will provide valuable insights and lessons for future governments, ensuring the audience feels informed and prepared for the future of Indonesian foreign policy.

This publication is the result of a collective effort, and we are deeply grateful to all who contributed. Our sincere appreciation goes to the authors for their dedication in writing and refining the manuscript to its final form. We also extend our thanks to the publisher for their collaboration with editors and authors throughout the production process, as well as to those who provided data, shared insights as informants, and assisted with document research, to bring this book to fruition. To everyone involved—from drafting and editing to publishing and distribution—we express our heartfelt gratitude. Lastly, we hope readers find this work insightful and valuable.

Jakarta, Indonesia
March 2025

Indriana Kartini
Khanisa

Contents

Editors and Contributors

About the Editors

Indriana Kartini is a Senior Researcher at the Research Centre for Politics, the National Research and Innovation Agency (BRIN), Indonesia. She was the Cluster Coordinator for Foreign Policy and International Issues in 2024. She obtained her Master of International Politics from the Faculty of Arts, The University of Melbourne, Australia in 2008. She received her Ph.D. from the Political Science Study Program, at the University of Indonesia in 2023. Her research interests are international politics, border studies, regionalism, and subregionalism. In 2017, she was awarded a Research Grant from the Sumitomo Foundation for a project on Japan's public diplomacy in Indonesia. She is the editor and contributor of academic books, including Potret Politik Luar Negeri Indonesia di Era Reformasi (Indonesian Foreign Policy's Portrait in the Reform Era) (Yayasan Pustaka Obor, 2019); Penguatan Konektivitas Lintas Batas dalam Kerja sama Ekonomi Subregional (Strengthening Cross Border Connectivity in Subregional Economic Cooperation (Yayasan Pustaka Obor, 2021).

Khanisa is a Researcher at the Research Center for Politics, the National Research and Innovation Agency (BRIN. She received her Bachelor's degree from the Department of International Relations (HI), Gadjah Mada University, in 2010 and completed her Master of Arts (International Relations) and Master of Diplomacy from the Australian National University in 2015. She focused her research on ASEAN and Southeast Asia issues.

Contributors

Athiqah Nur Alami Research Center for Politics, National Research and Innovation Agency, Indonesia (BRIN), Jakarta, Indonesia

Mohamad Zakaria Al-Anshori Indonesian Ministry of Foreign Affairs, Jakarta, Indonesia

Dewi Fortuna Anwar Research Center for Politics, National Research and Innovation Agency (BRIN), Jakarta, Indonesia

Gregorius Andika Ariwibowo Research Center for Area Studies, National Research and Innovation Agency, Indonesia (BRIN), Jakarta, Indonesia

Hamdan Basyar Research Center for Politics, National Research and Innovation Agency (BRIN), Jakarta, Indonesia

Tonny Dian Effendi The University of Muhammadiyah Malang, Malang, Indonesia

Faudzan Farhana Research Center for Politics, National Research and Innovation Agency (BRIN), Jakarta, Indonesia

Petrus K. Farneubun Cenderawasih University, Papua, Indonesia

Ahmad Helmy Fuady Research Center for Area Studies, National Research and Innovation Agency, Indonesia (BRIN), Jakarta, Indonesia

Fakhry Ghafur Research Center for Politics, National Research and Innovation Agency (BRIN), Jakarta, Indonesia

Saiful Hakam Research Center for Area Studies, National Research and Innovation Agency, Indonesia (BRIN), Jakarta, Indonesia

Ahmad Nuril Huda Research Center for Area Studies, National Research and Innovation Agency, Indonesia (BRIN), Jakarta, Indonesia

Awani Irewati Research Center for Politics, National Research and Innovation Agency (BRIN), Jakarta, Indonesia

Indriana Kartini Research Center for Politics, National Research and Innovation Agency (BRIN), Jakarta, Indonesia

Khanisa The Australian National University, Canberra, Australia

Johni R. V. Korwa Cenderawasih University, Papua, Indonesia

Muzzar Kresna Research Center for Area Studies, National Research and Innovation Agency, Indonesia (BRIN), Jakarta, Indonesia

Julia M. Lau the ISEAS-Yusof Ishak Institute, Singapore, Singapore

Defbry Margiansyah Research Center for Politics, National Research and Innovation Agency (BRIN), Jakarta, Indonesia

Dhurorudin Mashad Research Center for Politics, National Research and Innovation Agency (BRIN), Jakarta, Indonesia

Nicholas Metherall The Australian National University, Canberra, Australia

Yayan Ganda Hayat Mulyana Indonesian Ministry of Foreign Affairs, Jakarta, Indonesia

Prima Nurahmi Mulyasari Research Center for Area Studies, National Research and Innovation Agency, Indonesia (BRIN), Jakarta, Indonesia

Yusnarida Eka Nizmi Department of International Relations, Riau University, Pekanbaru, Indonesia

Amarulla Octavian National Research and Innovation Agency (BRIN), Jakarta, Indonesia

Irin Oktafiani Research Center for Population, National Research and Innovation Agency (BRIN), Jakarta, Indonesia

Yessi Olivia Department of International Relations, Riau University, Pekanbaru, Indonesia

Pandu Prayoga Research Center for Politics, National Research and Innovation Agency (BRIN), Jakarta, Indonesia

Rakhman Priyatmoko Research Center for Area Studies, National Research and Innovation Agency (BRIN), Jakarta, Indonesia

Tri Nuke Pudjiastuti Research Center for Politics, National Research and Innovation Agency (BRIN), Jakarta, Indonesia

Purwasandi Department of International Relations, Riau University, Pekanbaru, Indonesia

Agus R. Rahman Research Center for Politics, National Research and Innovation Agency (BRIN), Jakarta, Indonesia

Umi Oktyari Retnaningsih Department of International Relations, Riau University, Pekanbaru, Indonesia

R. A. Rizka Fiani Prabaningtyas Research Center for Politics, National Research and Innovation Agency (BRIN), Jakarta, Indonesia

Lidya Christin Sinaga Research Center for Politics, National Research and Innovation Agency (BRIN), Jakarta, Indonesia

Siswanto Research Center for Politics, The National Research and Innovation Agency (BRIN), South Jakarta, Indonesia

Made Selly Dwi Suryanti Cenderawasih University, Papua, Indonesia

Agus Trihartono Research Center for Area Studies, National Research and Innovation Agency (BRIN), Jakarta, Indonesia

Albert Triwibowo Department of International Relations, Parahyangan Catholic University, Bandung, Indonesia

Nostalgiawan Wahyudhi Research Center for Politics, National Research and Innovation Agency (BRIN), Jakarta, Indonesia

Humphrey Wangke Research Center for Politics, The National Research and Innovation Agency (BRIN), Jakarta, Indonesia

Bondan Widyatmoko Research Center for Area Studies, National Research and Innovation Agency (BRIN), Jakarta, Indonesia

Dian Wirengjurit The Indonesian Ambassador Extraordinary and Plenipotentiary to the Islamic Republic of Iran (2012–2016), Jakarta, Indonesia

Ganewati Wuryandari Research Center for Politics, National Research and Innovation Agency (BRIN), Jakarta, Indonesia

Meilinda Sari Yayusman Research Center for Area Studies, National Research and Innovation Agency (BRIN), Jakarta, Indonesia

Rahmi Yulia Department of International Relations, Riau University, Pekanbaru, Indonesia

R. R. Emilia Yustiningrum Research Center for Politics, National Research and Innovation Agency (BRIN), Jakarta, Indonesia

Chapter 1
Introduction

Indriana Kartini and Khanisa

Abstract In President Joko Widodo's first term, Indonesian diplomacy prioritized sovereignty, citizen protection, and economic engagement. In his second term, these priorities were reframed through the Ministry of Foreign Affairs' "4+1 priorities": advancing economic diplomacy, safeguarding Indonesians abroad, promoting sovereignty diplomacy, contributing to regional and global peace, and strengthening diplomatic infrastructure through human resource development, bureaucratic reform, and digitalization. A notable shift occurred as economic diplomacy—previously secondary—became the central focus, framed as TTI (trade, tourism, investment) diplomacy, directing diplomats to expand export markets, attract investment, and promote tourism. The book is structured into five sections: Jokowi and diplomacy; economic diplomacy; protection of citizens; sovereignty and national integrity; and diplomacy for peace and stability.

Keywords Foreign policy · Diplomacy · Indonesia · Joko Widodo

1.1 Background

Observing the journey of the two periods (2014–2019) and (2019–2024) of President Joko Widodo's (Jokowi) administration, we are reminded of his role in Indonesia's foreign policy performance. Purba (2022), a senior journalist for the Jakarta Post, views President Jokowi as "a latecomer in the foreign-policy arena." However, if we trace back Nawa Cita or the Nine Priority Agendas, which guided the Jokowi government's vision at the beginning of his leadership period, we observe that President Jokowi has significantly built a policy framework focusing on efforts to encourage economic growth, advance people's welfare, and enhance national independence. Therefore, President Jokowi seems more interested in realistic and achievable foreign

I. Kartini (✉) · Khanisa
Research Center for Politics, National Research and Innovation Agency (BRIN), Jakarta, Indonesia
e-mail: indr023@brin.go.id

I. Kartini and Kanisa (eds.), *Joko Widodo's Foreign Policy and Diplomatic Legacy*,
https://doi.org/10.1007/978-981-95-2085-5_1

policy goals to meet the interests of the Indonesian people, particularly those related to the economy and development.

Furthermore, under President Jokowi's leadership, Indonesia's foreign policy is practical and is known as "Down-to-Earth Diplomacy," a vision outlined in the Annual Press Statement of the Minister of Foreign Affairs upon President Jokowi's inauguration on October 20, 2014. In this statement, the Indonesian Minister of Foreign Affairs Retno Marsudi stated that "Indonesian Diplomacy will be connected with people's interests." Indonesian diplomacy will be "down-to-earth." During President Jokowi's administration, grounded diplomacy translated into a strategic approach yielding tangible results, such as economic diplomacy, sovereignty diplomacy through negotiations to resolve national boundaries, and the protection of Indonesian citizens abroad.

In the initial phase of President Jokowi's leadership (2014–2019), Indonesian diplomacy focused on three key priorities: maintaining sovereignty, protecting citizens, and promoting economic diplomacy. Subsequently, during the second phase of President Jokowi's tenure (2019–2024), the Indonesian Minister of Foreign Affairs operationalized Jokowi's vision and mission in diplomacy and foreign policy by introducing a strategic framework known as the "4 + 1 priorities." This framework emphasized the enhancement of economic diplomacy, the protection of Indonesian citizens abroad, sovereignty diplomacy, and Indonesian diplomacy for regional and global peace and stability, with an additional focus on strengthening diplomatic infrastructure (human resources, bureaucratic reform, and the digitalization of diplomacy). A notable difference between the diplomatic priorities of Indonesia during the two periods of Jokowi's leadership is evident. In the initial phase, economic diplomacy held a relatively lower priority, whereas in the latter phase, it assumed a primary position. Referred to as TTI (trade, tourism, investment) diplomacy, economic diplomacy underpins the second phase of Jokowi's leadership. Indonesian diplomats are tasked with actively seeking export markets, attracting investors, and promoting tourism to Indonesia.

Several challenges have arisen during the two periods of President Jokowi's administration. These include the global economic crisis, increasingly complex geopolitical dynamics, such as competition between major powers, climate change, the Covid-19 pandemic, and its ramifications for Indonesia, as well as issues like the refugee crisis and the situation in Palestine. Addressing these challenges necessitates a forward-thinking Indonesian foreign policy strategy.

1.2 Structure of the Book

The book is divided into five key segments: Jokowi and diplomacy; economic diplomacy; diplomacy for the protection of Indonesian citizens; diplomacy of sovereignty and national integrity; and diplomacy for regional and global peace and stability. Excluding the Introduction and Epilogue, the first group of chapters gives an overview of President Joko Widodo's administration vision that places greater emphasis on realistic and achievable foreign policy goals. This section is devoted to exploring the journey of foreign policy and diplomacy during Joko Widodo's two terms in office and how it achieved the Jokowi administration's goals. In the second chapter, Yayan G H Mulyana and Mohamad Zakaria Al-Anshori explore the evolution of Indonesia's foreign policy under President Jokowi by analyzing how his administration's pragmatic and results-oriented approach has redefined the nation's role on the global stage. The authors argue that the down-to-earth diplomacy was successfully conducted by President Jokowi to realize the Nawa Cita vision. It has proven to be able to meet the needs of the people and, at the same time, still demonstrate its leadership in the regional and international arena.

In Chapter 3, Ganewati Wuryandari, Athiqah Nur Alami, and R. A. Rizka Fiani Prabaningtyas discuss the growing influence of gender in Indonesian foreign policy under the presidency of Joko Widodo (2014–2024). They assert that in contrast to the previous government, Jokowi's government, with Retno L. P. Marsudi as a female foreign minister, has taken significant steps to implement gender-mainstreaming strategies as part of bureaucratic and institutional reform in the Ministry of Foreign Affairs.

In Chapter 4, Lidya Christin Sinaga and Tonny Dian Effendi highlight specifically the implementation of the GMF vision as Indonesia's maritime strategy. The authors argue that while Indonesia has succeeded in strengthening its maritime vision in ASEAN at the level of ideas and has received recognition from other countries and foreign institutions, from the practical level, it is still hampered by domestic obstacles, both from political and economic aspects, which means that this decade is not enough for Indonesia to gain momentum to improve the national economy from the maritime sector, especially maritime infrastructure and connectivity.

In Chapter 5, Julia Lau examines the legacy of Joko Widodo's pragmatic foreign policy after his two terms in office (2014–2019; 2019–2024) and looks ahead to Prabowo Subianto's likely conduct of foreign policy and statesmanship. The author then considers the opportunities there might be for Indonesia to enhance its reputation and record as a middle power. After a brief examination of some early foreign policy moves under the Prabowo administration, the author argues that there is still a distinct gap between ambition and reality where Indonesia's middle-power aspirations are concerned.

The book's second part focuses on economic diplomacy which became one of Indonesia's diplomacy key priorities. In the first period of Jokow's presidency, it was the third key priority, while in the second period, it became the first key priority. Given the importance of economic diplomacy as a step toward achieving economic prosperity and people's welfare, this part analyzes the various economic diplomacy measures taken by the Indonesian government on several strategic issues and its achievements. Ahmad Helmy Fuady, Bondan Widyatmoko, Muzzar Kresna, and Saiful Hakam's chapter leads with a discussion on the new initiatives undertaken during President Joko Widodo's administration to enhance Indonesia's economic and diplomatic engagement with Africa. From their analysis, the authors show that President Widodo's foreign policy toward Africa represents a pragmatic shift in diplomatic and economic strategies, emphasizing business opportunities and market expansion while seeking to overcome historical perceptions and build stronger bilateral relations.

In Chapter 7, Meilinda Sari Yayusman, Gregorius Andika Ariwibowo, Ahmad Nuril Huda, Prima Nurahmi Mulyasari, Rakhman Priyatmoko, Agus Trihartono examine the formulation of Indonesia's foreign policy in gastrodiplomacy during Jokowi's presidency terms and its implementation, including among Indonesian diasporas in Australia and the Netherlands. The authors suggest that while the goals of Indonesian gastrodiplomacy are clearly defined, they are set as "changeable," that is, from "promoting Indonesian cuisines and touristic destinations to international audiences" during Jokowi's first-term presidency, to "increasing the number of overseas Indonesian restaurants and spice export values" in his second term. This change was also followed by the shift of coordinating leadership of the program, from the Ministry of Tourism to a coordinating ministry, indicating the increasing seriousness from the side of the government to succeed in its gastrodiplomacy objectives.

The book's third part focuses on one of the most prominent priorities in Joko Widodo's foreign policy agenda, that is, the protection of Indonesians abroad. This part explores Indonesia's strategy for utilizing diplomacy to achieve this goal. The Ministry of Foreign Affairs has been cooperating with other ministries and governmental bodies in designing diplomatic tools that are suited to resolving specific issues.

In Chapter 8, Tri Nuke Pudjiastuti, R. A. Rizka Fiani Prabaningtyas, Athiqah Nur Alami, Faudzan Farhana, and Irin Oktafiani explore Indonesia's role as a norm entrepreneur in promoting human rights norms on TIPs handling and the challenges faced by the Indonesian government to realizing human rights norms in the TIPs issue and advancing protection for Indonesian Migrant Workers (IMW). The authors argue that in the two terms of President Joko Widodo's presidency, Indonesian foreign policy has actively promoted and made substantial progress in promoting human rights standards in addressing TIPs, particularly concerning Indonesian migrant workers. Indonesia's normative entrepreneurship, which was positioning itself as a leader in advocating for these norms at various levels, has demonstrated its contribution to addressing the issue of human trafficking affecting Indonesian migrant workers, emphasizing the importance of a human rights approach.

In Chapter 9, Yessi Olivia, Yusnarida Eka Nizmi, Umi Oktyari Retnaningsih, Purwasandi, and Rahmi Yulia focus on Indonesia's health diplomacy under Joko Widodo's administration. They analyze how the Indonesian government navigated the country amid the COVID-19 pandemic. The authors argue that Indonesia's foreign policy during the pandemic reflected Indonesia's status as a middle-income country where, on one side, Indonesia, like other developing countries, struggled to secure COVID-19 vaccines. From their analysis, the authors show how the foreign ministry plays a very strategic role in ensuring that Indonesia receives assistance in coping with COVID-19, starting from vaccines and supplies of medical equipment.

In Chapter 10, Albert Triwibowo examines the conduct of digital diplomacy by the Indonesian Ministry of Foreign Affairs on the issue of citizen protection, especially during the administration of Joko Widod. The author argues that the

Indonesian Ministry of Foreign Affairs has tried to be creative to face challenges and opportunities in implementing digital diplomacy regarding citizen protection.

The fourth part of the book focuses on efforts to maintain sovereignty and national integrity as one of the key priorities in the Jokowi two-term administration, both in the first and second periods. This part explores the issues of Indonesia's land boundaries with neighboring countries, which are still problematic, as well as issues in Papua related to national integrity. Indriana Kartini, Awani Irewati, and Agus R. Rahman's chapter leads with a discussion on border diplomacy at the Indonesia–Timor-Leste border, particularly in the Noel Besi–Citrana (Naktuka) segment, and the Indonesia–Malaysia border, especially in the Sebatik Island segment. Emphasizing a nation's territorial sovereignty as a key priority in the pursuit of national interests, the authors show that the Indonesian government has made various diplomatic efforts to finalize several problematic segments of land boundaries, such as the Naktuka segment on the Indonesia–Timor-Leste border and the Sebatik Island segment on the Indonesia–Malaysia border. Resolving land boundary problems in these segments remains a task and challenge that should be addressed by the new government under the leadership of Prabowo Subianto.

In Chapter 12, Johni R. V. Korwa, Nicholas Metherall, Petrus K. Farneubun, and Made Selly Dwi Suryanti examine how Jokowi's administration has also sought to protect Indonesia's sovereignty over Papua through strengthening relationships with neighboring Papua New Guinea (PNG) and discuss Jokowi's legacy in Papua within the broader context of Indonesia's national integrity. The authors argue that Indonesian diplomacy was directed toward containing the Papuan freedom movement and its political aspirations. Furthermore, Indonesia bolstered its bilateral ties with neighboring Pacific nations, particularly PNG, with the major objectives being, among other things, the suppression of Papuan freedom fighters and support for Indonesia's attempts to preserve its territorial integrity.

The 5th part of the book explores Indonesia's role in the global world and how Indonesia sets its diplomatic goals to achieve the national interest of having a peaceful and stable region. Several issues discussed in this part include the South China Sea, the Indo-Pacific region, Minilateralism as well as Multilateralism, and Indonesia's stance on Palestine. Amarulla Octavian's chapter leads with a discussion on South China Sea (SCS) issue that affect the stability of Indonesia's defense, security, and economy, whose territory includes and is adjacent to the SCS. The author argues that in responding to the US-China relations turbulence, Indonesia must skillfully "maneuver" in accordance with the Independent and Active foreign policy. The goal is none other than to ensure Indonesia's national interests, namely to safeguard state sovereignty, territorial integrity, and sovereign rights in the Natuna Sea in accordance with the applicable national and international laws and maintaining stability and order in the Southeast Asia region from the possibility of war between great powers.

In Chapter 14, RR Emilia Yustiningrum, Ganewati Wuryandari, Siswanto, Dewi Fortuna Anwar evaluates the Quad's ability to support Indonesia's national interests and a more equitable Indo-Pacific order by analyzing the opportunities and challenges that come with its evolution. By analyzing domestic and international factors pertaining to Indonesia's foreign policy, the authors argue that the transformation of the Quad has helped to change the perceptions of the Joko Widodo's government about the Quad from skeptical into becoming more open minded, which open opportunities to foster regional cooperation, stability, and resilience.

In Chapter 15, Dian Wirengjurit discuss on minilateralism and middle powers as major players in geopolitics and international relations and Indonesia's position in the midst of its dynamics. The author argues that Indonesia's involvement in minilateralism remains a preliminary attempt in Jokowi's diplomatic endeavor that are focused on economic growth, social welfare, and political independence. For Indonesia, which has vast and long experienced in ASEAN, minilateralism continues to be attractive, including in the cross-regional framework because of its specific objectives, especially in bringing minilateralism as an alternative and creative alignment.

In Chapter 16, Humphrey Wangke, Khanisa, and Pandu Prayoga examine how Indonesia's strategy on multilateralism was implemented in G20's presidency in 2022 and ASEAN's chairmanship in 2023. The question that will be the basis of the examination is, "how Indonesia has utilised its diplomatic capabilities and strategic vision to influence global and regional policies?" The authors argue that although President Joko Widodo is likely to have a more straightforward and pragmatic vision on foreign policy, Indonesia was still carrying its multilateral foreign policy by highlighting the importance of discussion and personal approach in smoothing its diplomatic strategy.

In Chapter 17, Nostalgiawan Wahyudhi, Hamdan Basyar, Dhurorudin Mashad, Fakhry Ghafur, and Defbry Margiansyah analyze why Jokowi's foreign policy toward Palestine is so prominent? What are the desired outcomes of this policy? What differentiates it from the foreign policy of Soekarno, Soeharto, and Susilo Bambang Yudhoyono (SBY)? The authors find that Indonesia's foreign policy toward Palestine is strongly influenced by the character of the regime in power and the existing domestic and international political context. Furthermore, it can be observed that Jokowi's policy places greater emphasis on political approach than that of Soeharto and SBY, yet it does not adopt a confrontational political approach as did that of Soekarno. In addition, it can be seen that Indonesia's foreign policy objectives toward Palestine are strongly influenced by the norms of international institutions, notably the UN, which advocates for a two-state solution.

The Epilogue gathers highlights and lessons learned from the issues discussed earlier. It will characterize Indonesia's diplomacy and its achievements during Joko Widodo's era and what distinguishes this era from previous ones.

References

Purba, Kornelius. 2022. Jokowi's Foreign Policy Comes Late, but Internationally Impactful. Retrieved from January 25, 2024, from https://www.thejakartapost.com/opinion/2022/11/25/jokowis-foreign-policy-comes-late-but-internationally-impactful.html.

Indriana Kartini is a Senior Researcher at the Research Centre for Politics, the National Research and Innovation Agency (BRIN), Indonesia. She was the Cluster Coordinator for Foreign Policy and International Issues. She obtained her Master of International Politics from the Faculty of Arts, The University of Melbourne, Australia in 2008. She received her PhD from the Political Science Study Program, at the University of Indonesia in 2023. Her research interests are international politics, international border, subregionalism, and geopolitics. She is the editor and contributor of academic books, including Potret Politik Luar Negeri Indonesia di Era Reformasi (Indonesian Foreign Policy's Portrait in the Reform Era) (Yayasan Pustaka Obor, 2019); Penguatan Konektivitas Lintas Batas dalam Kerja sama Ekonomi Subregional (Strengthening Cross Border Connectivity in Subregional Economic Cooperation (Yayasan Pustaka Obor, 2021)

Khanisa is a Researcher at the Research Center for Politics, the National Research BRIN. She received her Bachelor's degree from the Department of International Relations (HI), Gadjah Mada University, in 2010 and completed her Master of Arts (International Relations) and Master of Diplomacy from the Australian National University in 2015. She focused her research on ASEAN and Southeast Asia issues.

Chapter 2
Down-To-Earth Diplomacy: Connecting Diplomacy to the People and Translating Nawa Cita

Yayan Ganda Hayat Mulyana and Mohamad Zakaria Al-Anshori

Abstract This article aims to analyze the implementation of down-to-earth diplomacy that Indonesia carried out during one decade of the Joko Widodo—Jokowi—presidency. It assesses how diplomacy brings concrete deliverable that impacts the realization of Jokowi's Nawa Cita (Nine Priority Agendas). The analysis utilizes the observation of an insider plus practitioner of diplomacy. The study finds that down-to-earth diplomacy has been instrumental in helping the government materialize the Nawa Cita. It results in concrete deliverables in trade and investment, development, the protection of Indonesians overseas, securing sovereignty and territorial integrity, and contribution to the advancement of regional and global peace, stability, and prosperity. The study also finds that down-to-earth diplomacy has been executed in line with the country's active and independent foreign policy. It is, therefore, not transactional.

Keywords Down-to-earth diplomacy · Nawa Cita · Concrete deliverables · Active and independent foreign policy · Transactionalism

2.1 Introduction

The ascent of Joko Widodo, or Jokowi, into power, was hailed enthusiastically as he was Indonesia's first leader outside the political and military elite. He brought hope for a better Indonesia due to the absence of his burden of the past. However, unlike his predecessor, President Susilo Bambang Yudhoyono, who was considered enthusiastic, experienced, and capable in foreign policy, Jokowi's condition was the opposite. He was expected to implement an inward-looking and pragmatic foreign policy. However, over time, President Jokowi could show foreign policy directions that were not inferior to those of previous Indonesian leaders.

President Joko Widodo has significantly shaped Indonesia's foreign policy in two terms from 2014 to 2024. President Jokowi's administration has made substantial

Y. G. H. Mulyana (✉) · M. Z. Al-Anshori
Indonesian Ministry of Foreign Affairs, Jakarta, Indonesia

I. Kartini and Kanisa (eds.), *Joko Widodo's Foreign Policy and Diplomatic Legacy*,
https://doi.org/10.1007/978-981-95-2085-5_2

strides in redefining Indonesia's diplomatic priorities to achieve the nation's goals. His foreign policy framework had been intrinsically linked to the Nawa Cita or Nine Development Priority Agendas. The Nawa Cita was translated to the country's foreign policy pillars, especially those related to security, stability, and economic independence. The Nawa Cita has focused more on encouraging economic growth, improving people's welfare, and enhancing national autonomy. This strategic orientation led to a foreign policy that aims to deliver more tangible benefits for the people.

Toward that end, Jokowi promoted down-to-earth diplomacy as a practical approach to meeting the country's objectives. He seemed to instrumentalize the foreign policy dimension with domestic orientation. Down-to-earth diplomacy has been understood by many as an inward-looking kind of foreign policy (Andika 2016a). However, President Jokowi seemed successful in answering the underestimates of some observers about his inward-looking foreign policy and the country's regional and global role.

The down-to-earth diplomacy was clearly articulated in the 2015 Annual Press Statement by Minister of Foreign Affairs, Retno Marsudi. She emphasized that foreign policy must prioritize real benefits for the Indonesian people in an inclusive manner, especially in the socio-economic field. Diplomacy down to earth makes foreign policy more accessible and relevant to the public. The goal is to address socio-economic issues directly, ensuring diplomatic efforts generate concrete benefits for all levels of Indonesian society (Kemlu 2015a).

Such a diplomatic approach is strongly connected with people (Djumala 2019). Indonesia's foreign policy cannot be divided into domestic situations and regional and global aspirations. That approach continued as Jokowi's administration entered its second term (2019–2024), and these priorities evolved. The Indonesian Minister of Foreign Affairs introduced the "4 + 1 priorities" framework, which continued to emphasize economic diplomacy, the protection of Indonesia citizens abroad, sovereignty diplomacy, expanded Indonesia's role in promoting regional and global peace and stability, and alongside the strengthening of diplomatic infrastructure, including human resources, bureaucratic reform, and the digitalization of diplomacy.

This chapter explores the evolution of Indonesia's foreign policy under President Jokowi, analyzing how his administration's pragmatic and results-oriented approach has redefined the nation's role on the global stage. It is essential to understand how Jokowi's foreign policy approach became a tool to realize his people-oriented policy despite all the complexities of global diplomacy.

2.2 Diplomacy for the People

As stated earlier, the pro-people approach in foreign policy during Jokowi's administration is known as down-to-earth diplomacy. Term diplomacy can be defined as the art of creating peaceful relationships between countries, groups, or individuals (Boudreau 2023). President Jokowi wished to attain the Indonesian national goals

through down-to-earth diplomacy: sovereignty in its politics, independence in its economy, and distinct cultural character.

During the initial period of Jokowi's administration, he put the people's interest at the center of Indonesia's foreign policy to protect the Indonesian people's needs first and foremost. In other words, his primary foreign policy orientation was dedicated to serving the interests of its people, who mainly still grapple with economic matters. Foreign policy needs to be manifested in seizing benefits for the people. There has been a perception among the public that the works of diplomacy are elitist, having little relationship with the benefits and prosperity of the society. The scope of diplomacy is broad and covers various issues at the bilateral, regional, and multilateral levels (Djumala 2014).

Weatherbe (2016), Andika (2016b), and Wardhani and Dugis (2023) argue that apart from Jokowi's personal beliefs and experiences, Indonesia's proactive role in the global arena conducted by his predecessor, which is considered not to have a direct impact on the needs of the people, encouraged him to embrace down-to-earth diplomacy with a pro-people approach. Jokowi's foreign policy is generally implemented to address domestic economic and political problems, especially those relevant to domestic issues. His foreign policy was conducted to secure national development and domestic needs (Wardhani and Dugis 2023). Considering that foreign policy is about change and continuity, it is unnecessary to contrast President Jokowi with his predecessors heavily.

Some scholars, such as Priamarizki and Marzuki (2022), argue that President Jokowi was a domestic president due to his priorities on infrastructure development, bureaucratic reform, and other related domestic issues. President Jokowi instructed Indonesian diplomats to boost exports, attract investment, and increase services for Indonesian citizens abroad. His international initiatives, including being the host of regional and international meetings and his participation in international forums, merely aimed at supporting national economic development and the welfare of his people. His attendance at a virtual United Nations General Assembly, his efforts to acquire vaccines, and his other international endeavors could not be separated from domestic purposes not to boost Indonesia's international posture (Priamarizki and Marzuki 2022). Having focused more on domestic interests, President Jokowi's foreign policy was characterized as an inward-looking policy (Noor 2017). However, this notion is refuted by Lundry (2018) and Jemadu and Lantang (2021), saying that President Jokowi's international initiatives and leadership had remained in line with his political pragmatism and domestic targets.

Realizing the down-to-earth diplomacy, Indonesia had taken phenomenal steps, namely implementing harsh policies against illegal fishing and executing drug dealers. The country enforced laws and regulations at sea to overcome illegal, unreported, and unregulated (IUU fishing). Aggressive and strict measures were taken by the administration of Jokowi toward illegal fishing in Indonesia's territorial waters. Hundreds of foreign fishing vessels were destroyed and sunk as they operated illegally in the country's water. Jokowi also increased the use of capital punishment for those responsible for smuggling drugs. This policy was previously put on hold by his predecessor, President Yudhoyono (Lundry, 2018). The two above policies sparked protests from neighboring countries because their interests were affected or their citizens were convicted of the death penalty. Jokowi's policies to sink illegal fishing vessels and exercise the death penalty for drug traffickers realized his commitment to domestic constituents (Andika 2016a).

The law enforcement at sea and the execution of the death penalty, as the implementation of down-to-earth diplomacy, were principally in accordance with and did not violate the fundamental principles of Indonesia's independence and active foreign policy. Adhering to foreign policy principles guides the country in implementing the pro-people diplomacy approach. It is imperative to emphasize that independent and active foreign policy does not mean neutrality and detachment. Instead, the foreign policy principles promote *"an active role and engagement with all countries, guided by strategic autonomy and firm adherence to international law"* (Marsudi 2023). This principle also makes Jokowi's foreign policy more driven by economic pragmatism. Jokowi's "Down-to-Earth" diplomacy focused more on strengthening trade relations than playing a firm and multilateral role on the international stage (The Conversation 2024).

The country's strict policies on illegal fishing and drug traffickers for the sake of domestic and nationalistic goals raised questions about whether Jokowi had paid less attention to Indonesia's regional and international role. President Jokowi's leadership was considered to have neglected ASEAN. Since Jokowi first rose to power, ASEAN's central role has declined and disappeared from Jokowi's foreign policy (Rosyidin and Pattipeilohy 2020; Darwis and Putra 2022). However, President Jokowi's foreign policy does not neglect the importance of Indonesia's presence on the international stage. It cannot be denied that maintaining Indonesia's role on the global stage remains crucial for securing domestic benefits, particularly in terms of the economy. ASEAN continues to be a strategic organization for Indonesia to maintain its influence and strong image in the region. Additionally, fostering relations with neighboring countries in regions such as East Asia and the Indo-Pacific is also highly important (Nabbs-Keller 2020).

It cannot be denied that diplomacy, which prioritizes domestic interests, will attract not only public support but also responses from the global world. ASEAN is an example of this case. With the organization's vision of regional integration, Indonesia seemingly showed a foreign policy that tends to establish protectionism. However, that did not happen at all. President Jokowi's focus on bilateralism did not decrease his commitment to maintaining Indonesia's leadership in ASEAN (Lundry 2018).

In a nutshell, Indonesia's down-to-earth diplomacy and perceived post-ASEAN engagement would seemingly never relegate the importance of ASEAN (Sidiq 2022). Indonesia's leadership in ASEAN remained unchanged as Jokowi's predecessor had sought it. As its role as a peace broker in the 1980s Vietnam–Cambodia, the 2011 Thailand–Cambodia border conflict, and the establishment of ASEAN's Six Point Principles on the South China Sea in 2012, Indonesia took a similar approach in the era of President Jokowi. Indonesia has never left its interests to maintain security and stability in the region. Unsurprisingly, Indonesia took the initiative to mobilize a joint response to address a military coup in Myanmar in 2021, where ASEAN leaders agreed on the Five-Point Consensus (Sidiq 2022).

2.3 Diplomacy as a Crucial Instrument in Realizing the Nawa Cita

Jokowi came to the Indonesian presidency with a national development agenda known as Nawa Cita. It consists of nine development priorities. The Nawa Cita derived from President Jokowi's political, economic, and cultural vision of national sovereignty. The Nawa Cita was then translated into the National Mid-Term Development Plan (RPJMN). Based on the RPJMN, the Ministry of Foreign Affairs then made a Strategic Plan (Renstra) for foreign policy, followed by program formulation and implementation (UNDP 2015; Fitriani and Panduwinata 2018).

The Ministry of Foreign Affairs, as a bureaucratic machine for implementing foreign policy, has a vital role in realizing the Nawa Cita. In this regard, diplomacy is crucial to realize the Nawa Cita in the first and the second terms of Jokowi's presidency. Diplomacy encompasses all aspects of the Nawa Cita Development priorities, particularly on the agenda of ensuring the state's presence to protect the entire nation and enhance productivity and competitiveness in the international market. These objectives are articulated through three pillars of foreign policy priorities: maintaining Indonesia's sovereignty, enhancing the protection of Indonesian citizens, and intensifying economic diplomacy.

Those Indonesian diplomacy strategies aim to ensure that all diplomatic efforts are directed toward maximizing domestic benefits (Andika 2016a, b). Domestic benefits mean benefits for the people. The conduct of diplomacy in favor of the people was clearly reflected in the motto "Diplomasi untuk Rakyat" meaning Diplomacy for the People, enshrined in the 2015–2019 Ministry's strategic plan (Kemlu 2015b; Fitraini and Panduwinata 2018). Indonesia's diplomacy is reflected in efforts to protect and uphold democracy, development and good governance, encourage the socio-economic sector (including human resources, trade, and investment), and ensure both global and regional security as well as territorial sovereignty.

2.4 Diplomacy That Protects Indonesian Citizens

The problems facing Indonesian citizens abroad have become increasingly complex and diverse in their variants. These require the attention and role of the state in providing services and protection in a more comprehensive, precise and optimal manner. As mandated by the 1945 Constitution and national laws and regulations, the Ministry of Foreign Affairs (MOFA) is always committed to strengthening efforts to protect Indonesian citizens and legal entities abroad. This protection is conducted with the principle of partisanship and caring through three interrelated approaches: prevention, early detection, and rapid response treatment (Kemlu 2015a). The government's diplomatic protection model for Indonesian citizens abroad prioritizes legal protection efforts. In other words, when Indonesian citizens experience a problem or case, the Indonesian government provides legal assistance and protection according to the existing legal guidelines (Paramitaningrum et al. 2018).

The MOFA always faces high public expectations at home regarding the provision of maximum protection for all Indonesian citizens. The high public spotlight and sensitivity of protection issues urge the Government to provide further protection beyond normative legal and consular protection. The next challenge in managing protection issues includes overlapping national legislation in labor migration. This has resulted in duplication and inconsistency concerning Indonesian citizen protection. It also leads to differences in perspective between policy-making agencies.

There is still minimal public understanding of the importance of the safe migration process. Most Indonesian migrants work in the informal sector, spreading across Southeast Asia, East Asia, and the Middle East. The lack of access to information for Indonesian migrants in the informal sector frequently causes another challenge (Paramitaningrum et al. 2018). The limited resources of the Indonesian Missions abroad also hamper the optimal protection of Indonesian citizen protection abroad. The MOFA has built a system for handling cases of Indonesian citizens abroad by strengthening protection strategies, including strengthening institutional systems, strengthening protection diplomacy, and grand design for resolving cases (Kemlu 2015a).

Under President Joko Widodo, the protection of Indonesian citizens abroad has become a diplomatic priority, with the government taking proactive steps to address the challenges faced by its diaspora. For the last 10 years, there has been a marked increase in the number of legal cases resolved in favor of Indonesian citizens abroad, a reduction in the number of abuse cases against migrant workers, and more efficient crisis response mechanisms. The Indonesian Mission abroad's quick action during crises, such as the repatriation of citizens during the COVID-19 pandemic, has been widely praised domestically and internationally.

The diplomacy of protecting Indonesian citizens has shown extraordinary performance. The presence of the state can truly be felt by Indonesian citizens as a protector and servant of the people. Implementing down-to-earth diplomacy has provided safety and comfort for Indonesian citizens abroad. During the period 2014 to 2023, no less than 218,313 cases have been resolved, of which 360 Indonesians who were sentenced to death were saved. The Indonesian government has repatriated 18,022 people from conflict areas and also in dangerous conditions, such as at the beginning of the outbreak of the COVID-19 pandemic.

During the COVID-19 pandemic, more than 88 thousand Indonesian citizens have been provided with vaccine facilities. Indonesian citizens are often the targets of crimes, including crimes that demand ransom. In this case, no less than 56 Indonesian citizens have been rescued from hostage, including those that occurred in the Southern Philippines. The protection of Indonesian citizens abroad also includes how the rights of migrant workers can be fulfilled. In this case, more than Rp. 1 trillion of Indonesian citizens' financial rights have been successfully recovered.

In the context of this protection, the Indonesian government continues to improve its services, including by cooperating at the bilateral and regional levels. At the bilateral level, MOUs have been signed with other countries to maximize protection services, including a one-channel deployment system with Malaysia and Saudi Arabia. At the regional level, ASEAN cooperation has been established to address online scam crimes. All of this shows that down-to-earth diplomacy characterized by pro-people diplomacy has been truly realized based on the interests of the Indonesian people. This is because the constituents of implementing down-to-earth diplomacy are the Indonesian people, including migrant workers, who are struggling to contribute to the country's national income and the prosperity of the Indonesian people.

Indonesia also continues to establish a partnership concerning the protection of Indonesian citizens. For instance, in 2022, Indonesia signed a Memorandum of Understanding (MoU) on the Placement of Migrant Workers with Malaysia and Saudi Arabia. The protection of Indonesian citizens abroad is a vital component of President Joko Widodo's Nawa Cita vision, reflecting a broader commitment to people-centered diplomacy.

As Indonesia continues to grow and assert itself on the global stage, the protection of its citizens will remain a crucial pillar of its foreign policy. This commitment to citizen protection not only enhances Indonesia's global standing but also ensures that its citizens, no matter where they are in the world, feel the presence of their homeland. In this regard, the MOFA has built a system for protecting Indonesian citizens abroad, namely the Indonesian Citizen Care Portal and Safe Travel. The MOFA will continue improving services and protection for Indonesian citizens abroad through digital transformation, strengthening the human resources capacity in the MOFA and Indonesian Missions abroad.

2.5 Diplomacy That Promotes Democracy

Indonesia, as the largest democracy in Southeast Asia, has taken a leadership role in promoting democratic governance within the region. Through ASEAN, Indonesia has consistently advocated for the adherence to democratic principles, human rights, and the rule of law. The MOFA has been active in supporting democratic transitions in neighboring countries and has provided platforms for dialogue on democracy and good governance.

Indonesia was not silent when the military junta seized power undemocratically in 2021. Indonesia's various efforts, including holding the ASEAN Summit in Jakarta in 2021 to promote and restore democracy in Myanmar, have not shown significant progress. The uncooperative attitude of the Myanmar military government and China's role in "blessing" the military junta in Myanmar made Indonesia endorse the normative approach shown by ASEAN. President Jokowi shared the opinion with his predecessor that Indonesia should take leadership in promoting democracy and human rights in the ASEAN region (Jemadu and Lantang 2021).

The reasons why President Jokowi took the lead in the development and promotion of democracy are as follows. First, he was determined to present Indonesian Muslims as a moderate people. He was also determined to stem the increasing influence of militant Islam, which coincidentally was very critical of his government. Driven to campaign for moderate Islam, President Jokowi then controversially disbanded HTI and also the Islamic Defenders Front (FPI). President Jokowi also played a role in establishing the Indonesian International Islamic University (IIIU) in Depok through Presidential Decree Number 57 of 2016. The establishment of the university is expected to strengthen the character of Islamic culture, which is in harmony with modernity and democracy. With the university's existence, Indonesia is expected to become the hub of Muslim scholars worldwide.

The second reason is that President Jokowi was confident in managing the ethnic and religious diversity in Indonesia. He wanted to share his experience with the world, especially Islamic countries. That is why President Jokowi offered Indonesia the position of peace mediator between the conflicting parties in Afghanistan. He did this when he visited Afghanistan in 2017. The third is that President Jokowi sought Indonesian democracy to manage to generate impacts for economic development in Indonesia. Therefore, the 10th Bali Democracy Forum (BDF) used the theme "Democracy that Delivers" (Jemadu and Lantang, 2021) as President Jokowi understood that democracy could deliver economic impacts and benefits for the people.

President Jokowi committed to hosting the BDF as one of Indonesia's landmarks in advancing democracy and human rights in the Asia-Pacific and beyond. First held in 2008, the BDF has become the premier international forum for promoting democratic values and bringing together government leaders, civil society, and academics to share experiences and best practices. With the principles of inclusive and homegrown democracy, the forum has developed. Initially, the BDF was a government forum. Currently, it also includes civil society, media, student, and economy panels.

On the global stage, Indonesia has also actively promoted the principles of democracy in multilateral settings. Indonesia has been elected as a United Nations Human Rights Council member for the sixth time (Antara 2023). This opportunity has allowed Indonesia to champion democratic principles and call for the protection of civil liberties globally. The efforts to promote democracy in the regional or multilateral fora are also beneficial in strengthening Indonesia's democracy within the country. The down-to-earth diplomacy approach could be translated as an inward-looking foreign policy, as far as how foreign policy could serve the best interest of the people.

The Indonesian MOFA has made considerable progress in promoting democracy domestically and internationally. Indonesia's leadership in regional and global forums has bolstered its image as a democratic country committed to human rights and the rule of law. The BDF has emerged as a prominent forum for discussing democracy in the Asia-Pacific region. Indonesia's active participation in international organizations has strengthened its position on democratic governance issues.

2.6 Diplomacy for Development

In President Joko Widodo's administration, diplomacy has been deeply intertwined with Indonesia's national economic development agenda, aligning foreign policy with the country's domestic priorities (Setiawati and Roziqin 2024). This approach, rooted in the Nawa Cita, has emphasized the importance of leveraging international relationships to promote economic growth, improve infrastructure, and enhance the overall welfare of the Indonesian people.

Development cooperation is one of the strategic economic diplomacy instruments and a tool to increase engagement with partner countries. In addition to trade and investment cooperation, Indonesia's diplomacy also promotes development cooperation implemented, among others, through Technical Cooperation (KST) in the Asia-Pacific region and beyond. Aside from being a tool for economic diplomacy, development cooperation with developing countries is also intended to support Indonesia's sovereignty. Diplomacy for development is, for example, manifested through the country's Pacific Elevation vision, the holding of the Archipelagic and Island States Forum, and other regional and international cooperation.

The country's diplomacy is mainly directed toward national development. Diplomacy must contribute to national economic development and growth. During Jokowi's administration, Indonesia's economic growth has been impressive. Over the period from 2014 to 2023, Indonesia's GDP demonstrated significant growth. In 2014, the GDP was recorded at $890.815 billion. Although there was a slight decline in 2015 to $860.854 billion, the trend turned positive again in 2016 as the GDP rose to $931.877 billion, continuing its upward trajectory until 2019, when it reached $1,119.1 billion. In 2020, the GDP experienced a modest drop to $1,059.05 billion, likely due to global economic challenges such as the COVID-19 pandemic. Recovery began in 2021, with GDP climbing to $1,186.51 billion and maintaining this growth through 2023, culminating in a peak of $1,371.17 billion. Overall, Indonesia's GDP grew by an impressive 53.92% from 2014 to 2023 (World Bank 2024).

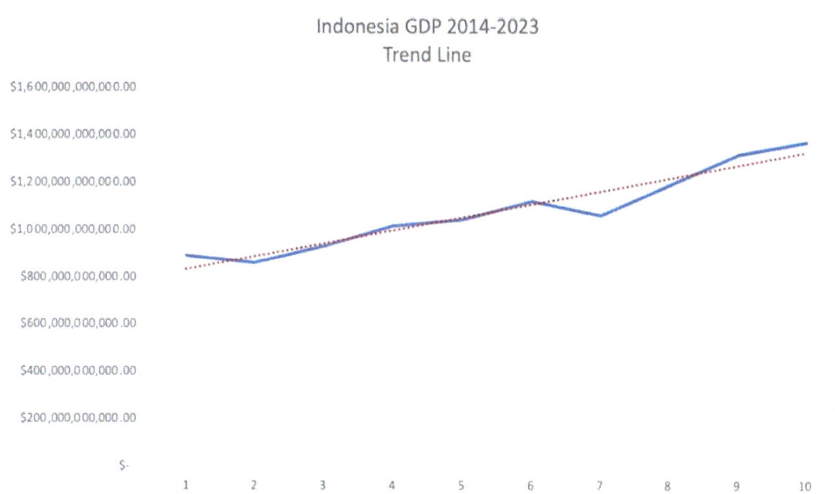

Source World Bank (2024)

Indonesia prioritizes economic partnerships over economic competition through several global free trade agreements while still paying attention to national interests. This is intended to secure the regional supply chain and flow of goods and services. Indonesia's economic diplomacy strives to encourage the realization of the ASEAN Economic Community as a center of growth for the region and the world. In addition, economic diplomacy seeks to position itself as a true partner amidst the constellation of major countries, especially with the PRC and the US, through membership in the Regional Economic Partnership (RCEP) and Indo-Pacific Economic Framework (IPEF). Economic diplomacy is also strengthened in the UN, G20, and ASEAN forums involving state-owned enterprises and the private sector.

Indonesian diplomacy has demonstrated its ability to build international consensus for post-pandemic economic recovery. This was done when it hosted the G-20 in 2022. Indonesia received much recognition from the international community. The agreed document, "G20 Action for a Strong and Inclusive Recovery," reflects the world's commitment to an inclusive and sustainable global economic recovery. In addition, Indonesia also initiated the establishment of the G-20 Pandemic Fund to strengthen global preparedness for future pandemics, especially for developing countries (Raharjo 2024).

During the G20 presidency, Indonesia also initiated many concrete deliverables G-to-G with 361 projects worth more than USD71 billion. Meanwhile, private projects recorded at more than USD 29 billion. During the ASEAN chairmanship, Indonesia initiated the ASEAN Indo-Pacific Forum (AIPF), which resulted in 93 cooperation projects worth more than USD 38.2 billion, of which USD 13.5 billion were in health, mineral processing, and private finance.

Indonesia's leading role in the region is crucial to economic growth. In 2015, ASEAN launched the ASEAN Economic Community (AEC) 2015 and adopted the AEC Blueprint 2025 and ASEAN Vision 2025. Through this cooperative framework, it is anticipated that all ASEAN member countries will be able to integrate further and strengthen their economies. As a result, the AEC will enable ASEAN and its member nations to emerge as significant players in the global market.

Several infrastructure collaborations were successfully carried out through bilateral diplomacy, such as with China and Japan. In 2015, Indonesia leveraged diplomatic relations to secure international financing and technical assistance for large-scale infrastructure projects, such as the Jakarta-Bandung High-Speed Railway, which was developed in collaboration with China through PT. KCIC. These projects were crucial in improving connectivity and supporting economic growth between the two big cities in Java. The cooperation between Indonesia and Japan resulted in significant investments in infrastructure projects, including the expansion of Jakarta's Mass Rapid Transit (MRT) system, especially for the development of the Phase 2 project.

The Jokowi administration's foreign policy reflected a strategic vision to address geopolitical and geoeconomic issues that link national security and economic prosperity. Indonesia's participation in the Belt Road Initiative underscored its commitment to leveraging international partnerships for the growth and prosperity of the Indonesian people (Mahadana 2023).

2.7 Diplomacy for Good Governance

Indonesia's diplomatic efforts have also played a crucial role in enhancing good governance within the country. Through international cooperation, capacity-building initiatives, and the adoption of global best practices, Indonesia's MOFA has contributed to the country's ongoing efforts to promote transparency, accountability, and law enforcement in governance.

Since 2014, Indonesia's participation in the Open Government Partnership (OGP) has been a critical avenue for enhancing transparency and accountability in public administration. Additionally, Indonesia has worked closely with global organizations/institutions like the World Bank, the United Nations Development Programme (UNDP), the United Nations Office on Drugs and Crime (UNODC) and the Asian Development Bank (ADB) to strengthen its governance capacity. Bilaterally, Indonesia has engaged with countries such as Singapore, Australia, South Korea, Estonia, Japan, and many others. These collaborations have led to various initiatives to improve governance reforms that promote open data, enhance the rule of law, and improve anti-corruption measures.

Concerning national good governance, Indonesia's Corruption Perception Index (CPI) for 2023 indicated a score of 34 out of 100, placing the country at 115th out of 180 nations. This score reflected no change from the previous year, maintaining a steady position that signals ongoing challenges in combatting corruption and enhancing national transparency. The trend over the past decade, from 2012 to 2023, showed fluctuations, with periods of improvement and decline. After reaching its highest point in recent years, around 2019, the CPI score has gradually decreased, pointing to persistent governance and anti-corruption issues that require attention (Transparency International 2024).

Indonesia was one of the 50 countries covered in the Business Ready (B-Ready) report 2024. The B-Ready replaces the Ease of Doing Business, which measures and compares business climates among countries. One of the things measured in the report is the ease of complying with regulations, the provision of public services, and the use of public services relevant to the business. Indonesia scored 63.98 on the regulatory framework, 63.44 on public services and 61.31 on operational services. Indonesia's scores on the regulatory and operational framework pillars were still below the average of 50 countries, while the score on the public services pillar was above average (Wildan 2024).

Indonesia had a partnership with the European Union in 2016 about cooperation in many fields, including capacity-building on good governance. The cooperation has been conducted through many working groups between Indonesia and the EU. Considering the existing challenges in implementing good governance in Indonesia, diplomacy has played an important role in fostering cooperation that helps increase the standard of good governance in Indonesia.

2.8 Diplomacy for Quality of Human Capital

Human capital development drives Indonesia's economic growth and social progress. The Nawa Cita has provided the guiding framework for national policies that enhance the quality of Indonesia's human capital. One of the key strategies to achieve this vision has been the active use of diplomacy to forge international cooperation that supports this vision. Through bilateral and multilateral engagements, Indonesia has forged partnerships that provide access to education, vocational training, and capacity-building initiatives.

Indonesia has partnered with countries like Germany, Japan, South Korea, and Singapore to establish vocational training centers and capacity-building programs that align with global industry standards. For example, the collaboration with Germany has focused on the dual education system, which combines classroom learning with hands-on training in industries. This model has been instrumental in equipping Indonesian workers with the technical skills needed to meet the demands of both domestic and international markets (Bappenas 2017).

So far, Indonesian diplomacy has always sought opportunities for cooperation to increase the capacity of Indonesian human resources. The EU supports Indonesia for 1000 initiative as a follow-up to the 7th EU–Indonesia Joint-Committee Meeting in 2023. This aligned with the EU's Green Deal, aiming to make Europe the first climate-neutral continent by 2024 (Tobing 2024). The Indonesia–Australia CEPA also provides opportunities for Indonesian youth and students to intern in Australian companies.

2.9 Diplomacy for Trade and Investment

The alignment of foreign policy with national economic goals has been a defining feature, mainly through trade and investment diplomacy. This approach has sought to optimize Indonesia's international relations to create tangible economic benefits at the domestic level, directly supporting the Nawa Cita's focus on national development. Amid major global economic competition, Indonesia still increased its trade and cooperation through several strategies.

Diplomatic efforts were also directed toward expanding the outbound investment and export opportunities. The outbound investment that Indonesia achieved consists of Infrastructure development works in Niger, the Philippines, as well as facilitating PT Telkom's investment in Hungary and the development of an instant noodle factory in Serbia. Indonesia also exported railway carriages to Bangladesh and Thailand. Indonesia also exported aircraft to Thailand, the Philippines, Senegal, and Mexico.

Foreign direct investment (FDI) has also been a critical component of Indonesia's economic diplomacy strategy under Jokowi's leadership. Recognizing the importance of a favorable investment climate, the government undertook significant regulatory reforms, including establishing the One-Stop Service (OSS) system. This initiative streamlined the investment process, making navigating Indonesia's regulatory landscape easier for foreign investors.

Indonesia's economic diplomacy has successfully expanded market access and attracted foreign direct investment (FDI). Indonesia has signed 27 international trade agreements, opening new market access in Indonesia's non-traditional markets in Latin America and the Caribbean, and Africa. The tangible benefits for the people were generated through diplomacy. Foreign direct investment increased from US$28 billion in 2014 to US$50 billion in 2023. Strategic sectors such as renewable energy, food, and health become the focus of this foreign investment, reflecting the success of results-oriented foreign policies (Raharjo 2024). Investment in the Asia-Pacific and Africa region in Indonesia in 2023 increased by 187% from 2014, with the investment value in 2023 reaching USD42.6 billion for 48,931 projects. The achievement of this investment had been focused on infrastructure and downstream industrial sectors.

In 2014–2022, the value of Indonesia's trade with the Americas and Europe continued to increase to USD 103 billion, with a trade surplus for Indonesia of USD 25.4 billion. In 2023, Indonesia's export value was recorded at USD54 billion. Meanwhile, investment from countries in the Americas and Europe to Indonesia in 2023 increased by 472% from 2014, with the investment value in 2023 reaching USD34.9 billion for 21,967 projects. Indonesia's total trade value with the Asia-Pacific and African regions continued to grow to USD 426.3 billion in 2022. Indonesia's economic trade in the Asia-Pacific region improved the balance of payments from a deficit of USD 11.3 billion in 2014 to a surplus of USD 30.8 billion in 2022. In 2023, Indonesia's export value to the Asia-Pacific region increased to USD 227.4 billion.

To increase trade value with traditional trading partners, Indonesia continues to optimize the use of existing economic and trade agreements, such as Indonesia–Australia Comprehensive Economic Partnership Agremment (CEPA) in 2020, Indonesia–Korea CEPA in 2023, and the Regional Comprehensive Economic Partnership (RCEP). In addition, Indonesia has also encouraged efforts to secure supply chains and market access from the threat of non-tariff barriers, such as Pillar II of the Indo-Pacific Economic Framework (IPEF) and the Forest Law Enforcement, Governance and Trade-Voluntary Partnership Agreement (FLEGT-VPA) in 2014. Indonesia has accelerated the negotiation process for the Indonesia–EU CEPA, which is essential in strengthening cooperation in the European Region.

Several promotional activities and new initiatives aimed at expanding non-traditional markets in the Asia-Pacific and Africa included the Indonesia–Africa Forum (IAF) and the Indonesia–Africa Infrastructure Dialogue (IAID). The Indonesia–Africa Forum (IAF) resulted in business deals amounting to USD 586 million (Kemlu, 2019). Many economic diplomacies have also been conducted in Latin America and the Caribbean (LAC), Europe, and South and

Central Asia, such as the Indonesia–Latin America and Caribbean Business Forum (INA-LAC), Indonesia–Europe Business Forums (IEBF), and Indonesia–South and Central Asia Business Forum (INASCA).

Indonesia has secured several market accesses in the region through the Indonesia–Chile CEPA, Indonesia–Mozambique PTA, Indonesia–Iran PTA, and Indonesia–Pakistan PTA. Several new trade agreements has also been negotiated with MERCOSUR, the Eurasian Economic Union, and several countries in Sub-Saharan Africa, including Mauritius, Tanzania, and Rwanda. With these efforts, it is hoped that entrepreneurs and BUMN can access new market niches to increase export value and open up potential for other business cooperation.

2.10 Diplomacy for Regional and Global Peace, Stability, and Prosperity

Indonesia's foreign policy has been deeply rooted in the principles of peace, stability, and prosperity regionally and globally. This commitment is reflected in the country's active participation in diplomatic initiatives that aim to create a more stable and prosperous international environment.

The maritime sector has emerged as a focal topic at the regional level in Southeast Asia and the Indo-Pacific. Through the IORA Summit in 2015, its deliverables have provided a solid foundation for strengthening maritime cooperation between the IORA member countries. The EAS Summit 2015 in Kuala Lumpur also adopted the EAS Leaders' Statement on Enhancing Regional Maritime Cooperation under Indonesia's initiative. Furthermore, Indonesia strived to maintain stability, security, and prosperity, particularly in the Indian and Pacific oceans, by establishing partnerships with ten world maritime countries.

Realizing that Indonesia must ensure the Indian and Pacific Oceans do not become arenas for competition over natural resources, territorial conflicts and maritime supremacy, the concept of Indo-Pacific Cooperation was developed. Presented by President Joko Widodo at the East Asia Summit (EAS) in November 2018 in Singapore, Indonesia, and other ASEAN members invited all partners to continue developing the "Indo-Pacific" cooperation concept.

Kidnapping cases in maritime areas remain a significant challenge for the Southeast Asia region. Under President Jokowi's administration, Indonesia has initiated the development of trilateral cooperation, namely Indonesia–Malaysia–the Philippines to maintain security in the Sulu Sea and its surrounding areas, as well as subregional meetings focused on countering terrorism. Moreover, Indonesia has also fostered development and humanitarian cooperation in conflict-prone areas such as the southern Philippines and Rakhine State.

As part of the international community, global peace and stability are always priorities. Support for Palestine in the conflict with Israel will always be at the forefront of Indonesia's foreign policy. Indonesia's alignment with Palestine has never withered. Instead, it was further strengthened in 2018 with the doubled contribution to UNRWA from Indonesia and education support. For Indonesia, the Two-State solution is the most visible solution. Peace efforts must continue within the corridor of the Two-State solution.

Under President Joko Widodo, Indonesia also contributed actively to the peace process in Afghanistan. Indonesia has been among the most significant contributors to UN Peacekeeping Missions (UNPKO). In 2018, Indonesia deployed a total of 850 peacekeeping personnel, of whom 22 were women. By November 2018, the total number of Indonesian peacekeepers deployed worldwide had reached 3,545.

2.11 Diplomacy That Upholds Sovereignty and Territorial Integrity

As the largest archipelagic state in the world, Indonesia faces unique challenges in defending its territorial integrity. With over 17,000 islands and extensive maritime boundaries, maintaining sovereignty over its land and sea is a complex task. This responsibility has been a central focus of Indonesia's foreign policy under President Joko Widodo, which prioritized strengthening the state's presence and defending national sovereignty.

Indonesia's diplomacy has committed to upholding Indonesia's sovereignty through diplomacy, legal frameworks, and strategic engagement. Indonesia has taken a firm stance in defending its maritime rights in the North Natuna Sea. Indonesia also pursued diplomatic channels to address incursions, engaging in high-level talks while emphasizing the importance of adhering to international law, particularly the United Nations Convention on the Law of the Sea (UNCLOS). These efforts have contributed to regional stability by preventing escalations and promoting peaceful dispute resolution.

Indonesia shares land and maritime borders with several countries, and resolving these borders is crucial for maintaining national sovereignty. During President Joko Widodo's tenure, Indonesia has successfully negotiated and concluded several border agreements with neighboring countries, including Malaysia, the Philippines, and Timor-Leste. One notable achievement was the conclusion of the maritime boundary agreement with the Philippines in 2014, which resolved long-standing disputes in the Celebes Sea. This agreement demonstrated Indonesia's commitment to resolving disputes through peaceful negotiation and diplomacy, setting an example for other nations in the region.

For the last 10 years, Indonesia's diplomacy has emphasized the importance of international law and multilateralism in resolving border disputes. By operating within the frameworks of ASEAN and other international bodies, the Indonesian Ministry of Foreign Affairs has ensured that the country's territorial interests are safeguarded while maintaining good relations with its neighbors.

Indonesia has consistently advocated for ASEAN centrality in addressing regional security issues, including territorial disputes. The country has played a crucial role in shaping ASEAN's approach to the South China Sea and other security challenges. Through ASEAN, Indonesia has promoted the acceleration of negotiations on the Code of Conduct in the South China Sea and has pushed for the peaceful resolution of disputes based on international law (Kemlu 2023). These efforts have reinforced Indonesia's position as a regional leader in promoting stability and security, essential for protecting its territorial integrity and reflecting a broader commitment to safeguarding the nation's sovereignty in all its forms.

During President Joko Widodo's administration, sovereignty diplomacy has been pursued, among other strategies, through increasingly intensive engagement with Pacific countries, both bilaterally and via regional organizations such as the Melanesian Spearhead Group (MSG) and the Pacific Islands Forum (PIF). In 2016, Indonesia successfully upgraded its status from observer to Associate Member at the 20th MSG Summit in Honiara, Solomon Islands. Strengthening cooperation and maintaining active participation in these organizations has served to underscore Indonesia's presence in promoting constructive narratives about the country, while simultaneously mitigating negative discourse related to its interests and sovereignty. Indonesia's diplomatic efforts have prevented the United Liberation Movement for West Papua (ULMWP), a group supporting Papuan separatists, from obtaining full membership in the MSG, as membership remains limited to sovereign states. Additionally, Indonesia has successfully countered Vanuatu's attempts to raise issues concerning Papuan self-determination and decolonization. PIF member countries have periodically reaffirmed Indonesia's sovereignty over Papua, as reflected in official communiqués, including that of 2023.

2.12 Beyond Transactional

In recent years, Indonesia's foreign policy has been widely recognized for its down-to-earth diplomacy—a term that reflects a pragmatic and people-centered approach to international relations. However, this approach should not be mistaken for transactionalism, prioritizing immediate gains or quid pro quo deals. Instead, Indonesia's down-to-earth diplomacy, particularly under the leadership of President Joko Widodo, has been rooted in principles, long-term strategic interests, and a commitment to global stability.

Meanwhile, transactional diplomacy refers to a pragmatic and often short-term approach to international relations, where countries engage in negotiations and agreements based primarily on mutual benefits and immediate interests rather than on long-term strategic alliances or ideological commitments (Bashirov and Yilmaz 2019). This approach contrasts with more traditional forms of diplomacy that emphasize enduring partnerships and shared values. Transactional diplomacy is often seen in contexts where countries prioritize immediate outcomes, such as trade negotiations or crisis response. Some scholars, such as Rosyidin dan Pattipeilhy (2020), have portrayed President Jokowi's foreign policy orientation inaccurately when he embraced down-to-earth diplomacy characterized by pro-people emphasizing the benefits for the people.

Down-to-earth diplomacy addresses real-world issues affecting ordinary people, such as economic development, security, and welfare, rather than engaging in abstract ideological battles. This approach aligned with President Widodo's Nawa Cita, which emphasized strengthening the state's presence, improving the quality of life for all Indonesians, and enhancing Indonesia's global role. At the 2024 Annual Press Statement, Foreign Minister Retno Marsudi affirmed that "Indonesia's diplomacy is well-measured, well-calculated, action-oriented, result-oriented. But at the same time, it upholds unwavering values and principles' (Kemlu 2024). This demonstrated that Indonesia's foreign policy has been more pragmatic than transactional, emphasizing adherence to international law and norms.

Indonesia has consistently advocated for global cooperation through international organizations such as ASEAN, the United Nations, and the G20. Rather than pursuing narrow national interests, Indonesia's diplomacy has sought to contribute to global stability and prosperity. For instance, Indonesia's leadership in the Bali Democracy Forum (BDF) demonstrated its commitment to promoting democratic values and human rights, even when no immediate benefit can be gained. Indonesia's active role in peacekeeping missions under the UN was another testament to its non-transactional approach. By contributing troops and resources to conflict zones worldwide, Indonesia has demonstrated a commitment to global peace and security, driven by principle than self interest.

During President Joko Widodo's administration, Indonesia's humanitarian diplomacy underscored its non-transactional approach. In global crises, the country has provided aid and support to affected nations, guided by solidarity and a sense of global responsibility rather than expectations of reciprocity. Indonesia's assistance to Palestine and disaster-stricken countries has reflected its commitment to humanitarian principles. Its diplomacy has prioritized building long-term strategic partnerships and contributing to global welfare rather than securing immediate benefits, aligning with Indonesia's national values and international responsibilities. For instance, Indonesia's support for Palestine is rooted in historical solidarity and a commitment to justice, despite yielding no direct economic or political gains.

2.13 Conclusion

Over the past ten years, Indonesia's international diplomacy under the leadership of President Joko Widodo has demonstrated significant achievements that strengthen Indonesia's position in the region and on the global stage. Focusing on the principle of active free foreign politics, Indonesia's diplomacy has been oriented toward real action as well as measurable results, as it reflects the objective of down-to-earth diplomacy.

Under the leadership of President Joko Widodo, Indonesia has strengthened Southeast Asia's position as the center of world economic growth and maintained regional stability amid ever-evolving geopolitical dynamics. Indonesia's diplomacy has not only rested on the national interest but also positively impacted global peace, stability, and prosperity. These achievements confirm Indonesia's position as a significant player on the international stage.

In President Jokowi's leadership, the principle of active freedom was tested by the increasingly rapid and dynamic developments due to technological advances and globalization. Over the past 10 years, the world has faced various challenges caused by geopolitical and geoeconomic dynamics, the pandemic, and the world recession. In such a situation, Indonesia has managed to ensure the implementation of independence and active foreign policy principles. The consistency of implementing Indonesia's independent and active foreign policy has proven to increase Indonesia's diplomatic influence. Referring to the Lowy Institute's Asia Power Index 2023, Indonesia's diplomatic influence is considered to have increased rapidly by 6.4 points and is ranked 7th out of 26 Asia-Pacific countries assessed.

In the future, the aforesaid Indonesian foreign policy principle will always remain relevant to navigate the country amid regional and global challenges. The down-to-earth diplomacy was successfully conducted by President Jokowi to realize the Nawa Cita vision. It has proven to be able to meet the needs of the people and, at the same time, still demonstrates its leadership in the regional and international arena. It is simplistic and inaccurate to say that President Jokowi's foreign policy is transactional.

References

Andika, M. T. 2016a. An Analysis of Indonesia Foreign Policy Under Jokowi's Pro-People Diplomacy. *Indonesian Perspective* 1 (2): 93–105. https://doi.org/10.14710/ip.v1i2.14284.

Andika, M. T. 2016b. The Role of Indonesia under Jokowi's Pro-People Diplomacy. Prosiding Seminar Nasional INDOCOMPAC. Jakarta, 2–3 Mei 2016.

Antara. 2023. Indonesia regains its seat in the UN Human Rights Council. Retrieved from August 3, 2024 from https://en.antaranews.com/news/295839/indonesia-regains-seat-in-un-human-rights-council.

Bappenas. 2017. Babak Baru Kerja Sama Indonesia-Jerman dalam Pendidikan dan Pelatihan Teknik dan Vokasi. Retrieved from August 4, 2024 from https://www.bappenas.go.id/id/berita/babak-baru-kerja-sama-indonesia-jerman-dalam-pendidikan-dan-pelatihan-teknik-dan-vokasi-qf5fe.

Bashirov, G., and I. Yilmas. 2019. The Rise of Transnationalism in International Relations: Evidence from Turkey's Relations with the European Union. *Journal of International Affairs*. https://doi.org/10.1080/10357718.2019.1693495.

Bodreau, Diane et al. (2023). Article Diplomacy. Retrieved from August 1, 2024 from https://education.nationalgeographic.org/resource/diplomacy/.

Darwis, and B. A. Putra. 2022. The Decline of Indonesia's Leadership Role in ASEAN: Systemic and Domestic Constraints of Jokowi's Foreign Policy Alterations. *Academic Journal of Interdisciplinary Studies* 11 (1). https://doi.org/10.36941/ajis-2022-0010.

Djumala, D. 2014. Getting Diplomacy Down-to-Earth. The Jakarta Post. Retrieved from 3 November 2024 from https://www.thejakartapost.com/news/2014/12/16/getting-diplomacy-down-earth.html.

Djumala, D. 2019. Diplomacy Becoming More Down-To-Earth. The Jakarta Post. Retrieved from July 28, 2024, from https://www.thejakartapost.com/academia/2019/08/31/diplomacy-becoming-more-down-to-earth.html.

Fitriani, Panduwinata. V. D. 2018. Analisis Kinerja Kementerian Luar Negeri Indonesia (2015–2018). CSIS Working Paper Series WPSINT-3/2018

Iannone, Aniello. 2024. Akankah Prabowo membawa kebijakan luar negeri Indonesia jadi lebih asertif? *The Conversation*. Retrived from 03 November 2024 from https://theconversation.com/akankah-prabowo-membawa-kebijakan-luar-negeri-indonesia-jadi-lebih-asertif-241580.

Jemadu, A., and F. Lantang. 2021. Contested Interpretation of Indonesia's International Leadership in the Foreign Policy of President Susilo Bambang Yudhoyono and Joko Widodo. *Journal of ASEAN Studies* 9 (22): 199–217. https://doi.org/10.21512/jas.v9i2.7478.

Kemlu. 2015a. Pernyataan Pers Tahunan Menteri Luar Negeri Republik Indonesia Tahun 2015, Retrieved from 2 November 2024, from https://123dok.com/document/yr1m77oq-kementerian-republik-indonesia-pernyataan-tahunan-menteri-republik-indonesia.html.

Kemlu. 2015b. Rencana Strategis Kementerian Luar Negeri Repulik Indonesia 2015–2019. Retrieved from 1 November 2024, from https://kemlu.go.id/akip/kementerian-luar-negeri/rencana-strategis-kemlu/26f5bd4aa64fdadf96152ca6e6408068?type=repository.

Kemlu. 2019. Pernyataan Pers Tahunan Menteri Luar Negeri Republik Indonesia Tahun 2019. Retrieved from July 20, 2024, from https://kemlu.go.id/portal/id/read/55/pidato/pernyataan-pers-tahunan-menteri-luar-negeri-ri-tahun-2019.

Kemlu. 2023. Indonesia's Initiative Accelerates South China Sea Code of Conduct Negotiations. Retrieved from July 29, 2024, from https://kemlu.go.id/portal/en/read/5153/berita/indonesias-initiative-accelerates-south-china-sea-code-of-conduct-negotiations.

Kemlu. 2024. Pernyataan Pers Tahunan Menteri Luar Negeri Republik Indonesia Tahun 2024, Retrieved from 31 July 2024, from https://kemlu.go.id/portal/id/read/5658/pidato/pidato-menlu-pada-pernyataan-pers-tahunan-menteri-luar-negeri-ri-tahun-2024..

Lundry, C. 2018. Assessing Indonesia's Foreign Policy under Jokowi. *Asia Policy* 13 (4): 30–35. https://www.jstor.org/stable/26533123.

Mahadana, S. J. 2023. Kebijakan Luar Negeri Indonesia di bawah Pemerintahan Jokowi: Perkembangan, Prioritas, dan Tantangan. *Advances in Social Humanities* 1 (10).

Marsudi, R. L. P. 2023. Indonesia's Non-aligned Foreign Policy Is Not Neutral. Retrieved from 02 November 2024, from https://thediplomat.com/2023/11/indonesias-non-aligned-foreign-policy-is-not-neutral/.

Nabbs-Keller, G. 2020. ASEAN Centrality and Indonesian Leadership in a Contested Indo-Pacific Order. *Security Challenges* 16 (3): 21–26. https://www.jstor.org/stable/26924335.

Noor, S. W. 2017. Indonesian Foreign Policy Under Three Years Of Jokowi's Administration. Retrieved from 21 July, 2024, from https://asc.fisipol.ugm.ac.id/2017/12/06/indonesian-foreign-policy-three-years-jokowis-administration/.

Paramitaningrum, R. V. Yustikaningrum, and G. Dian. 2018. *Jurnal Global & Strategis*. https://doi.org/10.20473/JGS.12.1.2018.17-37.

Priamarizki, A., and K. Marzuki. 2022. Inscribed in Stone: The Making of Jokowi's Foreign Policy Legacy. IDSS Paper 046. Retrieved 3 November 2024 from https://www.rsis.edu.sg/rsis-publication/idss/ip22046-inscribed-in-stone-the-making-of-jokowis-foreign-policy-legacy/.

Raharjo. B. 2024. 10 Years of Jokowi's Government, Strengthen RI's Position in the International Scene. Republika. Retrieved from 3 November 204 from https://en.republika.co.id/berita/skoa7p415/10-years-of-jokowis-government-strengthen-ris-position-in-the-international-scene.

Rosyidin, M., and C. H. Pattipeilhy. 2020. Regionalism under Challenge: Ideas and Joko Widodo's Foreign Policy Towards ASEAN, 20214-2019. *Journal of ASEAN Studies* 8 (2): 147–172. https://doi.org/10.21512/jas.v8i2.6596.

Setiawati, S., and A. Roziqin. 2024. New Trends of Indonesia's Foreign Policy Under President Joko Widodo in The First Period: Foreign Policy. *Technium Social Sciences Journal* 54: 249–262. https://doi.org/10.47577/tssj.v54i1.10554.
Sidiq, M. H. F. 2022. Cornerstone No More? The Changing Role of ASEAN in Indonesian Foreign Policy. The Diplomat. Retrieved from 3 February 2024 from https://thediplomat.com/2022/10/cornerstone-no-more-the-changing-role-of-asean-in-indonesian-foreign-policy/.
Tobing, R. 2024. Eropa Terbuka Lebar untuk Pendidikan dan Kerja Green Engineering. Retrieved from 4 November 2024 from https://seputarcibubur.pikiran-rakyat.com/ekonomi-bisnis/pr-1788734412/eropa-terbuka-lebar-untuk-pendidikan-dan-kerja-green-engineering?page=all.
Transparency International Corruption Perceptions Index 2024. Asia Pacific-Indonesia. Retrieved from 1 November 2024, from https://www.transparency.org/en/cpi/2024/index/idn.
UNDP. 2015. Converging Development Agendas: Nawa Cita, RPJMN and SDGs. Retrieved from 3 November 2024 from https://www.undp.org/indonesia/publications/converging-development-agendas-nawa-cita-rpjmn-and-sdgs.
Wardhani, B., and V. Dugis. 2023. Indonesian Foreign Policy under the Jokowi Administration: Variations on Traditional Ideas. *Issues and Studies* 59 (3), e2350005. https://doi.org/10.1142/S1013251123500054.
Weatherbee Donald, E. 2016. Understanding Jokowi's Foreign Policy. *Trends in Southeast Asia* 12.
Wildan, M. 2024. World Bank Rilis B-Ready, Indikator Kemudahan Usaha Pengganti EoDB. DDTC News. Retrieved from 4 November 2024 from World Bank Rilis B-Ready, Indikator Kemudahan Usaha Pengganti EoDB.
World Bank. 2024. Data GDP (current US$)-Indonesia. Retrieved from https://data.worldbank.org/indicator/NY.GDP.MKTP.CD?end=2023&locations=ID&start=2014&view=chart.

Yayan Ganda Hayat Mulyana is a career diplomat at the Indonesian Ministry of Foreign Affairs. He is currently the Ambassador of Indonesia to Sweden. He was Director-General/Head of Foreign Policy Strategy Agency, Ministry of Foreign Affairs of the Republic of Indonesia (MFA). He joined the MFA in 1993. Since then, he has assumed responsibilities in various divisions of the MFA, including Head of the Secretariat for Special Envoys of the President (2004–2006), and Head of the Education and Training Centre (2018–2022). He served as Assistant to the Special Staff of the President for International Relations at the Office of the President (2010–2014). Between June 2012 and June 2013, he was part of the Indonesian National Committee on the Vision of the Post-2015 Development Agenda. His first diplomatic posting was at the Permanent Mission of Indonesia to the UN in New York (2000–2004). The second posting was also at the Permanent Mission of Indonesia to the UN in New York, serving as the Political Coordinator of the Indonesian Team to the UN Security Council (2007–2008). He then served at the Indonesian Embassy in Singapore as the Coordinator of Information, Social and Cultural Affairs, also as the Spokesperson of the Embassy (2009–2010). He served as Indonesian Consul General for New South Wales, Queensland, and South Australia, based in Sydney from October 2014 to December 2017. He graduated cum laude from University of Padjadjaran, Department of International Relations (1992); pursued in the US two master programs under the Fulbright scholarship (1996–1998) at Ohio University and a doctoral program (2000–2004) at Madison University majoring political science, IR, and public administration. He is a contributor to various journals and newspapers, including the Jakarta Post, the Jakarta Globe, and the Diplomat.

Mohamad Zakaria Al-Anshori is a policy analyst at the Foreign Policy Strategy Agency, the Indonesian Ministry of Foreign Affairs. As a diplomat, he has been assigned to the Indonesian Embassies in Manila and Canberra. He holds a Master of Management (MM) in International Management from the University of Indonesia and a Master of Arts (MA) in Political Science from De La Salle University of Manila. He holds a PhD from Victoria University of Wellington. He was the head of the writing team for the book on Indonesian Health Diplomacy in America and Europe (2022) and co-authored a book with Tantang Razak and Heru Prayitno on BIMB-EAGA and Its Dynamic Process (2004). He is the Managing Editor of Jurnal Hubungan Luar Negeri (Journal of Foreign Relations), Ministry of Foreign Affairs. His research interests include issues on ASEAN, Indo-Pacific, global geopolitics and geoeconomics, global south, Islam, and foreign policy.

Chapter 3
Reforms in the Indonesian Ministry of Foreign Affairs Toward a More Gender-Inclusive Foreign Policy

Ganewati Wuryandari, Athiqah Nur Alami, and R. A. Rizka Fiani Prabaningtyas

Abstract This chapter examines gender inclusion in Indonesian foreign policy during Joko Widodo's administration (2014–2024). Expectations of a more progressive gender-mainstreaming agenda and policies within the framework of bureaucratic and institutional reform at the Ministry of Foreign Affairs (MoFA) have increased with the appointment of Retno L.P. Marsudi as the first female foreign minister. This chapter analyzes the ministry's efforts toward gender inclusion, assessing the implementation of gender reforms, identifying obstacles to further progress, and potential strategies in the future. It employs a qualitative case study and combines a number of data collection methods, including focus group discussions (FGD) and interviews with academics, diplomats, and officials from pertinent ministries. While significant advancements in gender reforms have been achieved under Marsudi's leadership, the study highlights ongoing challenges, including limited female representation, inadequate support for female diplomats, and persistent socio-cultural barriers.

Keywords Indonesia · Ministry of foreign affairs · Gender inclusive · Foreign policy

3.1 Introduction

Indonesia's foreign policy, which is based on the idea of "bebas and aktif" (independent and active), aims to play a proactive role in shaping international order to benefit its interests. Over the past decade of President Joko Widodo's government (2014–2024), gender has emerged as a significant, albeit evolving, aspect of its diplomatic strategy. This aligns with the country's efforts to increasingly respond to external and domestic commitments toward gender equality, including in foreign policy.

G. Wuryandari (✉) · A. N. Alami · R. A. R. F. Prabaningtyas
Research Center for Politics, The National Research and Innovation Agency, Indonesia (BRIN), Central Jakarta, Indonesia

I. Kartini and Kanisa (eds.), *Joko Widodo's Foreign Policy and Diplomatic Legacy*,
https://doi.org/10.1007/978-981-95-2085-5_3

The appointment of Retno Lestari Priansari Marsudi as Indonesia's first female Foreign Minister by the President of Joko Widodo (Jokowi) on October 27, 2014 certainly marked a symbolic shift toward greater gender inclusion in foreign policy. Her appointment reflects a breaking *glass ceiling* in terms of women's leadership within the Ministry of Foreign Affairs (MoFA), an institution that is viewed as masculine in the Indonesian context, where male dominance still tends to be dominant in state institutions in the political, defense, and security sectors. Her appointee also reflected the growing acceptance of women in leadership roles within the MoFA, not to mention answering the question "Where are the women?" in the field of international relations addressed by Enloe (1990).

During Marsudi's presence as foreign minister (2014–2024), representation and participation in women's diplomatic roles have increased, a trend likely to continue with ongoing efforts to promote gender socialization and institutionalization within the ministry. This may correspond to the ministry's attempts to empower women, as evidenced by the Ministry of Foreign Affairs' commitment to gender mainstreaming with the issuance of Minister of Foreign Affairs Regulation No. 21 of 2020. This policy integrates gender considerations into all aspects of the ministry's operations, from planning, implementation, monitoring, and evaluation. To implement its policy on gender mainstreaming, the ministry extended beyond the Jakarta headquarters to the Indonesian representative overseas. At its embassy in Seoul and general consulate in Johor Bahru (Malaysia), the Indonesian Ministry of Foreign Affairs carried out its first gender-mainstreaming pilot project.

Despite all forms of progress pointing toward Indonesia's foreign policy having a more gender-inclusive approach, it must be admitted that there are still enduring obstacles. The ministry must address issues like cultural, institutional, and budgetary barriers. Untackled of these problems will affect gender-mainstreaming implementation in the ministry. According to Moser and Moser (2005), there are three phases to the practice of gender mainstreaming: adopting the terminology of gender equality and gender mainstreaming; establishing gender-mainstreaming policies; and carrying out gender mainstreaming.

This chapter aims to explore the growing influence of gender in Indonesian foreign policy under the presidency of Joko Widodo (2014–2024). In contrast to the previous government, his government, with Retno L.P. Marsudi as a female foreign minister, has taken significant steps to implement gender-mainstreaming strategies as part of bureaucratic and institutional reform in the ministry of foreign affairs. Most studies on gender mainstreaming in development pay little attention to government institutions/ministries and instead concentrate on the experiences of international and regional organizations, such as United Nations agencies (Prugl and Lustgarden 2006), the European Union (Lombardo 2005; Hafner-Burton and Pollack 2009), and ASEAN (Alami 2018). Within this context, this chapter aims to analyze what reforms and measures have been undertaken in the Indonesian Ministry of Foreign Affairs to foster gender inclusivity in Indonesia's foreign policy, the progress at which gender mainstreaming has been implemented, and the challenges and strategies for the future. The strategy of inquiry for this study is a qualitative case study utilizing multiple methods of data collection, including interviews and Focus Group Discussions (FGD) with academicians, diplomats, and officials from relevant ministries.

3.2 International and Domestic Political Reforms Drive Indonesia Toward Gender Inclusion in Foreign Policy

3.2.1 International and Domestic Driving Factors

There is context for Indonesia's foreign policy, which must constantly to balance engagement on a global scale and domestic needs. While Indonesia advocates for international engagement, it treads carefully not to compromise its national interests. In the last several decades, the international community has increasingly emphasized gender equality as a cornerstone of development, peace, and security. Indonesia's membership in international and regional organizations like the United Nations (UN) and ASEAN, as well as its participation in global conferences, exposes it to these norms, creating pressure to align its domestic and foreign policies.

Internationally, the United Nations has issued several declarations and conventions for the advancement of gender equality and the empowerment of women. One of the central and most comprehensive documents is the 1979 Convention on the Elimination of All Forms of Discrimination Against Women (CEDAW). The Convention gives positive affirmation to the principle of gender equality by requiring States parties to take "all appropriate measures, including legislation, to ensure the full development and advancement of women, for the purpose of guaranteeing them the exercise and enjoyment of human rights and fundamental freedoms on a basis of equality with men" (Article 3) (United Nations Human Rights Office on the High Commissioner 1979).

In order to speed up its implementation, the United Nations Fourth World Conference on Women in Beijing in 1995 declared the Beijing Declaration and Platform for Action (BPFA). The Declaration urges governments and other actors, including international organizations, to adopt gender-mainstreaming policies as part of strategic goals and actions in twelve areas of concern for urgent action to accelerate the achievement of gender equality. Some of these areas are health, education, economics, environment, politics, and children-women issues. Each country must report its progress every five years. The document calls on governments and other actors to "promote an active and visible policy of mainstreaming a gender perspective into all policies and programs, so that, before decisions are taken, analysis is made of the effects on women and men, respectively" (paragraph 189) (UN Women 1995).

As a result, gender mainstreaming is not only an Indonesian commitment but also a global one that the UN oversees. Indonesia has demonstrated its commitment to bolstering efforts to protect and empower women by ratifying CEDAW by Law No. 7 of 1984, long before the Beijing conference, and by signing the BPFA in 1995.

The Indonesian government's decision to ratify CEDAW is based on the mandate of Article 27, Paragraph 2, of the 1945 Constitution, which addresses every citizen's right to decent employment and means of subsistence.

By taking these actions, Indonesia consents to be bound by its provisions and assumes responsibility for adhering to and incorporating the convention articles of CEDAW and the BPFA into national laws and policies. From 1995 onward, in every five years, Indonesia makes a country report to the Asia-Pacific Forum on the implementation of the BPFA, which is then reviewed in full by the UN Commission on the Status of Women (CSW), namely in 2000, 2010, 2015, and 2020 (Komnas Perempuan 2019).

Therefore, those two international documents promoting women's rights and addressing gender inequality have directly played a significant role in creating domestic pressures for the Indonesian government to pursue more inclusive gender equality. Indonesia's transition to democracy after the fall of Suharto in 1998 also created opportunities for women's increased participation in gender equality. Gender mainstreaming has become one of the national development strategies since the beginning of the reform era, especially in 2000. A number of state policies have subsequently been created to provide legal umbrellas for gender equality, including (1) President Instructions Number 9 of 2000 concerning Gender Mainstreaming in National Development and (2) Law Number 8 of 2016 concerning the protection of women with disabilities. Indonesia also adopted UN Resolution 1325 on the Women, Peace, and Security (WPS) agenda in 2014, when, by presidential decree Number 18 of 2014, the government launched the National Action Plan (NAP) for the protection and empowerment of women and children in times of social conflict. Moreover, the 2020–2024 National Medium Term Development Plan has sections to ensure that gender-responsive planning and budgeting are applied. All of these regulations signify that gender mainstreaming has been fully and systematically incorporated into Indonesian law. Gender mainstreaming is not a program but a strategy to achieve gender equality in various sectors, including foreign policy.

3.2.2 Institutional Reforms in the Ministry of Foreign Affairs

In accordance with the national development strategy on gender mainstreaming in Indonesia, the Ministry of Foreign Affairs undertook internal reform in 2001 to comply with the 2000 President's Instructions, which pushed government institutions to implement gender mainstreaming. Legal regulation was originally indicated internally at the MoFA in 2001 by the "Benah Diri" (Self Improvement) policy. This program, which was started by the former Minister of Foreign Affairs, Hasan Wirajuda, aims to reform organizational structures, reorganize Indonesian representatives overseas, and revitalize the diplomatic profession (Wuryandari 2009). MoFA has implemented a number of innovative measures to progress the diplomatic field, including creating more opportunities for women to be recruited as diplomats, involving them in the formulation and implementation of merit-based foreign policy, and allowing both couple diplomats to pursue their careers without having their profession sacrificed (FGD, July 25 2024). This rule is different from what was previously established. At that time, if a diplomat married another Indonesian diplomat, one of them had to resign.

MoFA created some new policies as part of its ongoing internal reform to better reflect its commitment to gender mainstreaming. In order to evaluate all ministry regulations and provide Indonesian representatives with an understanding of gender issues, the ministry formed a gender-mainstreaming task force in 2006. The ministry also unveiled a plan in 2013 to advance gender parity and empower women in foreign policy, and in 2015 it published Gender Responsive Planning and Budgeting Guidelines as a follow-up (Prajuli et al. 2021). In the same year, the Indonesian Minister of Foreign Affairs regulated placements for diplomat couples in foreign service. They will be placed in neighboring MoFA missions (Kompas 2015). Additionally, MoFA develops infrastructure that takes gender equality into account, such as lactation rooms and childcare centers. Gender-mainstreaming websites and communities of families with special needs children are two examples of new inventions that are being produced that reflect pro-gender policies. Additionally, a sisterhood of female diplomats is established with the goal of increasing their capacity through workshops, sharing of experience, and mentoring.

MoFA alignment toward gender issues is strengthened by the issuance of Minister of Foreign Affairs Regulation Number 21 of 2020 concerning Gender Mainstreaming's Guidance at the Ministry of Foreign Affairs. Though it underlines the ministry's taking a stronger stance on gender mainstreaming, in fact, it took twenty years to institutionalize gender mainstreaming after the Indonesian government issued a gender-mainstreaming regulation in 2000. In this policy, the ministry incorporates gender issues into every facet of its activities, ranging from planning, implementation, and budgeting to monitoring and evaluation of the implementation of gender mainstreaming in the ministry. In 2021, the ministry continued its internal reform to protect women by tackling the sexual harassment issue. In this regard, MoFA issued a circular from the Minister of Foreign Affairs regarding the prevention and handling of sexual harassment within the Ministry of Foreign Affairs (FGD, 30 August 2023). Additionally, MoFA created two pilot projects in Indonesian representation abroad, which are the Embassy of the Republic of Indonesia in Seoul and the Consulate General of the Republic of Indonesia in Johor Bahru, Malaysia, to expedite the adoption of gender mainstreaming (FGD, September 21 2022).

The Ministry of Foreign Affairs goes one step further in institutionalizing its gender-mainstreaming strategy by creating working groups since 2006. Currently, it has ten Gender-Mainstreaming Working Groups (*Pokja PUG*) in the ministry and one hundred thirty-two working groups in the Indonesian representatives' offices abroad. Among its tasks are (1) monitoring the implementation of gender mainstreaming; (2) implementing planning, preparation, implementation, monitoring, and evaluation of gender mainstreaming; (3) carrying out gender-responsive planning and budgeting; and (4) preparing the Gender-Mainstreaming Working Group report periodically, one time every six months.

At first glance, various formal legal and policy reforms issued by the Ministry of Foreign Affairs have contributed to a lasting impact on gender equality. The first is evidenced by the rising number of female diplomats in the past ten years. While women diplomats have traditionally been underrepresented, research suggests that in the era of Joko Widodo, the number of female diplomats entering the foreign service has significantly improved. During the New Order, the composition of male and female diplomats was quite unequal, with 114 female diplomats and 545 male diplomats (Prajuli et al. 2021). In the early years of the reform era, the number of female diplomats significantly improved to 549, while there were 646 male diplomats (Prajuli et al. 2021).

According to the Ministry of Foreign Affairs, since 2004, the proportion of male and female recruits to the diplomatic corps at MoFA has been about equal (FGD, June 14 2024). However, since 2010, more women have been admitted, with the exception of 2015, into each batch of diplomatic recruits (Fig. 3.1). As a consequence, the number of female diplomats in the ministry increased annually. In 2024, the total diplomats in the ministry are 1.849, with 1091 (59%) being male diplomats and 758 (41%) being female diplomats (FGD, 25 July 2024).

Second, these significant changes in female diplomat representation in the Ministry of Foreign Affairs have slightly transformed its pipeline process of leadership. It is evidenced in the evolving share improvement of the high-level executive leadership within the MoFA from 2015 to 2023. In fact, if we look at the pattern of ambassadorial appointments from 2015–2021 (Prihartini and Alami 2022), it does not show gender progress; with a total of 175 ambassadors appointed, there are only 20 female ambassadors (12.7%). In 2015, not a single woman was appointed as ambassador. Meanwhile, from 2016 to 2021, the percentage of women appointed as ambassadors has decreased. There were five female ambassadors in 2016, but that number dropped to three in 2017, two in 2018, and two in 2019. However, slowly

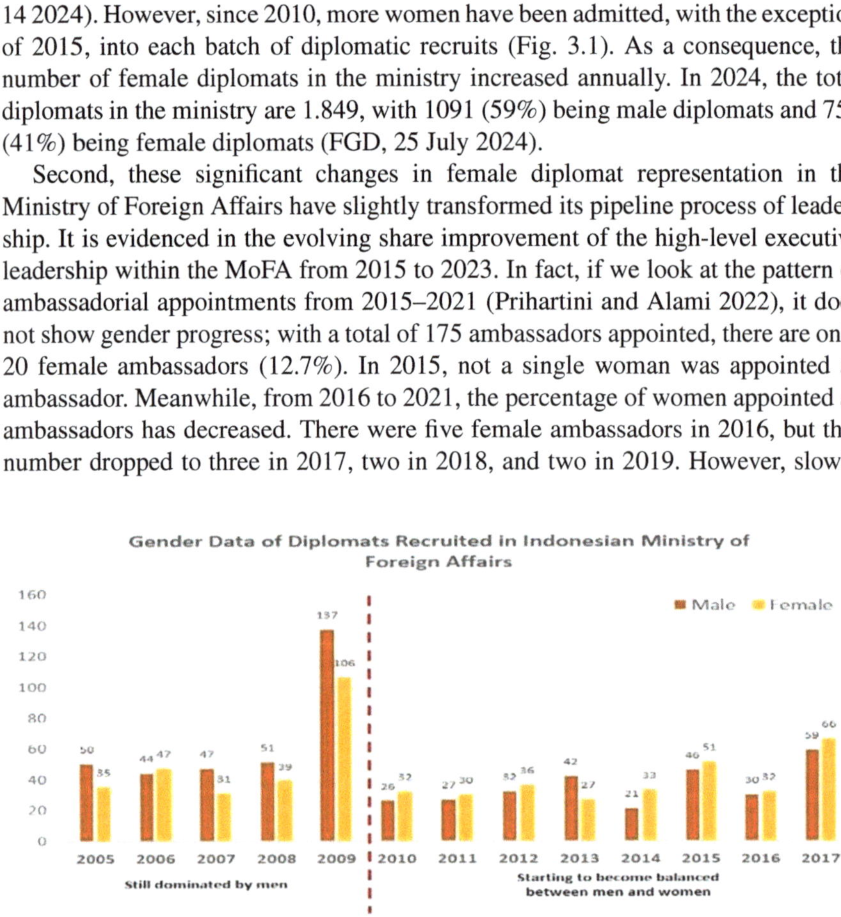

Fig. 3.1 Gender Data of Diplomats Recruited in Indonesian Ministry of Foreign Affairs. 2024. *Source* Data is processed from the Ministry of Foreign Affairs (FGD, July 25, 2024)

progress has been made in that direction from 2020 to 2023, with three, five, and nine female ambassadors.

Even though the trajectory of the ambassador post indicates the ministry's efforts to appoint more women to achieve gender parity, the percentage of female ambassadors is still behind that of male diplomats. Male ambassadors have consistently held a larger proportion of ambassador positions from 2018 to 2023 than female ambassadors, with the former accounting for 9.9% of all ambassadors in 2018, 11% in 2021, 11% in 2022, and 13% in 2023. In 2023, Indonesia was represented by forty-six male ambassadors and nine female ambassadors (Prajuli 2022).

Despite the trajectory of the ambassador post, which signifies the ministry's striving to appoint more to achieve gender balance, the share of women ambassadors is still lagging behind that of male diplomats. Its progress is being achieved, but at a slow pace. From 2018 to 2023, male ambassadors have continuously taken a major share of ambassadors compared to female ambassadors, which is 9.9% in 2018, 11.1% in 2021, 11% in 2022, and 13% in—(Chehab 2023). The composition of Indonesian ambassadors in 2023 consisted of 46 male ambassadors and nine female ambassadors (FGD, June 14 2024). The percentage of women holding ambassadorial positions in Indonesia in 2023 is less than the percentage of women holding ambassadorial positions worldwide (20, 54%), the G20 countries (23%), and several Asian nations, like the Philippines (26%) and Cambodia (25%) (Chehab 2023).

Moreover, according to research, the majority of women appointed as ambassadors tend to be stationed at the Indonesian embassy in nations with unique features, like recently opened diplomatic mission offices and Scandinavian nations with relatively high levels of gender equality. They have never held ambassadorial posts with as much geopolitical importance as those held by the US, China, Germany, Japan, and the UK (Tafonao et al. 2023; Prihartini and Alami 2022). In line with ambassadorial positions, women diplomats also remain underrepresented in high-level management positions in the ministry. In 2023, there were just one and eighteen female diplomats in Echelon I and Echelon II respectively, as opposed to eleven and thirty-one male diplomats in the same positions (FGD, July 14, 2023). Thus, these pictures demonstrated that Indonesia is still a long way away from ensuring gender parity at the top of diplomatic positions.

Third, increased opportunities for women diplomats and representatives in international forums. In addition to displaying its commitment to promoting gender parity, an internal reform of the Ministry of Foreign Affairs has facilitated the advancement of female diplomats by increasing their visibility and involvement in diplomatic arenas (Prajuli et al. 2021). Three young female diplomats at Indonesia's permanent mission to the United Nations, for example, have been assigned to deliver Indonesia's response on West Papuan issues raised by Pacific countries at the UN General Assembly (UNGA). They are Nona Gae Luna in the 70th session of the UNGA, Nara Masista Rakhmatia, and Ainan Nuran in the 72nd session of the UNGA. These numbers of Indonesian women diplomats holding important positions in the country's foreign policy and handling delicate matters like claims of human rights violations in West Papua in the multilateral forums indicate an increasing gender awareness and equality gains in Indonesian foreign policy.

3.2.3 Indonesia's International Role in Advancing Gender Issues

Under President Joko Widodo's administration, Indonesia has demonstrated a stronger global commitment to encouraging and actively participating in women's engagement in foreign policy, particularly with regard to peace and security. Indonesia officially endorsed the Women, Peace, and Security (WPS) 1325 agenda through the localization of the National Action Plan (NAP). In 2014, the government established the national action plan of WPS through the National Action Plans for the Protection and Empowerment of Women and Children during Social Conflicts in 2014–2019 (RAN P3AKS 2014–2019). In RAN P3AKS 2014–2019, the three main pillars of the WPS agenda are prevention, governance, and recovery (protection), and empowerment and engagement (participation). In 2015, the central government recommended ten local governments establish a Regional Action Plan (RAD P3AKS). Indonesia developed a second NAP (2020–2025) with a focus on broader issues like reducing the influence of misinformation and preventing violent extremism. Indonesia's Women, Peace, and Security (WPS) agenda is based on UN Security Council Resolution 1325, which was passed in 2000. This resolution established a framework for women's involvement in the maintenance and promotion of peace and security (Johnson-Freese and True 2021).

Indonesia played a pivotal role in winning the passage of UN Security Council Resolution 2538 on Women in Peacekeeping in 2020, demonstrating its advocacy in support of the global gender-mainstreaming agenda (FGD, July 25 2024). This resolution highlights the important role of women in peacekeeping operations. Indonesia is a significant contributor to UN peacekeeping operations, with a total of 2.715 soldiers, including 185 female peacekeepers in 2024. This is currently making the country the sixth-largest contributor globally (Widyawati 2024). The nation is one of the top 10 nations that committed to sending 4000 soldiers to various UN peacekeeping missions and dedicated to increasing its participation in peace operations (True et al. 2020).

Additionally, Indonesia is dedicated to raising the proportion of female peacekeepers and improving the training program's gender sensitivity. This demonstrates its belief that women are "agents of peace, agents of tolerance, and agents of prosperity" (Marsudi 2019). As agents of peace, women can be powerful in their role in peacekeeping operations. As argued by Indonesia's Foreign Minister Retno Marsudi, "[f]emale peacekeepers possess better situational awareness, can provide comfortable protection of civilians from sexual and gender-based violence, and are more easily acceptable in winning the hearts and minds of the local community" (Marsudi 2019).

As well as promoting gender inclusion in international peacekeeping, Indonesia has advocated for greater women's inclusion in peace negotiation processes. One concrete effort to address this challenge is the Indonesian Ministry of Foreign Affairs initiative to convene regional training on "women, peace, and security" in April 2019 (Marsudi 2019). The aim of this training of young female diplomats from ASEAN countries was to build the capacity of female diplomats as peace negotiators and mediators. To support the Palestinian cause, in the same year, Indonesia also organized training for women entrepreneurs in refugee camps to empower women to achieve sustainable development goals (FGD, July 25 2024). The country also organized numerous capacity-building efforts for Afghan women, including the Dialogue on the Role of Women in Building and Sustaining Peace and the Afghanistan-Indonesia Women's Solidarity Network in 2020 and the Workshop of the International Conference on Afghan Women Education: Education for Her, Progress for All in December 2023. At the regional level, Indonesia also demonstrates its commitment to increasing the role of women in peace negotiation processes. In 2019, Indonesia took a significant step by initiating the Southeast Asian Network of Women Peace Negotiators and Mediators (SEANWPNM) as a network of women negotiators and mediators in Southeast Asia (Indonesian MoFA 2020).

Due to Indonesia's above-mentioned pledges to promote gender equality agendas in its diplomacy efforts, Retno Marsudi's performance and leadership in the Ministry have received highly recognition on a national and international level. At the national level, in 2019 and 2021 she received the Women of the Year Award and the Indonesian Great Women Awards, respectively, for her leadership in the ministry (Hatta 2021). Minister Marsudi has also been awarded Outstanding Leadership in Foreign Diplomacy at the CNN Indonesia Awards 2024 (CNN Indonesia, 2024). At the international level, her breakthroughs in pursuing Agenda 2030 and sustainable development have resulted in Minister Marsudi's achievement as an Agent of Change from the UN Global and Partnership Forum (GPF) in 2017. The Award for Marsudi as Indonesia's Foreign Minister in 2024 further demonstrates the strong regard that Arabian nations have for Indonesia's steadfast leadership in the Palestinian issues. These accomplishments highlight how important it is for women to participate in leadership roles and also recognize their dedication to and role in advancing the gender equality agenda in the field of foreign services.

3.3 Gender-Mainstreaming Implementation: Evidence from the MoFA Representative Office

The Indonesian government under Jokowi has taken the case of Indonesian representative offices in South Korea and Malaysia for its position as a pilot project for a gender-responsive planning and budgeting program (*Perencanaan dan Penganggaran Responsif Gender-PPRG*). By taking the experiences of Indonesia's representative offices in Malaysia, particularly in the Indonesian Embassy in Kuala Lumpur and the Indonesian General Consulate in Johor Bahru, this section aims to investigate the practice of gender mainstreaming in the offices. In doing so, this section draws from Moser and Moser (2005) to investigate the practices through three stages of gender mainstreaming in institutions mentioned in the previous section.

3.3.1 Adopting the Gender-Mainstreaming Concept

The gender-mainstreaming concept involves the integration of gender perspectives and attention to gender issues at all levels of decision-making, program planning, and policy implementation (The United Nations 1997). Despite the widespread yet contested concept, gender mainstreaming has been widely adopted by international organizations, national governments, and civil society actors (True and Mintrom 2001; Hafner-Burton and Pollack 2009). Adopting gender mainstreaming involves articulating institutions' commitment and incorporating gender issues into organizational processes (Prugl and Lustgarden 2006).

It is recognized that there has been some commitment to implement gender-mainstreaming strategies at the Indonesian Embassy in Malaysia. One of the notable commitments is the appointment of Ms. Rossy Verona as the first female Deputy Chief of Mission (DCM) in Malaysia. During the discussion in the embassy, it was argued that the appointment of Rossy Verona indicated a change of mindset among policymakers in Jakarta on who should be in that strategic position.[1] The leadership commitment on gender equality has also been demonstrated through the issuance of the Decision Letter of the Head of the Republic of Indonesia Representative Office for Malaysia in Kuala Lumpur No. 200/RO/12/2021/01 concerning the Establishment of a Gender-Mainstreaming Working Group at the Indonesian Embassy in Kuala Lumpur in 2022, which was signed on December 28, 2021. However, the gender perspective has not been included in the 2020–2024 Indonesian Embassy Strategic Plan document in Kuala Lumpur. Nevertheless, it is understandable since the issuance of the document was in January 2020, which was before the enactment of Minister of Foreign Affairs Regulation Number 21 of 2020 on gender mainstreaming in the Ministry of Foreign Affairs.

Based on the monitoring and evaluation report, the commitment has been socialized to internal stakeholders at the Indonesian Embassy in Kuala Lumpur and its partners. However, this commitment has not been described in detail in the form of implementation guidelines, while the decree for the establishment of the gender-mainstreaming strategy was only issued on December 28, 2021, to provide direction for the implementation of tasks for the gender-mainstreaming working group at the Indonesian Embassy in Kuala Lumpur. The establishment of special units or divisions in institutions that handle women's and gender issues is also another way of adopting the gender-mainstreaming concept. The Indonesian embassy has established a working group for gender mainstreaming, consisting of 10 people, including the Ambassador as the director.

This working group's duties include, but are not limited to, providing its members with capacity-building exercises to increase their understanding of gender issues within the Embassy. It carried out training on gender-mainstreaming budget tagging as well as suitable case handling techniques. Despite all of this progress, gender mainstreaming still encounters challenges because of a lack of awareness of its significance and the diverse backgrounds of its staff, which originally came from the

[1] Focus Group Discussion Data, Malaysia, July 17 (2023).

Ministry of Foreign Affairs as well as other Ministries with various functions such as labor, immigration, and education.

3.3.2 Producing Gender-Mainstreaming Policies

After adopting gender-mainstreaming concepts among staffs in the Indonesian Embassy, the next stage is issuing gender-mainstreaming policies. Following up on the establishment of the working group on gender mainstreaming, the group members prepare a yearly work plan for implementing strategic gender-mainstreaming activities. The members also received budget tagging training on gender-mainstreaming initiatives, as well as technical help on how to become resource persons for related roles at the Indonesian Embassy. They also drafted a policy on gender mainstreaming, created a training plan on interviewing skills and advocacy for managing Indonesian female migrant workers, and created gender-based guidelines for running a shelter for these women. The working group also submits a report on the implementation of gender mainstreaming to the main coordinator of the working group in Jakarta twice a year.

In conducting their activities, they are supported by a general budget that is tagged in other activities or functions. The Embassy does not allocate a dedicated budget for gender-mainstreaming activities, so the activities are part of other general activities carried out by related functions at the Embassy (consular protocols, politics, economy, social-cultural, and education) whose activity targets are in line with the gender-mainstreaming targets (women, children, and other vulnerable groups). Also, there is no minimum percentage for the amount of budget that must be allocated to support gender-mainstreaming policies, depending on the commitment and political will of the head of representatives. Despite the unavailability of gender budgeting, our interview with the Consulate Staff at Johor revealed that this condition, to some extent, creates more flexibility in managing gender-mainstreaming programs.[2]

The Indonesian Embassy in Kuala Lumpur also developed sex-disaggregated data in the form of digital data containing personal data, including gender, in the passport service registration. By having separate data based on sex, this gender-mainstreaming policy could fulfill the needs of female migrant workers who stay at the shelter and provide logistical assistance according to their specific needs. This disaggregated data has been utilized to formulate policies on the activities of the Embassy. However, the 2021 report on the implementation of gender mainstreaming recorded the lack of gender analysis tools that are used in formulating policies in the Embassy. These efforts of gender-mainstreaming policies have highlighted the institutionalization of gender consideration in the Embassy's programs, indicating the tendency of integrationist approaches (Lombardo 2005) that incorporate gender or women's agendas in policies, but it may not challenge the existing frameworks.

[2] Interview Data, Malaysia, 20 July (2023).

3.3.3 Implementing Gender-Mainstreaming Strategies

The last stage of the gender-mainstreaming strategy is the implementation. According to Moser and Moser (2005), gender-mainstreaming implementation can be seen in two aspects: institutional and operational implementation. At the institutional level, implementation means the practice of gender mainstreaming in organizational culture and procedures. It is usually in the form of shared responsibilities among organizational staff and capacity-building programs. Meanwhile, the operational level is about addressing operational outcomes and the impact on gender mainstreaming. One way to handle the operational gender-mainstreaming implementation is by cooperating and collaborating with civil society in gender-mainstreaming policy.

In the experiences of the Indonesian Embassy in Malaysia, these practices are manifested in the context of citizen service and protection of Indonesian citizens in several ways, including case handling, provision of shelter and education services, and community participation. In handling cases, particularly for the distressed migrant workers who run away to the Embassy, the staff prioritize vulnerable groups such as Indonesian domestic migrant workers, including those who enter Malaysia with any improper documentation. They also utilize interview techniques that pay attention to the comfort of the victims and employ female staff to ask sensitive questions.

As part of implementing gender-mainstreaming strategies, the Indonesian representative provides some gender-sensitive facilities, such as nursing rooms and daycare in the Embassy, as well as shelters specifically for women in its Embassy in Kuala Lumpur and a separate shelter for women and men in its Consulate office in Johor. The availability of this separate shelter based on gender is enabled by the construction of a new building in Johor, while it is not possible to do the same in Kuala Lumpur because of the limited budget and infrastructure. In the Consulate office, it also develops quite strict standard operating procedures in the shelter, such as visitors not being allowed to enter without being accompanied by the staff members. Other basic services facilitated by the Embassy include the provision of education facilities, such as the establishment of Indonesian schools and Community Learning Centres (CLC) in some palm oil fields where most Indonesian migrant workers and their families work and live.

Another important effort to implement gender-mainstreaming strategies is by cooperating and collaborating with other stakeholders in doing some gender-related activities. The Indonesian Embassy in Kuala Lumpur has worked with Liber8 and the Malaysian Red Crescent Society to provide free vaccines for Indonesian citizens. The Embassy has also collaborated with some community organizations, such as Alfayat, Aisyiyah, EKUKI, Malino, and AOMI, to conduct social activities and provide access and legal assistance for migrant workers' repatriation to Indonesia.

The practices of the three stages of gender implementation highlighted the commitment of representative offices to gender mainstreaming, which is then manifested in its policies and implementation. Despite its limited budget and human resources, the representative offices have demonstrated their commitment to advancing gender equality in their institutions. Importantly, examining the practices of gender mainstreaming in Indonesia's representative offices highlights how the Ministry of Foreign Affairs has emerged as a gendered institution that has its own institutional gender norms and gendered division of labor (Aggestam and Towns 2019; Nicklasson and Towns 2022).

3.4 Gender-Inclusive Foreign Policy: Key Challenges and Strategies Ahead

While gender-mainstreaming programs and agendas were in place in both policies and practices during the presidency of Jokowi, some challenges remain. Three major concerns may affect the achievement of gender-mainstreaming objectives in the future. These challenges come from the implementation of existing policies and embedded socio-cultural norms in society.

3.4.1 Women's Meaningful Participation Beyond Nominal Representation

The institutionalization process in MoFA for gender mainstreaming has demonstrated progress in terms of gender representation. However, from the placement and promotion aspect, female diplomats remain posted to limited countries or in relatively new diplomatic missions. Since the progress is still low, it affects the overall performance of Indonesia's achievement of gender equality. Based on the Gender Gap Index 2024, Indonesia ranks 100 out of all 146 countries, which is even lower than some Southeast Asian countries such as the Philippines, Singapore, Thailand, Vietnam, and Lao PDR (Pal et al. 2024). Based on this index, while women's participation in the labor force is above average, it is more in professional and technical positions than in strategic roles such as managers, senior officials, or legislators. In the foreign policy realm, the existing representation has not yet manifested in high-level positions, such as ambassador appointments or high-rank structural positions.

The relatively large number of ambassadorial posts being given to political appointees also reduces the opportunities for female career diplomats to achieve an ambassadorial rank. Consequently, it may risk the success of gender-mainstreaming institutionalization efforts because it may lower the chance for female diplomats to have a fair and merit-based competition with other ambassador candidates, especially if there are not many female diplomats to choose from. Therefore, the inclusion of women in foreign policy and international politics requires a transformative change that focuses on improving socialization, increasing awareness of meaningful representation, and giving equal access and opportunities to women through a merit-based system.

The structural challenge within the ministry is how to further increase the representation of women in echelon 1 and echelon 2 positions. The presence of women in high-ranking and strategic positions would open possibilities for greater involvement of women as decision-makers and leaders in the process of foreign policy-making and practices. It should be acknowledged that recently there have been cases where some women have been increasingly involved in the handling of sensitive issues (Prajuli et al. 2021), indicating a recent transformation in MoFA toward a better environment for women in Indonesia's foreign policy. However, the homework is to build a critical mass of capable female diplomats so that these meaningful representations are not only confined to certain women figures who succeed in their career path as diplomats.

3.4.2 Provision of a Support System for Women Diplomats

The nature of a diplomat job is often perceived to be not women-friendly and family-friendly due to frequent job rotation, long working hours, placement abroad, and difficult career paths.[3] Moreover, women are also still burdened with domestic responsibilities. For women to have a career in diplomacy, there are patriarchal situations that cause differences in career progress compared to men in a similar career path. According to the experience of one Indonesian female diplomat, the crucial period to show their potential is after the first posting, as they would return to Indonesia and start positioning themselves in the ministry to build their career path.[4] At the same time, this period is usually also the productive age for having children. There is limited intervention for female diplomats to be able to anticipate the risk of this career gap. This situation explains why, during the selection process for high-ranking positions, there are not enough female diplomats who are considered eligible and registered as options for the Minister to choose. Consequently, male diplomats are easier to gain leadership recognition.

Another issue with the support system for diplomats emerged when they were posted in countries with strict regulations prohibiting diplomats from bringing along their domestic helpers.[5] It has become an issue, particularly for female diplomats who have children since the cost of daycare is usually quite expensive. Moreover, there is no childcare allowance to cover this cost. Consequently, those challenges to building a support system during their assignment abroad may cause difficulties in adapting for female diplomats and thus can affect their performance. In the long run, it may hinder women's potential to obtain strategic posts. Therefore, providing support for female diplomats requires a deep understanding of the benefits of gender-responsive policies for both women and men in MoFA.

[3] Interview with diplomat, July 25 (2024).

[4] Interview with female ambassador, August 28–29 (2023).

[5] Interview with female ambassador, August 28-29 (2023).

3.5 Dealing with Social-Cultural Barriers

Women's empowerment is crucial to pave the way for their ability to make meaningful involvement in foreign policy decision-making processes and practices. Our interviews with female ambassadors (August 2023) find that although women have fought for high-ranking positions, it does not necessarily mean they have the power to influence the policies. The main problem is rooted in socio-cultural challenges, which have shaped certain mindsets about women's role in public and private spaces (Prajuli et al. 2021), which can affect the personal motivation of women to make greater participation in foreign policy.

Therefore, gender socialization at home and in the community is an important part of gender-mainstreaming efforts in everyday life to shape the personal encouragement for women to participate in a political realm that is traditionally considered to be a masculine environment. Socialization from the smallest unit of the community would make women understand their capabilities to be anything, which encouraged many women after political reform in 1998 to enter the Ministry of Foreign Affairs. This is different from what happened before 1998, when the number of women in the Ministry of Foreign Affairs was very small, as there was still a strong view that diplomats are a man's job. This strongly correlates with the ideology of state *ibuism* during the New Order era, which shaped the social construction of womanhood within the patriarchal system in Indonesia (Ida 2001). Therefore, gender socialization should be passed down from generation to generation to be able to break the patriarchy rooted in society.[6]

3.6 Anticipating the Change of Administration: Sustaining Commitments

During the leadership of Retno Marsudi in MoFA, regulations and supporting guidelines and programs have been made to shape the corridor for gender mainstreaming and awareness within the MoFA. This was also supported by initiatives and innovations made by the Gender-Mainstreaming Working Group. Besides, Minister Marsudi has been acknowledged and appreciated nationally and globally for her leadership capabilities and good reputation while serving her role. Recognition of Marsudi's leadership role on gender issues, along with her outstanding performances in other diplomatic affairs such as the Palestine issues (Indonesian MoFA 2024), has demonstrated that female leadership matters for both advancing gender-mainstreaming agendas and serving as an exemplary model of female capabilities to perform as a competent diplomat in the traditionally male-dominated field of foreign affairs.

Now, the question is whether the next leadership in the ministry would find the importance of maintaining and even advancing the progress and legacies of

[6] Interview with IR scholars, August 6 (2024).

Indonesia's first female foreign minister. Meanwhile, the leadership in MoFA is greatly determined by the country's presidential leadership. Idiosyncratic factors in Indonesia's leadership are instrumental in directing the orientation and priorities of Indonesia's foreign policy (FGD, July 19 2024).

On the one hand, some may have confidence that Retno's policies on gender mainstreaming have set a quite strong foundation for continuing the implementation of gender mainstreaming in MoFA regardless of changes in administration or within Kemlu themselves (FGD, July 25 2024). This may be supported by the fact that the composition of female diplomats in the ministry is continuously rising to the extent of a comparable ratio with the male diplomats. On the other hand, while gender socialization has improved the practices of gender mainstreaming within MoFA, both in the headquarters and representative offices abroad, gender norms have not been fully institutionalized in the MoFA. In fact, besides the existing regulation, the socio-cultural transformation within MoFA is also a crucial factor that would ensure the continuation of the gender-mainstreaming agenda in Indonesia's foreign policy regardless of the changing administration. However, there remains misunderstanding and hesitance to consider gender as a concept that may involve the active participation of both men and women. Moreover, some scholars also see the potential for a more masculine approach from the successor of Jokowi's presidency.[7]

To anticipate the continuation of the gender-mainstreaming agenda in Indonesia's foreign policy, some strategies should be adopted. The first strategy relates to the efforts to strengthen the institutional framework. There are four institutional aspects that are crucial in maintaining the sustainability of gender mainstreaming in MoFA: strategic plans, monitoring and evaluation mechanisms, allocated budget, and reward system. A strategic plan can be designed in the form of a clear and targeted road map that becomes a guideline for MoFA in keeping the gender-mainstreaming agenda on track and making progress. Furthermore, the road map should be equipped with a more effective monitoring and evaluation mechanism beyond the submission of regular reports. In the Philippines, for example, the Department of Foreign Affairs holds a regular evaluation meeting yearly, which invites all Gender and Development (GAD) focal points in each representative office abroad. This kind of meeting is not only aimed at monitoring the progress of the GAD plan and projects but could serve as an avenue for knowledge sharing of best practices, better communication, and coordination among the GAD focal points.

In terms of gender-responsive budgeting, there needs to be political will at the national level to regulate the budget allocation of national and local governments to support the gender-mainstreaming agenda at all levels. As explained in the practices of gender mainstreaming in Indonesia's representative offices in Malaysia, the implementation of gender-related programs utilizes the existing budget resources by tagging system due to the absence of specific allocation. While the tagging budget system provides flexibility to some extent, there remain concerns about sustainability in the long run. Therefore, the designation of a certain percentage of the total government budget for the gender-mainstreaming agenda will help technical ministries and

[7] Interview with IR scholars, August 6 (2024).

the units within them to be able to run more optimal, targeted, and measurable gender-mainstreaming programs. Indonesia could learn from the Philippines, which regulates all departments and agencies to allocate a minimum of 5% of their total annual budget for financing gender programs.

Furthermore, the availability of financial resources for the gender-mainstreaming agenda would open the room for an innovative approach to encouraging parepartments and agencies to give their best commitments to the programs, such as by creating a reward system. Acknowledgment and rewarding contributions of representative offices and units at MoFA that have succeeded in meeting all the gender indicators during the implementation of gender-mainstreaming strategies is expected to motivate them to put in more effort and thus help sustain the execution of gender-mainstreaming plans and projects.

Besides institutional strategies, MoFA should design a strategy to improve human resources and advance gender equality awareness. There must be innovative ways to bring gender mainstreaming closer to more MoFA staff. During the leadership of Retno Marsudi, there have been some innovations that can be potentially developed as gender-mainstreaming flagship programs. The first innovation can be done by making the sisterhood project a more institutionalized and structured program. Recently, the Sisterhood in MoFA remains a fluid and informal forum that depends on the leadership figure of a senior female diplomat as the initiator. In fact, the Sisterhood has the potential to increase gender awareness through its activities and projects.

Another initiative is the mentoring program, which connects senior diplomats with both male and female junior diplomats, as well as between female senior and junior diplomats. The mentorship is not only an attempt to bridge the hierarchy and seniority between male–female and senior–junior diplomats but also provides opportunities for widening networks among female diplomats.

Furthermore, gender socialization programs and activities should also engage more male colleagues for a better understanding of the benefits of gender equality for both men and women in MoFA and in the conduct of foreign policy. Socializing gender equality should not rely on ad hoc events or temporary programs, but it should be more regular and institutionalized. Integrating gender in the diplomat educational curriculum, for example, is not only carried out in the junior diplomatic training (Sekolah Dinas Luar Negeri/Sekdilu) but also for every career diplomat educational level at mid-career diplomatic training (Sekolah Staf Dinas Luar Negeri/Sesdilu) and the senior diplomatic course (Sekolah Staf dan Pimpinan Kementerian Luar Negeri/Sesparlu).

3.7 Conclusion

This chapter has demonstrated policies and practices of gender-inclusive foreign policy during the presidency of Jokowi, highlighting the Ministry of Foreign Affairs as a gendered institution. During Jokowi's administration, the gender-mainstreaming agenda in the MoFA became more significant. The appointment of Retno Marsudi as the first female foreign minister (2014–2024) plays a significant role in the advancement and inclusivity of gender mainstreaming in MoFA. Not only her high-ranking position in MoFA, but Retno also makes use of her identity as a woman to promote gender-related visions and programs.

Retno's concerns on gender reforms in MOFA are not only at the headquarters in Jakarta but also among Indonesia's representatives abroad. During her leadership, Retno introduced a pilot project on gender mainstreaming to be implemented by Indonesia's representatives in its embassy in Seoul and its consulate general in Johor Bahru. Furthermore, Indonesia is also active in the global promotion and advancement of gender mainstreaming.

However, challenges in advancing and promoting the gender equality agenda, such as limited meaningful representation, an inadequate support system for female diplomats, and socio-cultural obstacles, should be addressed. Neglecting these challenges may affect the successful implementation of domestic gender mainstreaming and our rank in the Gender Gap Index. Specifically, women's representation in foreign policy is part of political empowerment, in which Indonesia ranks the lowest among other indicators in the Global Gender Gap Report 2024.

To conclude, ensuring the sustainability of the gender-mainstreaming agenda despite the changing administration is an important task. Sustainability is not only on internal MoFA but also on maintaining Indonesia's role in the global forum by utilizing gender issues as a new arena for greater global engagement.

References

Aggestam, Karin, and Ann E. Towns. 2019. The Gender Turn in Diplomacy: A New Research Agenda'. *International Feminist Journal of Politics* 21 (1): 9–28.

Alami, A. N. 2018. The Implementation of Gender Mainstreaming in ASEAN. *IKAT: The Indonesian Journal of Southeast Asian Studies* 1(2):185–207.

Chehab, S. 2023. *Women in Diplomacy Index, Research Analysis.* Anwar Gargash Diplomatic Academy, 1–17.

CNN Indonesia. 2024. Menlu Retno Marsudi Raih Outstanding Leadership in Foreign Diplomacy. Retrieved from August 16, 2024, from https://www.cnnindonesia.com/internasional/20240814183113-106-1133089/menlu-retno-marsudi-raih-outstanding-leadership-in-foreign-diplomacy.

Enloe, C. 1990. *Bananas, Beaches, and. Bases: Making Feminist Sense of International Politics.* University of California Press, xiii + 244

Hafner-Burton, E. M., and M. A. Pollack. 2009. Mainstreaming Gender in the European Union: Getting the Incentives Right. *Comparative European Politics* 7: 114–138.

Hatta, R. T. 2021. Retno Marsudi Raih Penghargaan Khusus dalam Anugerah Perempuan Hebat Indonesia 2021. Liputan 6. Retrieved from August 16, 2024, from https://www.liputan6.com/global/read/4538301/retno-marsudi-raih-penghargaan-khusus-dalam-anugerah-perempuan-hebat-indonesia-2021.

Ida, R. 2001. The Construction of Gender Identity in Indonesia: Between Cultural Norms, Economic Implications, and State Formation. *Masyarakat, Kebudayaan Dan Politik, XIV* 1: 21–34.

Indonesian Ministry of Foreign Affairs (MoFA). 2020. Indonesia Initiates the Establishment of the Southeast Asian Network of Women Peace Negotiators and Mediators (SEANWPNM) to Advance the Agenda of Women, Peace, and Security in the Region. Retrieved from July

30, 2024 from https://kemlu.go.id/portal/en/read/2019/berita/indonesia-initiates-the-establishment-of-the-southeast-asian-network-of-women-peace-negotiators-and-mediators-seanwpnm-to-advance-the-agenda-of-women-peace-and-security-in-the-region.

Indonesian Ministry of Foreign Affairs (MoFA). 2024. Indonesian Foreign Minister Receives Award from Arab Countries for Consistency in Supporting Palestine. Retrieved from August 16, 2024, from https://kemlu.go.id/portal/en/read/5843/berita/indonesian-foreign-minister-receives-award-from-arab-countries-for-consistency-in-supporting-palestine.

Johnson-Freese., J and J. True. 2021. Women, Peace, and Security: Competitive Edge for Australia and the US in the Indo-Pacific. *Pacific Forum, Issues and Insights. Working Paper* 7 (21):1–12.

Komisi Nasional Anti Kekerasan Terhadap Perempuan (Komnas Perempuan). 2019. *Laporan Independen Lembaga Nasional Hak Asasi Manusia tentang 25 Tahun Pelaksanaan Kesepakatan Global Beijing Platform for Action (BPFA+25) di Indonesia.* Komnas Perempuan, vii+112

Kompas. 2015. Kebijakan Baru kemenlu: Sesama Diplomat RI Kini Boleh Menikah. Retrieved from July 15, 2024 from https://nasional.kompas.com/read/2015/01/21/21375721/Kebijakan.Baru.

Lombardo, E. 2005. Integrating or Setting the Agenda?Gender Mainstreaming dalam the European Constitution-Making Process. *Social Politics: International Studies in Gender, State, and Society* 12 (3): 412–432.

Marsudi, R. L. P. 2019. Creating a Global Network of Female Peacekeepers. *The Jakarta Post.* Retrieved from July 15, 2024 from https://www.thejakartapost.com/academia/2019/04/09/creating-global-network-of-female-peacekeepers.html.

Moser, C., and A. Moser. 2005. Gender Mainstreaming since Beijing: A Review of Success and Limitations in International Institutions. *Gender & Development*, 13 (2):11–22.

Niklasson, B., and A. E. Towns. 2022. Introduction: Approaching Gender and Ministries of Foreign Affairs. *The Hague Journal of Diplomacy* 17: 1–31.

Pal, K. K., Piaget, K., and Zahidi, S. 2024. Global Gender Gap Report 2024, the World Economic Forum. Retrieved from July 2, 2024 from https://www.weforum.org/publications/global-gender gap-report-2024/.

Prajuli, W. A., R. V. Yustikaningrum, and D. N. Amurwanti. 2021. How Gender Socialization is Improving Women's Representation in Indonesia's FOREIGN AFFAIRS: Breaking the Ceiling. *Australian Journal of International Affairs* 75 (5): 527–545. https://doi.org/10.1080/10357718.2021.1893653.

Prajuli, W. A. 2022. Data 2015–2021: Pengangkatan duta besar di Indonesia belum setara gender. *The Conversati*on. Retrieved FROM July 5, 2024 from https://theconversation.com/data-2015-2021-pengangkatan-duta-besar-di-indonesia-belum-setara-gender-173663.

Prihartini, E. S., and N. A. Alami. 2022. Data Bicara: Keterwakilan perempuan Indonesia sebagai duta besar kurang dari 6%, strategi pengarusutamaan gender perlu diperkuat. *The Conversation*:1–5. Retrieved from June 18, 2024 from https://theconversation.com/data-bicara-keterwakilan-perempuan-indonesia-sebagai-duta-besar-kurang-dari-6-strategi-pengarusutamaan-gender-perlu-diperkuat-197033.

Prugl, E., and A. Lustgarten. 2006. Mainstreaming Gender in International Organizations. In *Women and Gender Equity in Development Theory and Practice: Institutions, Resources, and Mobilization*, eds. Jane S. Jaquette and Gale Summerfield, 53–70. Duke University Press.

Tafonao, W., S. Gea, and Z. Erma. 2023. Legal Review Regarding Appointment of Ambassador in Indonesia Based on Gender Equality. *Legal Brief (*Institute (Institut Hukum Sumberdaya Alam) 12 (1):137–147. https://doi.org/10.35335/legal137.

True, J., I. H. Gayatri, and Veronika. 2020. Indonesia's UN Security Council Drive for Inclusive Peace and Security. *The Interpreter.* Retrieved from June 30, 2024 from https://www.lowyinstitute.org/the-interpreter/indonesia-s-un-security-council-drive-inclusive-peace-security#:~:text=Indonesia's%20Foreign%20Minister%20Retno%20Marsudi%20argued%20in%20a%20Jakarta%20Post,minds%20of%20the%20local%20community.

True, J., and M. Mintrom. 2001. Transnational Networks and Policy Diffusion: The Case of Gender Mainstreaming. *Oxford University Press* 45 (1): 27–57. https://doi.org/10.1111/0020-8833.00181.

United Nations Human Rights Office on the High Commissioner. 1979. Convention on Elimination of All Forms of Discrimination Against Women, New York. Retrieved from July 20, 2024 from https://www.ohchr.org/en/instruments-mechanisms/instruments/convention-elimination-all-forms-discrimination-against-women.

United Nations. 1997. *Report of the Economic and Social Council for 1997*, A/52/3.

UN Women. 1995. *Beijing Declaration and Platform for Action*. United Nations. Retrieved from July 15, 2024 from https://www.un.org/womenwatch/daw/beijing/pdf/BDPfA%20E.pdf.

Widyawati, S. 2024. United Nations Thanks to Indonesia for its Contribution to Peacekeeping, United Nations Press Release, 11 July 2024. Retrieved June 5, 2024 from https://indonesia.un.org/en/273728-united-nations-thanks-indonesia-its-contribution-peacekeeping#:~:text=Today%2C%20Indonesia%20provides%202%2C715%20uniformed,currently%20serving%20in%20peace%20operations.

Wuryandari, G. 2009. Politik Luar Negeri Indonesia dan Restrukturisasi Kelembagaan. In *Ganewati Wuryandari*, ed. Format Baru Politik Luar Negeri. Indonesia, 61–104. Lembaga Ilmu Pengetahuan Indonesia.

Focus Group Discussions (FGD)

FGD Jakarta, September 21, 2022
FGD Jakarta, August 30, 2023
FGD Jakarta, June 14, 2024.
FGD, Jakarta, July 19, 2024.
FGD Jakarta, July 25, 2024.
FGD Malaysia, July 17, 2023.

Interview

Interview Data, Malaysia, July 20, 2023.
Interview with female ambassadors, August 28–29, 2023.
Interview with diplomat, July 25, 2024.
Interview with IR scholars, August 6, 2024.

Ganewati Wuryandari is a researcher at the Research Center for Politics, National Research and Innovation Agency (BRIN) in Indonesia. She received her PhD in Asian Studies from the University of Western Australia in Perth in 2006 and a research professorship in the field of international relations in 2022. She has been involved in the ASEAN Regional Forum (ARF) since 2020 as an Expert and Eminent Persons (EEP) of the Republic of Indonesia and is now active as a Secretary General of the International Federation of Social Sciences Organization (IFSSO) and a member of the BRIN Social Humanities Ethics Clearance. Her areas of interest in academia are international relations, particularly as it relates to Indonesian foreign policy, Indo-Pacific, gender and security, and border issues. "Indonesia's engagement in Climate Change Negotiations: Building National Resilience" (in Uday Chatterjee and Rajib Shaw, et al., *Climate Change, Community Response, and Resilience*, Elsevier, 2023), *Indo-Pasifik dalam Politik Luar Negeri Indonesia*: D*imensi Geopolitik, Geostrategi & Geoekonomi* (Andi, 2024), and *Politik Luar Negeri Indonesia Terhadap Tiongkok* (BRIN Press, 2024) are some of her most recent academic works.

Athiqah Nur Alami is a researcher at the Research Center for Politics, the National Research and Innovation Agency (Badan Riset dan Inovasi Nasional-BRIN), Indonesia. She completed her PhD from the Department of Southeast Asian Studies, National University of Singapore. Her main interests are Indonesia's foreign policy, labor migration, and gender in international relations, particularly to apply a gender lens in analyzing both migration and foreign policy and diplomacy issues. She has been part of the research team on Indonesia's foreign policy for the last ten years. Her team has produced several publications, including a book on "Foreign Policy and Energy Security Issues in Indonesia" in Springer in 2017. Currently, Athiqah and her team have been conducting research on Indonesia in the Indo-Pacific, gender in Indonesia's foreign policy, and education for displaced children.

R. A. Rizka Fiani Prabaningtyas is a researcher at the Research Centre for Politics, National Research and Innovation Agency (Badan Riset dan Inovasi Nasional/BRIN), Indonesia. She graduated from the Gadjah Mada University, Indonesia, with a bachelor's degree in international relations. She obtained her master in International Relations (Advanced) from the Australian National University in 2018. In BRIN, she is a member of two research teams, namely Indonesia's foreign policy and forced migration. Her research interest includes international relations, particularly Indonesia's foreign policy, forced migration, and gender in IR. Currently, her research discusses issues on refugees in Indonesia, education for displaced children, gender in Indonesia's foreign policy, and Indonesia's foreign policy on the Indo-Pacific.

Chapter 4
Evaluating Maritime Vision in Jokowi's Foreign Policy: The Implementation of Global Maritime Fulcrum Vision

Lidya Christin Sinaga and Tonny Dian Effendi

Abstract As an evaluation of a decade of President Joko Widodo's foreign policy, this article specifically examines how the Global Maritime Fulcrum (GMF) vision is implemented as Indonesia's foreign policy. As known, in the era of President Joko Widodo, Indonesia for the first time had the white paper on Indonesian Maritime Policy and on Maritime Diplomacy. Therefore, in evaluating the extent to which the maritime vision, namely the GMF, has been embodied in the Jokowi's foreign policy, we can measure from two indicators, which are the idea level and the practical implementation level. First, the idea level is the extent to which Indonesian maritime diplomacy is able to make it the "center" in the context of leadership in the maritime sector in the region. Referring to this definition, there are several diplomatic achievements, for example in IORA or in ASEAN in the form of the ASEAN Outlook on Indo-Pacific. The idea level includes the aspect of "recognition" from other countries and foreign institutions. Second, the practical implementation level means to what extent Indonesia implements the targets of its maritime diplomacy. As known, Indonesian maritime diplomacy in the GMF scheme views the importance of inter-regional connectivity or maritime connectivity, both from a domestic perspective and Indonesia's regional/global needs in its geostrategic position in the region. In this context, the presence of International Hub Port (IHP) is relevant and important because its presence basically shows the existence of global aspirations in the GMF vision by developing world class seaports which are expected to become important international navigation traffic axes. By the presence of these ports, Indonesia can gain the greatest possible benefits from global trade shipping activities that pass through Indonesian waters. However, this paper argues that while Indonesia has succeeded in strengthening its maritime vision at the level of ideas, but from the practical level, it is still hampered by domestic obstacles, both from political and economic aspects. This contrast between international and domestic achievements within the GMF framework underscores a paradox: Indonesia has

L. C. Sinaga (✉)
Research Center for Politics, The National Research and Innovation Agency, Indonesia (BRIN), Jakarta, Indonesia

T. D. Effendi
The University of Muhammadiyah Malang, Malang, Indonesia

I. Kartini and Kanisa (eds.), *Joko Widodo's Foreign Policy and Diplomatic Legacy*,
https://doi.org/10.1007/978-981-95-2085-5_4

made visible progress on the international stage, yet domestic maritime initiatives face slower or even stagnant progress.

Keywords Global maritime fulcrum · Foreign policy · Indonesia · Joko Widodo · ASEAN

4.1 Introduction

When Joko Widodo assumed office as the seventh President of the Republic of Indonesia in October 2014, a notable and significant development was the introduction of a new vision termed the Global Maritime Fulcrum (GMF). This vision is particularly noteworthy as President Joko Widodo is the first leader to explicitly define Indonesia as a maritime nation (Liow and Shekhar 2014). Historically, while Indonesia initiated the Djuanda Declaration that established the framework for an archipelagic state, the focus of development has predominantly favored land-based initiatives. Consequently, the emergence of the GMF vision under President Joko Widodo represents a renewed aspiration for a development strategy centered on maritime principles. Furthermore, it is essential to highlight that the GMF was articulated within the framework of Indonesia's foreign policy from its inception. On November 13, 2014, during the East Asia Summit (EAS) held in Naypyidaw, Myanmar, President Joko Widodo unveiled the five pillars of the GMF, which encompass maritime culture, marine resource management, maritime connectivity and infrastructure development, maritime defense capabilities, and maritime diplomacy. According to Nelson and Sulaiman (2015), this vision has the potential to serve as "the core of President Joko Widodo's foreign policy leadership." Following the 2014 EAS Summit, this initiative evolved into a maritime agenda that President Joko Widodo actively promoted, both domestically and in multilateral forums, to attract investment for the enhancement of maritime infrastructure and connectivity. At the 2014 APEC Meeting, President Widodo formally presented his five-year plan, which included the construction of 24 seaports and deep-sea ports across Indonesia, such as Kuala Tanjung Port (Sumatra Island), Tanjung Priok (Java Island/Jakarta), Tanjung Emas (Java Island), Makassar (Sulawesi Island), and Bitung (Sulawesi Island). President Jokowi's intention to maintain the policies of the preceding administration is evident, particularly in the initiative to develop International Hub Ports, specifically Kuala Tanjung and Bitung. Numerous analyses have been conducted regarding the GMF, particularly concerning Indonesian maritime diplomacy during President Widodo's tenure. These studies predominantly concentrate on the resolution of maritime boundary disputes (Pratomo and Kwik 2020; Andika 2017), Indonesia's maritime diplomacy through the Indian Ocean Rim Association or ASEAN (Indrayani et al. 2019; Supriyanto 2017), and the nation's power to realize the global maritime fulcrum (Saragih et al. 2018). However, there is a noticeable lack of analysis regarding maritime diplomacy in the context of maritime connectivity, particularly concerning the commitment of the Joko Widodo administration to the development of the International Hub Port (IHP). The significance of the IHP warrants discussion, as its establishment fundamentally reflects the global aspirations embedded in the GMF vision, aimed at developing international standard seaports that are anticipated to serve as crucial axes for international navigation traffic. The "global" dimension of maritime connectivity development pertains to how these ports enable Indonesia to maximize benefits from global trade shipping activities traversing its waters. This chapter commences by outlining how Indonesia's maritime diplomacy within the GMF framework perceives the importance of this initiative.

As an assessment of a decade of foreign policy under President Joko Widodo, this paper focuses specifically on the implementation of the GMF vision as Indonesia's maritime strategy. In evaluating the extent to which the maritime vision, namely the GMF, has been embodied in the Jokowi's foreign policy, we can measure from two indicators, which are the idea level and the practical implementation level. First, the idea level is the extent to which Indonesian maritime diplomacy is able to make it the "center" in the context of leadership in the maritime sector in the region. Second, the practical implementation level means to what extent Indonesia implements the targets of its maritime diplomacy. In this case, this paper focuses on evaluating the achievement of welfare and connectivity target for Indonesian maritime diplomacy.

As known, Indonesian maritime diplomacy in the GMF scheme views the importance of inter-regional connectivity or maritime connectivity, both from a domestic perspective and Indonesia's regional/global needs in its geostrategic position in the region. In this context, the presence of IHP is relevant and important because its presence basically shows the existence of global aspirations in the GMF vision by developing world class seaports which are expected to become important international navigation traffic axes. By the presence of these ports, Indonesia can gain the greatest possible benefits from global trade shipping activities that pass through Indonesian waters. However, this paper argues that while Indonesia has succeeded in strengthening its maritime vision in ASEAN at the level of ideas and has received recognition from other countries and foreign institutions, but from the practical level, it is still hampered by domestic obstacles, both from political and economic aspects, which means that this decade is not enough for Indonesia to gain momentum to improve the national economy from the maritime sector, especially maritime infrastructure and connectivity.

4.2 Two-Level Evaluations of the Global Maritime Fulcrum

In 1988, Robert D. Putnam developed a model to explain the negotiation process in the development and practice of diplomacy. Rather than being trapped in the debate about which one is more dominant or starting earlier between domestic and international politics, Putnam proposes two processes of negotiation that coincide and are related to each other. For Putnam, domestic and international are strongly linked and interdependent. Therefore, in explaining the state's behaviors, Putnam underlines the critical roles of the theory, integrating both the domestic factors that influence international affairs and vice versa. He argues:

> The politics of many international negotiations can be conceived as a two-level game. At the national level, domestic groups pursue their interests by pressuring the government to adopt favorable policies, and politicians seek power by constructing coalitions among those groups. At the international level, national governments seek to maximize their ability to satisfy domestic pressures while minimizing the adverse consequences of foreign developments. Neither of the two games can be ignored by central decision-makers, so long as their countries remain interdependent, yet sovereign (Putnam 1988).

Putnam developed the two-level games model as a metaphor for domestic-international interactions. According to him, a national political leader has to deal with two games. In the "Level I" at the international game, he discusses with his diplomats and foreign policy advisors, while in the "Level II" domestic game, he is surrounded by party members, parliamentary members, spokespersons for domestic agencies, interest groups, political advisors, and others. In these two games, the leader must negotiate with different groups: the international and domestic audiences.

Although he acknowledges the equal importance of the negotiation at each level, Putnam emphasizes the critical role of domestic politics in determining the win-set in negotiation. In domestic politics, the power and preferences of the significant actors provide principles, frameworks, alternatives, and evaluations critical to determining negotiation strategies in international politics. According to Putnam, those domestic actors could be bureaucratic agencies, interest groups, social classes, and public opinion. Regarding public opinion, the positive public perception of foreign policy supports the government in negotiations and diplomacy. In other words, the two-level theory seems focused on explaining Level I by exploring the dynamic at Level II.

Furthermore, some scholars use this theory to explain foreign policy by exploring the negotiation in domestic politics. Lisowski (2002) argues that California's energy crisis and Byrd-Hagel Senate Resolution became critical factors behind President Bush's hawkish decision to repudiate the Kyoto Protocol. Chung (2007) explains that political preferences, economic priorities, risk assessment, historical memories, and societal institutional constraints are among several factors that explain China's bargaining behavior in border disputes. In the case of the Anglo-Irish Peace process case, Trumbore (1998) argues that the public's preference, the issue's intensity, and the public's ratification power all determine public opinion that influences the negotiation. Public opinion at domestic also a critical factor influencing Israel's negotiation on the Israeli–Palestinian conflict (Shamir and Shukaki 2005). In China, the Chinese Communist Party uses the two-level games to encourage both support and participation of the domestic public in the public diplomacy agenda (Fallon & Smith 2022). Based on the situation in Iran, Bjola, and Manor propose the concept of domestic digital diplomacy, where the government uses social media to influence negotiation at the domestic level (Bjola & Manor 2018). The concept of two-level games explains the narration in Iran's domestic politics that influence its diplomacy with the United Kingdom (Kaussler 2011). In the case of Japan, Schoppa expands Putnam's approach by enlarging elite involvement in the two-level game negotiations (Schoppa 1993).

Since the two-level game theory links domestic and foreign policy, it also provides insights into policy evaluation at home and abroad. The theory inspires a more nuanced and comprehensive evaluation since it highlights the interplay between domestic and international factors. When foreign policy and diplomacy are strongly linked to negotiation at home, it implies that the evaluation of the negotiation abroad is also linked to the evaluation of bargaining activity at home. Thus, the two-level game theory insights encourage us to consider how the GMF is evaluated at the international and domestic levels.

This study assesses the GMF from different but linked levels by evaluating the policy's implementation in diplomacy and infrastructure development. The GMF is a vision encompassing domestic and foreign policy. GMF has five pillars: maritime diplomacy represents its foreign policy vision, and the other four pillars focus on domestic maritime development. Since the vision was declared, Indonesia has been actively engaged with maritime diplomacy at regional and international forums. At the same time, it focuses, among others, on maritime infrastructure and connectivity at the domestic level. Therefore, a two-level evaluation provides a more comprehensive and holistic understanding of the GMF's progress.

4.3 Level I Evaluation: Maritime Diplomacy

As a vision, the GMF provides direction for foreign policy and domestic developments centered on Indonesia's maritime identity. In 2017, the Indonesian government established official documents that detail the visions, projections, plans, and guidelines of the GMF through Indonesia's Maritime Policy. In this policy, maritime diplomacy is one of the five pillars of the GMF. Furthermore, the government established the Maritime Diplomacy White Paper in 2019. However, maritime diplomacy began before these two policy documents were established.

Indonesia's maritime diplomacy under Widodo's presidency is conducted at bilateral, regional, and international levels. At the bilateral level, Indonesia signed about 31 maritime cooperations (joint statement, letter of intent, and memorandum of understanding) with some countries from 2015 to 2021. These cooperations encompass many areas of maritime with Southeast Asia (Malaysia, Philippines, Singapore), East Asia (China, Japan, South Korea), Europe (Denmark, France, Germany, Netherlands, Norway, Poland), Australia, India, the United Arab Emirates, the United Kingdom, the United States, and international organization (UNDP). In the 2014 annual speech, Foreign Minister Retno Marsudi stated that in the past nine years, Indonesia had solved six border agreements with neighboring countries, including ZEE delimiting borders with Vietnam, the Indonesia–Malaysia agreement of maritime border at Sulawesi Sea, and the southern Malacca Strait (MoFA 2024).

At the regional level, Indonesia actively initiates and promotes maritime cooperation among the countries in the Indian Ocean, Southeast Asia, and the Pacific region. In the Indian Ocean, Indonesia co-founded the Indian Ocean Rim Association (IORA). Within this organization it plays a critical role in Fisheries Management and the Working Group on the Blue Economy. In 2017, Indonesia hosted the IORA Leader Submit and successfully initiated the Jakarta Concord and IORA Action Plan 2017–2021, while it also served as chair in 2022 and as Secretary General in 2024. Moreover, Indonesian diplomats hold a crucial position at the Blue Economy and Fisheries Management and Director for Public Diplomacy and Communication. In 2016, with some maritime countries, Indonesia had the 1st Maritime Local Government Network International Conference, resulting in the Wakatobi declaration (Coral Triangle Initiative 2016). Indonesia also engaged with the Pacific Island Forum as the post-forum dialogue partner, where in 2019, the organization established the Declaration on Preserving Maritime Zone in the Face of Climate Change-related Sea Level Issues (MOFA 2019; Pacific Island Forum 2021). In 2023, when Indonesia chaired ASEAN, it successfully launched the ASEAN Maritime Outlook (AMO). This document emphasizes the region's strategic importance due to its geographic location at the crossroads of intercontinental and interoceanic trade and incorporates Indonesia's maritime vision into the regional framework. Consequently, the AMO underscores Indonesia's regional leadership and illustrates its efforts to advance the region in line with its GMF vision.

At the international level, Indonesia engages with the pivotal maritime multilateral cooperation organization, the International Maritime Organisation (IMO). Indonesia joined the organization in 1961, held the position of External Auditor (2020–2023), and has been a continuous member of the IMO Council since 1983. Moreover, Indonesia also facilitated inter-regional maritime cooperation. In 2017, Indonesia initiated the Archipelagic and Island States (AIS) Forum, and in 2018, it hosted the first AIS ministerial meeting, resulting in the Manado Joint Declaration. Indonesia hosted four of five ministerial meetings, and in 2023, it hosted the AIS Forum (Indonesia Baik) (Fig. 4.1).

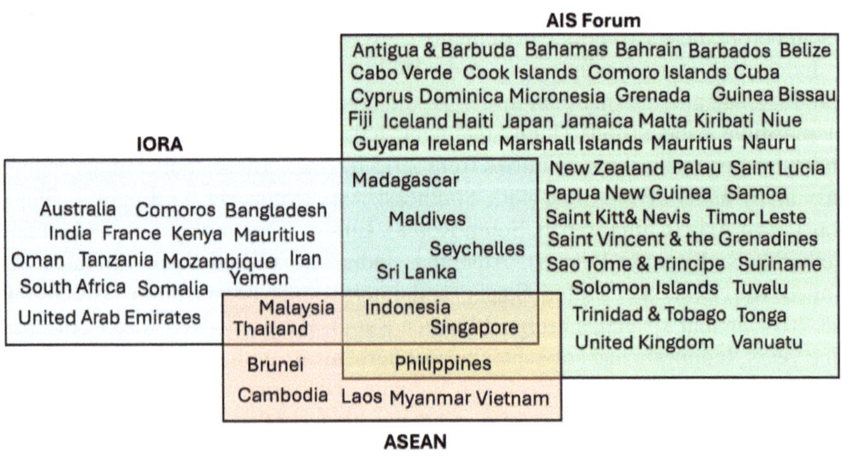

Fig. 4.1 The member states of IORA, ASEAN, and AIS Forum

Indonesia's maritime diplomacy in bilateral and multilateral relations illustrates some critical points. First, Indonesia is a vital norm builder in maritime cooperation. It initiated some agreements, such as the Wakatobi Declaration of Coral Triangle Initiative in 2016, the Jakarta Concord of IORA in 2017, the Manado Declaration of AIS Forum in 2018, and the Maritime Outlook of ASEAN in 2023. Second, Indonesia plays a critical role as a facilitator of regional maritime cooperation. It hosted meetings, conferences, and summits on maritime cooperations, particularly for IORA and AIS. Third, Indonesia bridges inter-regional cooperations in maritime. Indonesia is both the initiator and administrator of the AIS Forum, where the members of this organization are cross-regional. It encompasses Europe (the United Kingdom, Ireland, and Cyprus), Africa (Guinea-Bissau and Madagascar), South Asia (Sri Lanka, Maldives), Southeast Asia (Indonesia, Singapore, and the Philippines), Caribbean, South America (Suriname, Guyana, Trinidad and Tobago) and Pacific and Oceanian countries. Moreover, Indonesia's significant position in ASEAN, IORA, and AIS Forum provides an opportunity to enhance its leadership in regional maritime cooperation while pursuing its maritime national interest as the fulcrum of global maritime. The positive trend of Indonesia's diplomatic influence is captured by the Lowy Institute which reported that in 2023 and 2024, its diplomatic influence is the highest among five other ASEAN member states (Fig. 4.2).

However, there are two critical points besides some of the achievements of the maritime diplomacy of Indonesia. First, it seems more regional rather than global fulcrum. While the GMF positioned Indonesia as the center of global maritime affairs, the last ten years of maritime diplomacy achievements are primarily at the regional level. This situation may reflect two possibilities. First, there is inconsistency in

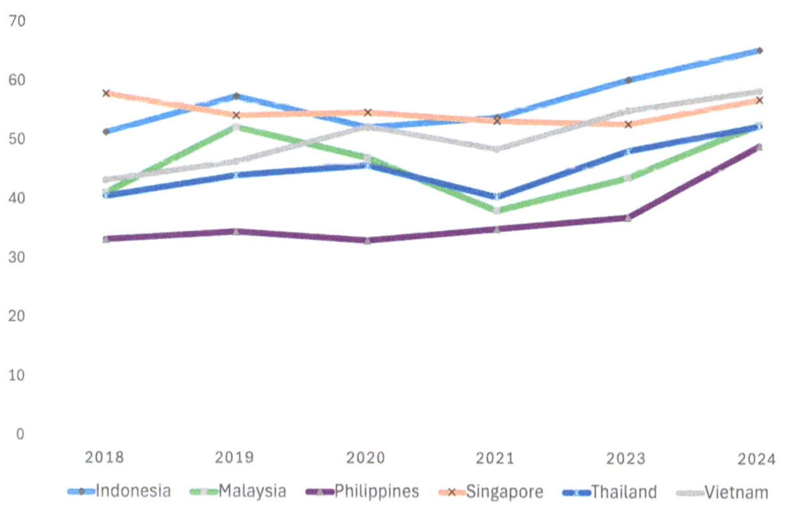

Fig. 4.2 Diplomatic influence of five ASEAN member states, 2018–2024 *Source* Lowy Institute (2024)

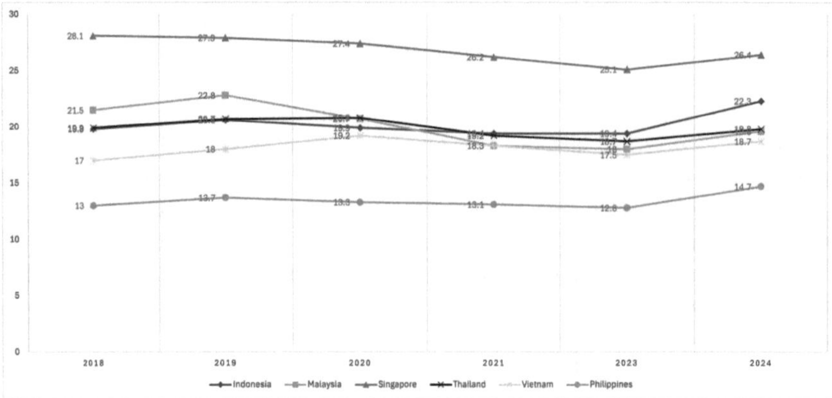

Fig. 4.3 The comprehensive power of five ASEAN member states, 2018–2024 *Source* Lowy Institute (2024)

the focus on maritime vision. The achievement mostly happened in the first term of Widodo's administration. Like some scholars and critics, the GMF was slightly sound in the second term of Widodo's administration, raising questions about the commitment to maritime development, including maritime diplomacy. GMF in the second term of Second, the vision is limited by Indonesia's maritime capability. Indonesia's position as the middle power may reflect its influence on the regional rather than a global level. Lowy Institute reports Indonesia's comprehensive power is grouped as the middle power, and from 2018 to 2024, the trend of Indonesia's comprehensive power increased, even passing Malaysia in 2022. However, it is still below Singapore (Fig. 4.3).

Second, is the diplomatic achievement in line with maritime security? An interesting situation is when we compare maritime diplomatic achievement with maritime security: military development and Illegal Unreported Unregulated Fishing (IUUF). Regarding military development, Indonesia has the highest score for the long-range maritime force projection compared to other ASEAN countries. However, this score stagnated from 2020 to 2023 and decreased in 2024 (Fig. 4.3). In the IUUF case, the Global Initiative Against Transnational Organized Crime reported that Indonesia experienced increasing score trends in 2023 compared to 2019, indicating a decreasing performance (Fig. 4.4). The report shows that in 2023, Indonesia ranked 6 of 152 countries, 3 of 20 Asian countries, and even 2 of 9 East Indian Ocean countries on IUUF. This contradictive situation shows that the maritime diplomacy achievement is not linear with the IUUF (Fig. 4.5).

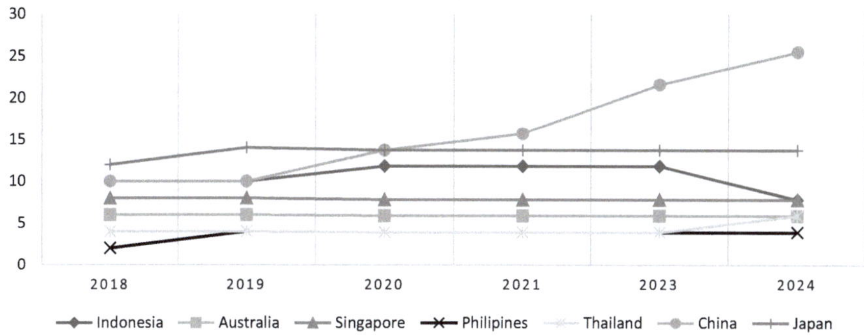

Fig. 4.4 Long-Range Maritime Force Projection of Five ASEAN Member States, China, Japan, and Australia, 2018–2024 *Source* Lowy Institute (2024)

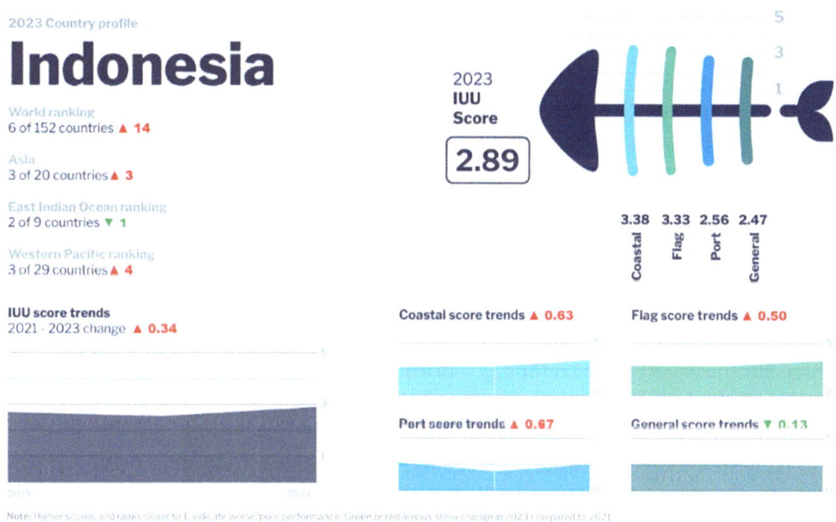

Fig. 4.5 Indonesia's Country Profile for IUU Fishing Index Risk, 2023 *Source* The Global Initiative Against Transnational Organized Crime (2023)

4.4 Level II Evaluation: Maritime Infrastructure and Connectivity

Achievements at the diplomatic level, as outlined in the preceding section, warrant a comparison with assessments at the domestic implementation level. A critical area for evaluation pertains to the execution of maritime connectivity and infrastructure development within the framework of the Global Maritime Fulcrum vision.

Since its inception, the "sea toll" program has emerged as a cornerstone of this maritime connectivity initiative. This program aims to establish effective sea connectivity through regular and scheduled shipping routes from West to East Indonesia. Guided by the principle that "ship promotes trade," the program seeks not only to mitigate price disparities between Indonesia's western and eastern regions but also to demonstrate the state's commitment to reaching remote, underdeveloped, and border areas (T3P). Consequently, while it was launched within the context of Indonesian foreign policy, its initial implementation was predominantly focused on domestic needs.

Nevertheless, the advancement of the "sea toll" initiative, which embodies the maritime connectivity development pillar of the GMF vision during Jokowi's administration, has yet to yield significant regional or international effects. The "sea toll" program is intended to enhance national connectivity, facilitating links between domestic ports and global trade routes. However, designated ports as regional and global hubs have not fully realized their potential functions. This situation has been acknowledged, prompting proactive measures such as the Decree of the Minister of Transportation No. 901 of 2016. This decree was established to prepare for the implementation of the ASEAN Economic Community in early 2016, which is expected to boost regional trade volume, including increased ship visits and maritime freight to Indonesian ports, and outlines the development of international routes such as Dumai-Malacca (Malaysia), Belawan-Penang, and Bitung-General Santos. The government has thus far focused solely on developing the infrastructure and physical facilities of the port, neglecting the cargo aspect, which results in a lack of sustainability considerations.

The Indonesian Maritime Policy, established in 2017, has fundamentally highlighted the necessity of developing maritime infrastructure, particularly ports that serve as trade hubs, alongside the provision of international standard transportation facilities and supporting services for trade activities in Indonesia's peripheral regions, with a focus on Eastern Indonesia. Bitung and Sorong have been identified for development as significant hub ports in Eastern Indonesia, while Saumlaki is noted for requiring enhancement due to its considerable geostrategic importance. Prior to the issuance of this KKI, the government enacted Presidential Regulation (Perpres) No. 3 of 2016, which pertains to the acceleration of National Strategic Projects, serving as a framework for hastening the development of international hub ports. This regulation specifically mentions the advancement of the Kuala Tanjung international port and the Bitung international hub port. These initiatives aim to integrate Indonesia into the global logistics system and contribute to the realization of the nation's vision of becoming a maritime fulcrum. Additionally, the white paper on maritime diplomacy, ratified in 2019, asserts that through national development acceleration programs, Indonesia seeks to responsibly broaden access to international contributions, thereby enhancing connectivity and facilitating smoother distribution and transportation within the country. Consequently, an evaluation of the development of the International Hub Port (IHP) in Indonesia, particularly in Bitung and Kuala Tanjung, is deemed pertinent.

The government's designation of Kuala Tanjung appears to be more contentious compared to Bitung Port. A careful examination of Presidential Decree 3/2016 reveals a distinction in the classification of Kuala Tanjung as an international port, while Bitung is referred to as an international hub port. Notably, Presidential Decree No. 32/2011, which pertains to the Master Plan for the Acceleration and Expansion of Indonesia's Economic Development (MP3EI), identifies Kuala Tanjung as the international hub for Indonesia's Western Region, whereas Bitung serves this role for the Eastern Region. Subsequently, Presidential Decree No. 26/2012, which outlines the Blueprint for the National Logistics System Development Policy, further solidified the status of Kuala Tanjung Port, granting it the additional designation of an international hub for the western region. The Minister of Transportation's Decree Number 414 of 2013 established two primary ports as international hubs: Kuala Tanjung Port in Batubara Regency, North Sumatra, and Bitung Port in Bitung City, North Sulawesi. However, the introduction of Presidential Decree No. 3 of 2016 has altered the status of Kuala Tanjung Port. The subsequent Decree of the Minister of Transportation KP 901 of 2016, which supersedes KP 414/2013, designates Tanjung Priok Port as the international container hub port.

The selection of Kuala Tanjung and Bitung as IHP is indeed justified when considering their geographical positioning. These ports are situated in the outermost regions, in contrast to Tanjung Priok and Tanjung Perak Ports, which are located more inland. Furthermore, Kuala Tanjung Port's proximity to the Malacca Strait, a vital trade route linking Southeast Asia with Europe and the Middle East, enhances its strategic importance. In terms of capacity, Kuala Tanjung Port can handle up to 60 million TEUs (twenty-foot equivalent units) annually, making it the largest port in Western Indonesia, surpassing Tanjung Priok Port (Pelni.co.id 2022). Consequently, it is anticipated that this port will attract significant international cargo traffic within the Malacca Strait. The Kuala Tanjung Multipurpose Terminal (KTMT) commenced operations in 2019 (Pelni.co.id 2022), although it had been operational since October 2018, accommodating various international vessels. Notably, several cruise ships have docked at Kuala Tanjung Port, bringing tourists from Malaysia, Singapore, and Australia, who have visited Lake Toba and Medan, the capital of North Sumatra province. Additionally, in terms of export activities, the Taipei-based Wan Hai Line vessel, with a capacity of 50,000 tons, has made two visits to Kuala Tanjung Port to transport regional products to China on December 27, 2018, it carried 250 TEUs of CPO derivative products, and on February 14, 2019, it transported 261 TEUs of the same products. However, the volume of activities conducted thus far does not reflect the port's substantial capacity. The level of port services at Kuala Tanjung remains relatively low, which is not commensurate with the significant investments made in its construction (Ombudsman.go.id 2023).

Bitung Port serves as a crucial entry point for trade in the Pacific Ocean. It is anticipated that Bitung Port will emerge as a key access point for the Philippines and East Asia, given the significantly reduced distance to various destination ports when utilizing Bitung (refer to Table 4.1). Nevertheless, the full potential of Bitung Port remains hindered by certain limitations.

Table 4.1 Comparison of travel distance from Bitung port and other ports to destination countries

Bitung Port	Port of Destination Country						
	Darwin	Kaohsiung	Hong Kong	Shanghai	Busan	Tokyo	Los Angeles
Via Tanjung Priok Port and Singapore (Nautical Mile)		3526	3365	4142	4408	3429	9574
Bitung Direct (Nautical Mile)	1094	1346	1423	1901	2113	2220	6651
Distance Difference		2180	1942	2241	2295	1209	2923
Duration Difference		346,46	336,11	349,11	351,46	364,24	378,76
Day Difference		14,44	14,00	14,55	14,64	15,18	15,78

Source Salim, Z., Mychelisda, E., Rahmayanti, A.Z. (2015). Bitung sebagai Pelabuhan Hub Internasional. *Warta Pengkajian Perdagangan*, III (8)

It is evident that geographical factors alone do not account for the disparities in port capacity relative to operational activities. Several domestic factors contribute to the sluggish development of IHP as Indonesia's gateway to becoming the Global Maritime Fulcrum.

Firstly, there is a notable absence of economic hubs in proximity to IHP. According to the *Sislognas*, the main port's location should ideally be integrated with economic hubs (such as special economic zones and industrial areas), Logistics Hubs, and International Port Hubs. The alternative main ports identified in the *Sislognas* include Sabang, Belawan, Kuala Tanjung, Batam, Jakarta, Surabaya, Banjarmasin, Balikpapan, Makasar, Bitung, Kupang, Sorong, and Biak. However, the lack of adequately developed economic hubs remains a significant obstacle to the optimal functioning of IHP in facilitating global connectivity. The establishment of these economic hubs is crucial, as it directly impacts sustainability and operational regularity. The sustainability aspect hinges on the availability of cargo to be transported; if the hub port can facilitate the movement of goods to and from export/import destinations, international trade routes will remain active. Consequently, this sustainability influences regularity, ensuring that vessels operate on a consistent schedule. This regularity, in turn, affects shipping costs; for instance, transporting goods from Batam to Singapore may be more expensive than from Singapore to Tanjung Priok, despite the latter being a longer distance, due to the established regularity of the Singapore-Tanjung Priok route. As noted by a participant in the Focus Group Discussion/FGD (September 13, 2019): "It is unrealistic to expect that connecting the eastern region by land will automatically stimulate industrial growth; industry must be established first, followed by the necessary infrastructure." This perspective is particularly valid, given that 80% of the industrial sector remains concentrated on the island of Java (Jaya 2019).

The implementation of the agreement to enhance international crossings along the Dumai-Malacca (Malaysia), Belawan-Penang (Malaysia), and Bitung-General Santos (Philippines) routes has unfolded as follows. The Indonesian government focused solely on the development of port infrastructure and physical facilities, neglecting the cargo aspect, which undermines the sustainability of the initiative. Nevertheless, there exists significant potential for regional connectivity development, contingent upon the presence of mutually beneficial interests among the involved countries (Interview 2019). These interests are primarily linked to trade activities (cargo) and the movement of individuals, particularly in the tourism sector. This remains a considerable challenge for Indonesia. The stagnation observed on the Bitung-Davao route represents a significant setback for Indonesia, highlighting a lack of preparedness regarding the types of goods to be traded. This route is integral to the ASEAN Connectivity Master Plan and the BIMP EAGA Blueprint. On April 30, 2017, President Joko Widodo and President Rodrigo Duterte officially inaugurated the Ro-Ro (roll on-roll off) shipping service, marking the commencement of the Davao-Bitung trade route at Kudos Port in Davao. A subsequent welcoming ceremony for the Ro-Ro vessel took place at Bitung Port on May 2, 2017. However, following the inaugural voyage, no additional trips occurred due to insufficient export cargo (Kompas, 13 July 2019). This situation is believed to stem from the fact that Bitung and Davao essentially offer similar products (Transcript of FGD 2019). This raises significant concerns, especially considering the long-standing relationship established between Bitung and Davao since the initiation of sister city cooperation in 1993, under the leadership of Mayor S.H. Sarundajang in Bitung and Duterte in Davao (bumn.go.id 2019). The issue of overlapping commodities should have been addressed from the outset.

On May 17, 2019, Olly Dondokambey, the Governor of North Sulawesi, announced plans to reinstate the Bitung-Davao route by introducing an extended route that includes Davao-Bitung-Ho Chi Minh (Vietnam)-Labuan (Malaysia)-Bitung-Ternate-Davao. Subsequently, on July 11, 2019, the MV Baltic Summer, a vessel with a capacity of 250 TEU that had previously serviced the Bitung-Davao route, arrived at Bitung Port after departing from Labuan, East Malaysia. However, this vessel did not transport any goods to the Philippines due to prohibitively high shipping costs (Kompas, 12 July 2019). In August 2023, during a visit by Philippine Consul General Lolita B Capco to Governor Olly Dondokambey, discussions were held regarding enhancing maritime connectivity between Bitung and Davao (Antaranews 2023). It was not until February 2024 that the North Sulawesi Provincial Government celebrated the inaugural voyage of the SITC Container Line cargo ship MV. SITC Batangas Voy.2403 N, which carried various agricultural and fishery products from North Sulawesi intended for export to several East Asian nations, including Japan, South Korea, and China. The Bitung Special Economic Zone (KEK) has been in place since May 2014, established under Government Regulation Number 32 of 2014. This initiative was proposed by the North Sulawesi Provincial Government and is closely

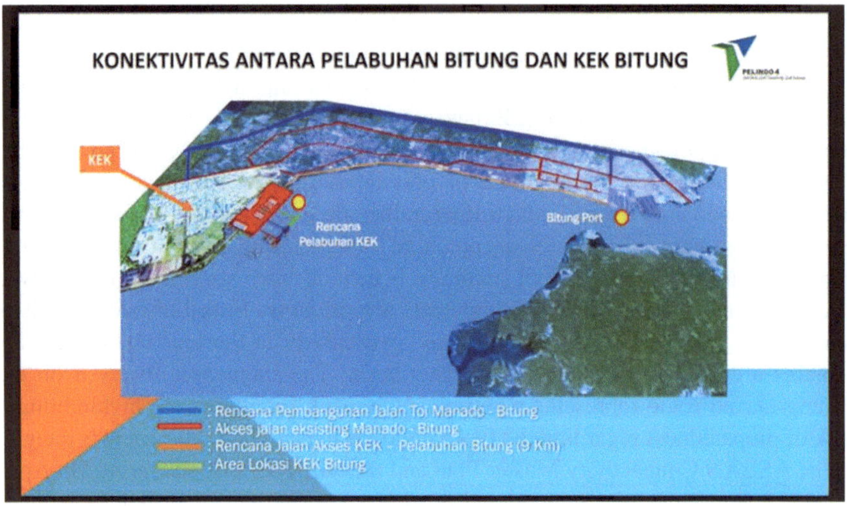

Fig. 4.6 Connectivity between Port Bitung and Bitung Special Economic Zone. *Source* Administration Office of Bitung SEZ

linked to the geoeconomic and geostrategic advantages of the Bitung KEK's location. The geoeconomic benefits include its strategic position as a hub for growth and a distribution center for goods and logistics in eastern Indonesia, international access particularly to the BIMP-EAGA, the Australia-Indonesia Development Area (AIDA), East Asia, and the Pacific, as well as its proximity to the IHP development plan (Fig. 4.6).

A memorandum of understanding (MoU) has been executed between the North Sulawesi Provincial Government and China Communications Construction Company (CCCC), the parent entity of CRBC. Additionally, on June 14, 2019, a memorandum of agreement (MoA) was established between PT Membangun Sulut Hebat (MSH), the manager of the Bitung Special Economic Zone and a regional government-owned enterprise (BUMD) of the North Sulawesi Provincial Government, and China Road and Bridge Corporation (CRBC). The Bitung Special Economic Zone is designated to occupy an area of 534 hectares, with potential expansion up to 2500 hectares, focusing on core industries such as coconut processing and its derivatives, fish and seafood processing, as well as the pharmaceutical sector and logistics. Despite being officially operational since April 1, 2019, and the signing of an agreement by China to develop the Bitung Special Economic Zone, land acquisition remains a significant challenge hindering the swift realization of the Special Economic Zone.

The second issue pertains to the lack of coherent government policies. As previously highlighted in the discussion regarding the ongoing debate over the status of Kuala Tanjung port within the framework of the Minister of Transportation's policies, this reflects a deficiency in the consistency of policy development at the governmental level, leading to the emergence of disjointed policies that fail to provide adequate support. Following the implementation of Presidential Regulation No. 81/2018 aimed at accelerating the development of Kuala Tanjung, the Coordinating Minister for Maritime Affairs, Luhut Binsar Pandjaitan, in February 2019, designated seven ports to serve as international hubs. These ports include Belawan/Kuala Tanjung (North Sumatra), Tanjung Priok (DKI Jakarta), Kijing (West Kalimantan), Tanjung Perak (East Java), Makassar (South Sulawesi), Bitung (North Sulawesi), and Sorong (West Papua). Luhut's vision is to establish an integrated seaport system through these seven hubs, which will subsequently connect with the domestic sea toll route. Cargo from smaller ports will be transported to these hub ports via feeder ships before being shipped directly to their international destinations. Luhut aims to diminish reliance on Singapore, which has traditionally served as a transshipment port for Indonesian exports, by offering more competitive transshipment rates than those available in Singapore (Azka 2019). However, it is crucial for Indonesia to recognize that a direct comparison with Singapore is not feasible (Transcript of Interview 2019). Singapore excels in providing value-added services, consistently enhancing its capacity and service quality each year, while Indonesia continues to grapple with fundamental port requirements, such as outdated cranes. The perspective of the private sector regarding the initiative to establish these seven international hubs is particularly noteworthy.

> Hub ports serve as critical gateways for exports, imports, and international transhipment. Competing with established ports such as Singapore and Tanjung Pelepas presents significant challenges due to geographical disadvantages, unless a port is developed in Batam. Although the government has proposed the establishment of seven hub ports, it is improbable that these will achieve international status, as major shipping lines tend to favor Singapore, Tanjung Pelepas, or Laem Chabang. If we consider the logistics of these seven hub ports, one must question the fuel requirements for transportation and the intended destinations. Singapore, Tanjung Pelepas, and Laem Chabang primarily serve their surrounding regions. It is important to note that international hub ports do not primarily cater to the cargo of the host country; for instance, only 20 percent of the cargo handled in Singapore is destined for Singapore itself, while the remaining 80 percent is for other nations, including the Philippines, Brunei, Vietnam, and Cambodia. If we aim to establish an international hub port in Bitung, we must consider the distribution channels for the goods. Will the focus be on the Philippines? Indonesia's potential lies in becoming a gateway hub for export and import cargo. (Transcript of FGD 2019)

The evolving government policies that do not reference the National Logistics System contribute to a lack of clarity regarding the government's focus and priorities, while simultaneously raising concerns about its dedication to the development of an IHP. For instance, in the case of the Kuala Tanjung IHP, the concurrent development of the Belawan New Port suggests that the Kuala Tanjung IHP may struggle to compete with this established port. In the eastern region of Indonesia, while awaiting the operational launch of the Bitung IHP, the government is revitalizing several key ports, including Makassar Port. Since December 2015, under the management of PT. Pelindo IV, Makassar Port has been facilitating direct exports, allowing shipments to reach their destination countries without the need to transit through Tanjung Perak Port in Surabaya, Tanjung Emas Port in Semarang, or Tanjung Priok Port in Jakarta. This development is particularly significant given the absence of public ports in eastern Indonesia capable of handling direct export or import activities (Wangke 2019). However, the initiative to establish Makassar New Port as an IHP in eastern Indonesia raises significant questions about the future of the Bitung IHP, especially as both ports aim to attract cargo from across Sulawesi and its neighboring regions. The ambiguity in government policy inevitably fosters competition among these major ports.

Lastly, the complexity associated with the "Old Routes, Old Players" presents a significant challenge in the advancement of the IHP. A pivotal question in the development of IHP is how to rejuvenate IHP Kuala Tanjung and Bitung in light of the established appeal of Tanjung Priok and Tanjung Perak Ports, which benefit from their long-standing routes and proximity to industrial hubs. On paper, IHP Kuala Tanjung and Bitung appear to be at a disadvantage compared to these two prominent ports. This raises concerns regarding the willingness of Main Liner Operators (MLOs) to invest in Bitung or Kuala Tanjung, or to relocate their operations and containers from Surabaya or Tanjung Priok to these ports. Acknowledging the realities of the shipping industry reveals that infrastructure alone is not the sole concern; despite being viable in theory, practical implementation remains challenging. Moreover, it is essential to recognize that established players in the market may feel threatened by the emergence of new hub ports. A private sector representative shared insights during a focus group discussion, stating:

> When we initiated the Bitung-General Santos route for exports to the West Coast, we encountered competition from the existing Bitung-Surabaya-Singapore-West Coast route. They were willing to reduce prices to the point that our new route became unviable, only to subsequently increase their prices again. Recently, there was an effort to revive the Bitung-Davao route, initially aimed at General Santos, but it did not gain significant traction. I consistently advise the Philippine Ambassador in Indonesia and the Coordinating Minister to engage with Tanto and Temas first, extending offers to them before considering new entrants. If they are not included in the discussions, they may perceive a threat and could potentially withdraw from the competition. (Transcript of FGD 2019)

The presence of competing ports in proximity to IHP presents a significant challenge. It is well established that 90 percent of international trade occurs via maritime routes, with 40 percent of this trade transiting through Southeast Asian waters. This raises the question of Indonesia's share in this trade, a query complicated by the existence of ports such as Tanjung Pelepas, Singapore, and Laem Chabang. While Batam has the potential to serve as a significant hub for cargo, its growth is hindered by Singapore's dominance.

To address this issue, it is essential to foster collaboration between the private sector's shipping operations and the government's trade promotion initiatives. However, it is important to acknowledge that the previously adopted paradigm of "ships follow the trade" has resulted in disparities, not only in port infrastructure but also in pricing. This approach suggests that shipping companies will transport goods to locations where they are available, often disregarding the adequacy of port facilities. This trend is evident both domestically and internationally; for instance, despite Vietnam and Cambodia lacking advanced port facilities, they still attract large vessels for cargo transport. This situation invites some criticism of the government, as significant investments have been made in port development without corresponding increases in shipping activity (Transcript of FGD 2019).

In response to the disparities in port infrastructure and pricing, the government has shifted its focus toward an Indonesia-centric approach. This strategy aims to implement the "ships promote trade" model, which seeks to establish new shipping and trade routes. Consequently, the government's encouragement of this paradigm should be viewed as a necessary complement to the existing "ships follow the trade" model, which remains crucial for the private sector's operations.

4.5 Conclusion: A Paradox of the GMF

The Global Maritime Fulcrum (GMF) is fundamentally rooted in reviving Indonesia's identity as a maritime nation. As a national doctrine, it embodies both domestic and international ambitions. Maritime diplomacy is central to Indonesia's international vision, positioning the country as a significant global player in maritime affairs. Domestically, connectivity is key, inspiring initiatives like the sea toll and the development of international port hubs. To fully assess Jokowi's maritime vision under the GMF, it's essential to consider both domestic and international contexts. Here, Putnam's concept of the "Two-Level Game" offers valuable insight into the interaction between domestic and international dynamics, though this study only partially applies it as a framework to emphasize the need for a dual-level evaluation of the GMF.

While Indonesia has made strides in maritime diplomacy, the situation for domestic maritime development is more complex. Internationally, Indonesia has shown leadership as a regional maritime norm-setter by initiating frameworks like the Jakarta Concord, Wakatobi Declaration, and ASEAN Maritime Outlook. It has also positioned itself as a prominent actor in maritime cooperation. However, significant domestic challenges remain, such as the lack of an economic hub near the Indonesian Hub Ports (IHP), incoherence policies, and limited appeal of the new IHPs compared to established ports. Establishing the IHP as the backbone of maritime connectivity requires a holistic approach encompassing social, cultural, economic, and political dimensions. Although infrastructure may be developed swiftly, garnering community support, fostering a shifting paradigm from "ships follow the trade" to "ships promote the trade," and achieving effective collaboration among government agencies all pose substantial challenges. These obstacles highlight that human and ideational factors often prove more difficult to address than material infrastructure issues. This contrast between international and domestic achievements within the GMF framework underscores a paradox: Indonesia has made visible progress on the international stage, yet domestic maritime initiatives face slower or even stagnant progress. A comprehensive evaluation of Jokowi's maritime vision in foreign policy requires considering each pillar of the GMF. Future studies may consider these questions: How are marine and human resource development, maritime security, law enforcement, ocean governance, economic development, spatial management, and maritime culture aligned with maritime diplomacy? Does maritime diplomacy under the GMF represent and support these other pillars, and are they, in turn, reinforcing diplomatic efforts? Furthermore, the paradox presented in this study leads to an essential question: How has Indonesia achieved notable international success in maritime diplomacy amid significant domestic challenges?

References

Andika, M. T. 2017. Indonesia Border Diplomacy Under The Global Maritime Fulcrum. *Ritsumeikan International Affairs* 15: 1.

Antaranews. 2023. *Gubernur Sulut-Konjen Filipina Membahas Konektivitas Bitung dan Davao.* Retrieved October 20, 2024, from https://www.antaranews.com/berita/3670674/gubernur-sulut-konjen-filipina-membahas-konektivitas-bitung-dan-davao.

Bjola, C., and I. Manor. 2018. Revisiting Putnam's two-level game theory in the digital age: Domestic digital diplomacy and the Iran nuclear deal. *Cambridge Review of International Affairs* 31 (1): 3–32. https://doi.org/10.1080/09557571.2018.1476836.

BUMN.GO.ID. 2019. *Pengusaha Sambut Rute Manado-Davao City.* Retrieved October 19, 2024, from http://bumn.go.id/angkasapura1/berita/0-Pengusaha-Sambut-Rute-Manado--Davao-City.

Chung, C. P. 2007. Resolving China's Island Disputes: A Two-Level Game Analysis. *Journal of Chinese Political Science* 12 (1): 49–70. https://doi.org/10.1007/s11366-007-9001-7.

Fallon, T., and N. Smith. 2022. The Two-Level Game of China's Public Diplomacy Efforts. In *China's International Communication and Relationship Building*, ed. Xiaoling Zhang and Corey Schultz, 63–78. Routledge.

Indrayani, I., C.K.B. Pratita, and P. Halim2019. Indonesian Maritime Diplomacy, Through The Indian Ocean Rim Association, In Realising The Goal To Become The World Maritime Axis. *International Journal of Innovation.*

Jaya, W. K. 2019. *Sea Toll as a Catalyst for Equal Development and Welfare.* The National Seminar "Towards Sustainable Maritime Connectivity in Indonesia". Jakarta: ERIA, Ministry of Transportation, and LPEM FEB Universitas Indonesia.

Jaringan Dokumentasi dan Informasi Hukum, Kementerian Kementerian Perhubungan Republik Indonesia. Keputusan Menteri Perhubungan Nomor: KP 414 2013 Tentang Rencana Induk pelabuhan Nasional. Retrieved October 19, 2024, from http://jdih.dephub.go.id/assets/uudocs/kepmen/2013/KP_414_TAHUN_2013.pdf.

Jaringan Dokumentasi dan Informasi Hukum, Kementerian Kementerian Perhubungan Republik Indonesia. Keputusan Menteri Perhubungan Republik Indonesia Nomor KP 901 Tahun 2016 tentang Rencana Induk Pelabuhan Nasional. Retrieved October 19, 2024, from https://supplychainindonesia.com/wp-content/files/KM_KP_901_Tahun_2016.pdf.

Kaussler, B. 2011. British-Iranian Relations, The Satanic Verses and the Fatwa: A Case of Two-Level Game Diplomacy. *British Journal of Middle Eastern Studies* 38 (2): 203–225. https://doi.org/10.1080/13530194.2011.581819.

Kompas. 2019a. *Disiapkan, Insentif di Jalur Bitung-Davao.*

Kompas. 2019b. *Ekspor-Impor dari Bitung Perlu Solusi.*

Liow, J. C., and V. Shekhar 2014. Indonesia as a Maritime Power: Jokowi's Vision, Strategies, and Obstacles Ahead. *Brookings.* Retrieved October 18, 2024, from https://www.brookings.edu/articles/indonesia-as-a-maritime-power-jokowis-vision-strategies-and-obstacles-ahead/.

Lisowski, M. 2002. Playing the Two-level Game: US President Bush's Decision to Repudiate the Kyoto Protocol. *Environmental Politics* 11 (4): 101–119. https://doi.org/10.1080/714000641.

Nelson, B., and Y. Sulaiman. 2015. The Implications of Jokowi's Global Maritime Axis. *Strategic Review.* Retrieved September 19, 2024, from https://www.stratfor.com/the-hub/implications-jokowis-global-maritime-axis.

Ombudsman.go.id. 2023. *Layanan Pelabuhan Kuala Tanjung Sepi, Kepala Ombudsman Sumut: Prihatin dan Sia-sia Besarnya Investasi Untuk Pembangunan.* Retrieved October 28, 2024, from https://ombudsman.go.id/perwakilan/news/r/pwkmedia--layanan-pelabuhan-kuala-tanjung-sepi-kepala-ombudsman-sumut-prihatin-dan-sia-sia-besarnya-investasi-untuk-pembangunan-.
Pelni.co.id. 2022. *Terminal Point Kuala Tanjung.* Retrieved October 28, 2024, from https://agencies.pelni.co.id/berita/detailberita/terminal-point-kuala-tanjung
Peraturan Presiden Republik Indonesia Nomor 3 Tahun 2016 Tentang Percepatan Pelaksanaan Proyek Strategis. Retrieved October 19, 2024, from https://www.djppr.kemenkeu.go.id/uploads/files/Perpres%203%202016.pdf.
Peraturan Presiden Republik Indonesia No.32 Tahun 2011 Tentang Masterplan Percepatan dan Perluasan Pembangunan Ekonomi Indonesia. Kementerian Koordinator Bidang Perekonomian.
Peraturan Presiden Republik Indonesia No.26 Tahun 2012 Tentang Cetak Biru Pengembangan Sistem Logistik Nasional. Kementerian Koordinator Bidang Perekonomian.
Pratomo, E., and J. Kwik. 2020. *Good agreements make good neighbours: Settlements on maritime boundary disputes in South East Asia.* Marine Policy: Elsevier.
Putnam, R. D. 1988. Diplomacy and Domestic Politics: The Logic of Two-Level Games. International Organization, 42(3), 427–460. https://www.jstor.org/stable/2706785.
Salim, Z., E. Mychelisda, and A. Z. Rahmayanti. 2015. Bitung sebagai Pelabuhan Hub Internasional. *Warta Pengkajian Perdagangan* 3 (8): 1.
Saragih, H. M., R. Siregar, and E.A.B. 2018. Indonesian national power to achieve the global maritime fulcrum. In *Proceedings of the 5th International Conference on Community Development (AMCA 2018).* https://www.atlantis-press.com/proceedings/amca-18/25901695.
Schoppa, L. J. 1993. Two-level games and bargaining outcomes: Why gaiatsu succeeds in Japan in some cases but not others. *International Organization* 47 (3): 353–386. https://doi.org/10.1017/S0020818300027995.
Shamir, J., and K. Shukaki. 2005. Public Opinion in the Israeli-Palestinian Two Level Game. *Journal of Peace Research* 42 (3): 311–328. https://doi.org/10.1177/0022343305052014.
Supriyanto, R. A. 2017. The Indian Ocean and Indonesia's Global Maritime Fulcrum: Relevance to ASEAN. In *ASEAN and the Indian Ocean: The Key Maritime Links*, ed. S. Bateman, R. Gamage, and J. Chan, 52–61. S Rajaratnam School of International Studies.
Transcript of FGD. Jakarta: September 13, 2019.
Transcript of Interview. Singapore: August 14, 2019.
Trumbore, P. F. 1998. Public Opinion as a Domestic Constraint in International Negotiations: Two-Level Games in the Anglo-Irish Peace Process. *International Studies Quarterly* 42 (3): 545–565. https://doi.org/10.1111/0020-8833.00095.
Wangke, H. 2019. *Diplomasi Indonesia dan Pembangunan Konektivitas Maritim.* Yayasan Pustaka Obor Indonesia.

Lidya Christin Sinaga is a researcher at the Research Centre for Politics, National Research and Innovation Agency (BRIN). Her research interests are Indonesia–China Relations and ethnic Chinese Indonesians. She obtained her master's degree from School of History and International Relations-Flinders University, Australia in 2016. She is the editor and contributor of several academic books, including Refleksi Satu Dekade Belt and Road Initiative (Yayasın Pustaka Obor Indonesia, 2024); "Projecting A Global Identity as a Maritime Nation in the New Capital City" in The Road to Nusantara (ISEAS, 2023); Six Decades of Indonesia-China Relations: An Indonesian Perspective (SPRINGER, 2018).

Tonny Dian Effendi is a lecturer in the Department of International Relations at the University of Muhammadiyah Malang, Indonesia. He was a visiting research fellow at the Japan Institute of International Affairs and as a visiting scholar at esteemed institutions such as the Universidade do Minho in Portugal, the Institute of International Relations at National Chengchi University in

Taiwan, and the Center for Chinese Studies at the National Central Library in Taiwan. He was also a recipient of the Graduate Student Research Training program at the Center for Asia–Pacific Area Studies, Academia Sinica, Taiwan. Currently, he is a member of the Research Center for Advanced Science and Technology Open Laboratory for Emergence Strategies (ROLES) at the University of Tokyo. His research interests encompass East Asian international relations, diplomacy, diaspora studies, and Chinese Indonesian studies.

Chapter 5
Indonesia's Efforts in Building a Middle-Power Identity: An Unfinished Project

Julia M. Lau

Abstract Indonesia is a middle-power country in Asia that is beginning to be recognized for its influence at the regional and global levels. It is increasingly studied by scholars as a case of a middle power. This chapter examines how Indonesia, under the Joko Widodo (2014–2024) administration, established its identity as a rising middle power but argues that this project is incomplete. Examining the public record of the Widodo administration and its foreign ministry's activities and statements, scholarly studies of Indonesia as a middle power whose status is contested, and how other powers and regional neighbors have responded to Indonesian foreign policy, the chapter argues that there is still a distinct gap between ambition and reality where Indonesia's middle-power aspirations are concerned. In what capacities does Indonesia operate as a middle power and what are the prospects for its evolution under the administration of President Prabowo Subianto?

Keywords Indonesia · Middle power · Foreign policy · Joko Widodo · Prabowo Subianto

5.1 Introduction

This chapter discusses how Indonesia, during the Joko Widodo administration (2014–2024), began self-identifying and establishing its identity as a rising middle power in a more concrete way than during his predecessor Susilo Bambang Yudhoyono's tenure (2004–2014), when the term was first introduced into the Indonesian foreign policy lexicon. International relations (IR) and area studies scholars have increasingly studied Indonesia as a case of a middle power in the past decade, adding to the concept's utility when it comes to defining Indonesia's status and role in the current international and regional order. Nevertheless, this project—of firmly

J. M. Lau (✉)
the ISEAS-Yusof Ishak Institute, Singapore, Singapore

I. Kartini and Kanisa (eds.), *Joko Widodo's Foreign Policy and Diplomatic Legacy*,
https://doi.org/10.1007/978-981-95-2085-5_5

securing Indonesia's self-identity, status, and recognition as a middle power—remains incomplete in late 2024 at time of writing, as former president Widodo hands over power to his successor, Prabowo Subianto.

Finding Indonesia's niche as a middle power and establishing consistent recognition from the international community thus remains the task of Prabowo and Prabowo's successors to complete, given the gap between Indonesia's middle-power aspirations and reality. This task is further complicated by geopolitical uncertainties including US–Sino rivalry, regional tension in the South China Sea and the unresolved Myanmar crisis, as well as unending war in Ukraine and the tensions in the Middle East. On its part, Indonesia has endeavored to keep its ties to both great powers on an even keel, as it upholds its long-standing foreign policy tenet of *bebas aktif* (free or independent and active). The US is an important security partner while China continues to invest in and trade with Indonesia. The new president has vowed to maintain *bebas aktif* as a guiding principle in the conduct of his administration's foreign policy (Intan 2024).

The next section will examine scholarly discussions and analyses of Indonesia's middle-power role and status, especially those that were written during Widodo's presidency, to first determine what kind of middle-power Indonesia is regarded as in the present time. Following that discussion, the chapter will empirically consider whether Indonesia has truly acquired the identity of a "middle power"—what Wendt describes as "relatively stable, role-specific understandings and expectations of self" (1992, p. 397)—and if its relational ties with other nation-states reveal that they treat Indonesia as such. One way is by looking at the public record of President Widodo's overseas visits and foreign policy accomplishments, together with his administration and foreign ministry's activities and statements. Whether certain ideas and projects have failed to take off, even as others have succeeded, is one factor affecting the strength of Indonesia's middle-power claims. The world's reaction to Indonesia's diplomacy and whether regional neighbors and other powers consider and acknowledge Indonesia as a middle power (by calling it such, for instance) are additional indicators, if one takes a constructivist view (Wendt 1992).

Examining the legacy of Joko Widodo's pragmatic foreign policy after his two terms in office (2014–2019; 2019–2024) and looking ahead to Prabowo Subianto's likely conduct of foreign policy and statesmanship, this chapter then considers the opportunities there might be for Indonesia to enhance its reputation and record as a middle power. After a brief examination of some early foreign policy moves under the Prabowo administration, the chapter argues and concludes that there is still a distinct gap between ambition and reality where Indonesia's middle-power aspirations are concerned.

5.2 The Scholarly Debate: What Is a Middle Power and Is Indonesia One?

One 2003 attempt to define "traditional" versus "emerging" middle powers by Jordaan (2003) argued that "all middle powers" had foreign policy behavior aimed at stabilizing and legitimizing the world order. By this definition, Indonesia certainly fits the mold of a status quo middle power, meaning it generally abides by and supports in rhetoric and action upholding the rule of international norms and law, and does not challenge the existing global order in any significant way. This does not mean that it cannot rhetorically criticize the existing order, however, and former president Widodo, like some of his predecessors, had criticized on several occasions the existing international order for failing to deliver equitable growth. In Indonesia's case, it has benefited from relative stability in the world order as it seeks higher growth and continues to deal with a wide range of domestic challenges, as its position in current economic rankings testifies.

Jordaan notes that middle powers also "typically adopt an activist style," interfering in "global issues beyond their immediate concern" (2003, p. 167). Focusing on conflict reduction with "like-minded" states, middle powers work through multilateral institutions and channels to achieve their goals in these issue areas. While back in 2003, Indonesia did not make Jordaan's list of middle powers, these descriptions certainly fit the Indonesia of today. Another notable point Jordaan made was that in examining middle-power behavior, one had to allow for "a degree of agency, in terms of leadership by specific individuals and the exercise of a choice as to whether to become more involved in a certain foreign policy issue (i.e. niche diplomacy)" (Jordaan 2003, p. 173). This reminder about the role of agency and leadership is an important one as we consider Widodo's legacy in foreign policy and how that might change under the new president, who comes in with a Realist sensibility, a different approach toward positioning Indonesia in the world order, and a strong personal interest in world politics (Lau 2024).

A little more than a decade after Jordaan's analysis, scholars began more seriously examining Indonesia as a possible country case when deliberating the concept of "middle power." In their article on security strategies of "middle powers in the Asia–Pacific," for instance, Emmers and Teo (2015) examined Indonesia alongside South Korea as a case of a middle power which lacked "great-power capabilities but still aim(ed) to influence the regional security environment." The authors acknowledged the usual quantitative and qualitative classifications of middle powers—the former "ranking them according to…gross domestic product (GDP), GDP per capita, population and military expenditure"—and the latter involving "diplomacy and foreign policies." Using their own metric of "functional" versus "behavioural" measurements, Emmers and Teo argued that middle powers could target "resources to address a specific issue" (that is, "niche diplomacy") or promote "broad behavioural norms of multilateralism and conflict management" (2015, p. 187). In their piece, the authors listed Thailand, Japan, South Korea, Russia, Indonesia, and Vietnam as "middle powers" using a slight modification of another author's indicators measuring economic, military, and development factors (sorted into great, middle, and small power levels).

In their analysis, Indonesia had "relatively low level of resource availability" compared to other regional middle powers like Australia and South Korea but faced (at the time) a "relatively benign" and "low-threat" strategic environment (Emmers and Teo 2015, p. 195–196). Thus, they described Indonesia's performance of middle-power status as relying on "confidence building through multilateral" dialogue and assuming a "mediating or facilitating role" to manage regional challenges, or in sum, acting as a "behavioural" middle power based on "moral authority." Multilateralism and network diplomacy were thus Indonesia's key tools (Emmers and Teo 2015, p. 199). By comparison, South Korea had high resource availability and a high threat environment, leading it to use a "functional regional security strategy"—that is, allying with a great power, the US, and implementing various measures to preserve its own security interests (Emmers and Teo 2015, p. 201)..

The world and the Asia-Pacific, including Southeast Asia, have changed since Emmers and Teo's analysis of middle powers a decade ago, as has Indonesia itself. Arguably, Indonesia's strategic environment now faces slightly elevated levels of threat or is at least not as "relatively benign" as it used to be. This is in large part because China is now much more assertive about issues including the South China Sea dispute and intent on building up influence vis-à-vis smaller countries in what it sees as its own backyard, and this takes place against worsening tensions in the US–China rivalry.

Indonesia has also got richer, meaning it has slightly more resource availability, although it is not yet an advanced economy and the odds are against it reaching that status by 2045, despite its leaders' hope and rhetoric. Nevertheless, Emmers and Teo's description of Indonesia being more behavioral than functional will likely still hold true until President Prabowo can significantly improve its military capabilities. Given Indonesia's limitations in hard power, it should continue to display a strong preference to use multilateralism or regionalism in its foreign policy toolkit, despite the personal inclinations of the new president. It is difficult at this juncture to imagine an Indonesia willing to directly confront larger powers except in dire or potential conflict situations—such as rebuffing the Chinese coast guard's recent incursions into the waters surrounding the Natuna islands (AMTI 2024)—or seeking a formal alliance with one great power to balance against another, given Indonesia's long-standing commitment to non-alignment and autonomy.

Teo's updated theorizing on middle powers (Teo 2022a) drills more deeply into how middle powers may take up distinctive key roles in international politics, thus carving for themselves a niche that others cannot easily take over. She challenges the neorealist emphasis on material power when analyzing middle powers. In her "differentiation-based framework," middle powers engage in "relative, relational, and social power politics" and context is important. She, like many scholars who study middle powers, acknowledges how multilateralism and building like-minded coalitions are important tools of middle powers but notes the limitations of using this to distinguish them from others in the international order (Teo 2022a, p. 8, 10). She argues that "ideational and social aspects are equally important" as they determine how a middle power's behavior and interests "are received by others" (Teo 2022a, pp. 11–12).

If the above logic is applied to Indonesia, a country whose self-description as a middle power and whose neighbors including Australia already refer to it as such, its population size (fourth most populous in the world) and increasing economic and geopolitical weight has launched it firmly into middlepowership. We can certainly see a focus in Indonesian diplomacy that clearly includes the ideational and social aspects. Indonesia's foreign minister (Menlu) and ministry (Kemlu), and its key officials or office bearers, are strong champions of certain social or normative ideals when diplomatic rhetoric or action is required.

Today, Indonesia clearly "fulfils…requirements of being a middle power" if one assesses its "material capabilities," when this is defined merely as "economic size," as Teo (2022b, pp. 153–154) points out in her chapter analyzing Indonesia as an "awkward" example of a middle power. Indonesia is the only Southeast Asian country in the G-20 grouping, the world's fourth most populous nation-state and "the tenth largest economy in terms of purchasing power parity" (World Bank 2023). This "awkwardness" is evident when a different criterion, using "composite indices spanning the economic, military and social sectors" to measure Indonesia's middle power clout, is applied and when Indonesia is compared to its peer competitors in the middle-power category. Compared to Australia and South Korea, for instance, Indonesia's rankings for wealth per capita, military capabilities, infrastructure, technological capability, and human development, etc., lag far behind (Teo 2022b, p. 155). Teo (2022b) also analyzed Indonesia's self-description as a middle power and whether this corresponded with "other actors' perception" of it, finding inconsistencies including the alternative label of "aspiring great power" to some parts of the Indonesian political and strategic community (Teo 2022b, p. 155–156).

5.3 Recognition with Limitations

What the various scholarly works referenced above illustrate is that the term "middle power" has experienced a slight renaissance in recent IR literature, despite qualms about its utility as a concept, which all the scholars note. As discussed, Teo (2022a) doubles down on the idea that middle powers can and do perform certain distinctive functions in the world order. In Indonesia's own rhetoric and from its neighbors and partners' references to its development, Indonesia has indeed *arrived* as a middle power.

However, how Indonesia wields this power is constrained by material realities and limitations, including a limited budget for military modernization and establishing sufficient outposts across the world with which to project diplomacy. For instance, by one measure of global diplomacy (but with data for only 66 countries), Indonesia ranks 23rd, after Poland and Greece, with 130 diplomatic posts in 95 cities (Lowy Institute 2024). For comparison, Switzerland is ranked 19th with 141 posts and South Korea 13th, with 187. The same ranking lists, in the middle-power category, Indonesia at ninth—just below Singapore's eighth position for its influence, after examining the diplomatic networks of 66 countries and territories in Asia, the Organisation for Economic Co-operation and Development (OECD) and the Group of 20 (G20).

Given the above, Indonesia's way forward as it entrenches itself as a middle power likely lies in the ideational or normative and social aspects of its power, until it is sufficiently wealthy materially to acquire a true measure of hard power alongside its influence.

5.3.1 President Widodo's Foreign Policy Record and Legacy

In his two terms as president, what did former president Widodo achieve on the foreign policy front? In his first term, many observers had criticized his focus on domestic issues and reluctance to attend important international summits including the United Nations General Assembly, dismissing him as a foreign policy lightweight and his country as ineffective in resolving or influencing the outcome of several pressing issues in Asia (Pattiradjawane 2017). In terms of strategic vision for Indonesia's foreign policy, Widodo's record is mixed at best. Commentators tend to praise his foreign minister and her ministry rather than the president for whatever achievements Indonesia registered in terms of diplomatic success. Some critics debated whether Widodo failed to be even-handed in maintaining strategic autonomy, reflected as equidistance where possible, in managing the US–China rivalry, suggesting that he might have tilted too closely to Beijing in his pursuit of Chinese investment for his infrastructural projects (Dzakwan 2023).

For one, the much-promoted concept of a "global maritime fulcrum" (GMF or *Poros Maritim Dunia*) that Widodo launched near the beginning of his presidency did not successfully take off (Arifia 2020). There is some merit in the GMF concept, given Indonesia's core identity as an archipelagic state and the pressing concerns in the region on the South China Sea. As outlined by Widodo when he first became president, the five areas under this concept are essential to Indonesia's core security and territorial concerns: maritime diplomacy, maintaining Indonesia's territorial integrity, maritime safety and sovereignty; safeguarding the country's national resources and exclusive economic zone (EEZ); intensifying defense diplomacy; and diminishing maritime rivalries among major powers and promoting peaceful resolution of territorial disputes (Widodo 2014). However, this concept languished as Widodo became more focused on domestic political concerns toward his second term as president and gave most of his attention to economic diplomacy and attracting foreign investment. At this point, there is no indication that President Prabowo is interested in reviving the GMF concept.

Nevertheless, Indonesia during the Widodo era managed to push forward with its efforts to gain recognition for its conception of the regional order, as embodied in the ASEAN Outlook on the Indo-Pacific or AOIP (Agastia 2020). Scholars like Agastia view Indonesia's success in promoting the AOIP concept as being evidenced by how dominant actors in the Indo-Pacific region have generally accepted it. Agastia uses role theory—where states are actors that "conceive their own role" and assessed by how they perform such roles—to explain that Indonesia's push for the AOIP reflects Widodo's "role conception…of Indonesia as a middle power" (2020, p. 294). In the ASEAN context, the AOIP is listed together with regionally accepted international declarations and agreements including the UN Charter, the Treaty of Amity and Cooperation in Southeast Asia, and the UN Convention on the Law of the Sea, reflecting that it has gained a certain degree of acceptance at the normative level. That at the ASEAN Summit in 2024 (which Widodo skipped, nominating his vice president to attend) produced a statement reiterating the importance of the AOIP concept in an ASEAN-centered regional architecture supports this development (ASEAN 2024). The challenge, however, would be for this commitment to show some concrete progress after an unclear course of action despite Indonesia's initial advocacy for the concept (Lin 2024).

Toward the end of his second term, the former president was active in visiting foreign counterparts to seek investment for his mega project to build a new capital city, Nusantara, in East Kalimantan. This augmented his image as "salesman in chief," where he earlier encouraged his diplomats to sell Indonesia to the world (Sheany 2018). On the positive front, Widodo succeeded in attracting investment in many infrastructural projects from the Chinese and other foreign entities, including the high-speed railway from Jakarta to Bandung. This is likely to continue under Prabowo's administration and studies are already being conducted to see if the Jakarta–Bandung rail line can be connected to Surabaya (Faridz 2024) while Prabowo is expected to seek Chinese investment in other projects such as a "sea wall" for Jakarta. More critically, however, such a strong focus on seeking foreign investment over other foreign policy concerns had opened Widodo to criticism that he was overly transactional in his conduct of economic diplomacy.

Importantly, Widodo left office with ties to Indonesia's closest neighbors, Malaysia and Singapore, on a positive slate, after resolving key bilateral disputes over issues ranging from defense cooperation to airspace. President Prabowo looks set to continue this legacy of maintaining strong ties with both; he will receive Singapore's PM Wong in Jakarta in early November and had visited Malaysian PM Anwar Ibrahim in the interim between winning the election and being inaugurated as president.

There were other areas which showed potential, including the elevation of Indonesia's relationship with Australia to a Comprehensive Strategic Partnership in 2018, about midway through Widodo's presidency. His visit to Sydney and to Port Moresby in Papua New Guinea were arguably among the rarer occasions on which Widodo displayed statesmanship and a friendly exchange with his Australian counterpart PM Albanese (Lau 2023). There is a chance that Indonesia can build up its relations with Papua New Guinea and the smaller Pacific Island nation-states, in the coming years.

Another area for growth in Indonesia's foreign policy is its role in the Global South, which was articulated as a desire for leadership but not well executed during Widodo's presidency. While speaking in Bali at the Indonesia-Africa Forum (IAF), Widodo expressed commitment to "being part of the global solution" and "defending the interests of the Global South" but seemed to focus on business deals arising from the forum and the economic side of things rather than normative or other forms of leadership (Shibata 2024).

Whether former president Widodo left a distinctive mark on Indonesian diplomacy is debatable. One Indonesian scholar argues that Widodo shifted Indonesia's perspective on the world order and credits him for criticizing "the liberal international order by emphasising democracy's impacts in terms of delivering equitable development" (Umar 2023, p. 1460). Noting that Widodo has criticized the liberal international order on multiple occasions, Umar argues that he has "focused more on democracy's material benefits for economic development and Indonesia's political and economic sovereignty" (Umar 2023, p. 1468 and 1473).

For Umar, Widodo also "engaged with alternative conceptions of democracy" provided by countries outside the liberal international order, like China but "continued to seek foreign direct investment" and "upheld its commitment to multilateralism" (Umar 2023, p. 1474). At the same time, former Menlu Retno "linked democracy to equitable development and a just global financial system" post-COVID-19 pandemic (Umar 2023, p. 1472). While all of the above may be factually accurate, this rhetorical criticism against the world order and Widodo's consistent seeking of closer trade and investment ties with a non-democratic great power (China) may have undermined existing regard for Indonesia's role as a middle power that upholds international rules of law or at the minimum, sent conflicting signals to even Indonesia's partners. What is more, Indonesia's domestic politics and the backsliding in its democracy and transparency rankings during the latter half of Widodo's presidency make for uncomfortable reminders and dampen any claims Indonesia may make in speaking for developing democracies going forward.

5.4 Indonesia as a Muslim Middle Power

One of Indonesia's distinct qualities is that it is a majority Muslim middle power, the only one in the world at present that is recognized as such. This allows Indonesia's leaders and diplomats to champion certain causes that Western and non-Muslim middle powers may shy away from. While Widodo himself did not seem to make too much of this situation, since the outbreak of hostilities following the Hamas attack on Israel on 7 October 2023, we have seen such championing especially on critical issues like the Israel–Gaza conflict, especially by former Menlu Retno Marsudi. Her soaring pro-Palestinian rhetoric, especially in the earlier months of the conflict, included mention of Indonesia's humanitarian ideals, solidarity with the Palestinian cause, and ideology supporting Indonesia's own anti-colonial and pro-Global South positions (Grehenson 2024).

Ibu Retno's trenchant calls on her peers in many countries to seek a ceasefire and peaceful end to the ongoing Gaza-Israel conflict and her emotive tweets on X were unusually strong. These rhetorical moves were accompanied by her tireless diplomacy in many in-person and phone meetings with foreign counterparts on raising humanitarian efforts for the Palestinians and calling for ceasefire. Part of the impetus was domestic, given the outrage sparked among Indonesians of all faiths by the crisis in Palestine (Lau 2023). However, the facts speak to a hard limit as to how much Indonesia can influence or sway the outcome, given the multiple players entrenched in the conflict and Indonesia's geographical distance from the center of the violence in Gaza and now Lebanon.

It will be interesting to see if the new foreign policy actors under Foreign Minister Sugiono and his three deputy ministers takes the same strident line rhetorically and whether Prabowo orders a more robust and active response by the foreign ministry or even the Indonesian military (as peacekeepers) to help the Palestinian cause. While there is a high likelihood that the Prabowo administration will continue to rhetorically and materially support the Palestinian cause, with Prabowo mentioning Palestinian independence during his inaugural speech as president, limited resources and a litany of other pressing priorities, domestic and foreign, will still mean that Indonesia is more of an observer than a strong player in this crisis' resolution.

5.5 The Foreign Minister's Role

On a separate tack, Jost et al. (2024) have examined the role of foreign policy advisers' impact on a country's foreign policy. Looking at the deliberations of American presidents and their foreign policy advisers from 1947 through 1988, they argue that "advisers' predispositions regarding the use of force shape state behavior through the counsel advisers provide in deliberations." While Indonesia has not been placed in a position to decide on the use of force in the past decade, at least not vis-à-vis waging any war against any party, this reminder about the importance of the president's foreign policy advisers is an apt one.

For the decade of his presidency, Widodo's choice of career diplomat Retno Marsudi as Menlu has seen an energetic foreign minister who was tireless in visiting her counterparts to discuss a range of key issues. The Ukraine crisis in 2022 followed by the Gaza conflict in 2023 significantly impacted her agenda. Correspondingly, the relevant Lowy Institute's Asia Power Index, which tracked defense and diplomatic interactions among 25 regional countries for 2022, showed how Indonesia was third "most active diplomatic player" that year, after China and Japan (Patton and Sato 2023).

Given that Prabowo has dropped the long tradition of appointing a technocrat who has risen through the ranks of Kemlu as foreign minister, choosing a close confidant from his inner circle, there is a possibility that Kemlu might have difficulty adjusting to its new minister. Whether Sugiono is the right person to push forward with Indonesia's drive toward true middle-power status and consolidate its foreign policy agenda will partly depend on his ability to compel the staff and diplomats to cohere to a clear vision.

While it is too early to tell how Sugiono will lead as Menlu, among his first moves was attending the BRICS Summit in Kazan, Russia in October 2024, where Indonesia was announced as a partner (not a full member)—36 countries sent representatives. Ostensibly the new president could not attend, as his Cabinet was undergoing their orientation program during the summit date. Kemlu's message on the BRICS summit noted that Indonesia would "voice a message of peace" (Jakarta Post 2024), which is ironic since Russia is waging a war in Ukraine and has been since early 2022. Sugiono's office stated that his presence at this summit reflected Indonesia's "commitment to playing an active role in international forums and strengthening ties with all countries" (Jakarta Post 2024). Soon after, Russia and Indonesia conducted their first joint naval exercise in the waters near Surabaya in early November 2024.

Both gestures have alarmed some observers, who question if the Prabowo administration is already shifting Indonesia's foreign policy by seemingly throwing its lot in with Russia and China—which Prabowo will visit on his first official tour overseas as president—rather than the US and Europe. Prabowo will visit the US after China, on the same tour. Those who are positive about the latter development suggest that Indonesia can gain ideas and build up its military capacity by interacting with the Russians in military drills (*CNA* 2024). If the new Indonesian government does not intend to destabilize its relationships with the US and western European countries, however, it will need to quickly balance its approach and send other signals to reassure its other partners that it truly is autonomous.

There is potential for Indonesia to consider strengthening its ties with other middle powers. One possible area of growth lies with the European Union (EU) as a security actor (as opposed to Indonesia forging closer bilateral relationships with individual EU member countries). The EU desires to form stronger security relationships in the Indo-Pacific in general; while Indonesia has generally focused on gaining military acquisitions and hardware from countries including France and Italy, a growing prospect for cooperation has emerged on defense initiatives, especially on transnational challenges including organized crimes and maritime security (Lau 2024a). In this regard, multiple security dialogues between the EU and Indonesia have taken place over the past decade (Tempo 2024).

5.6 Conclusion: Redefining "Middle Power" for the Twenty-First Century?

After examining the previous president's record on foreign policy and highlighting some areas of challenge or possible improvement for his successor, this chapter concludes by asking whether Indonesia might be able to redefine the nebulous concept of "middle power" as it consolidates its rise this century. First, under the Yudhoyono and Widodo presidencies, Indonesia tried to position itself as a responsible power that plays by international rules and aspired to be a "bridge builder" on issues like Israel–Gaza and Russia's war in Ukraine. President Prabowo has also used the language of bridge building in his presidential campaign and later statements on Indonesia's foreign policy aspirations.

Second, Indonesia is now a G20 member and aspiring to join the ranks of the OECD and BRICS nations. This can be linked to the above concept of bridge building, where Indonesia, as a developing country from the Global South, can represent other developing countries to ensure more inclusivity and diversity when finding solutions to global problems. Yet, while joining more multilateral institutions where Indonesia's voice can be heard among great and wealthy powers can increase Indonesia's soft power and influence, there is a risk that other ASEAN member states might view Indonesia as being uncertain about its commitment to ASEAN. Such an outcome would ironically weaken Indonesia's clout and credibility in ASEAN. The new foreign minister and president would need to weigh those risks against their desire to appear even-handed vis-à-vis the great powers and to maintain strategic autonomy.

Third, as discussed in a previous section, Indonesia remains the world's largest Muslim-majority middle power. It thus has the chance and some credibility to act as interlocutor or spokesperson on certain issues, representing the Global South or other Muslim nation-states, in a way that other middle powers, which mostly hail from Europe or Northeast Asia, may not. While no formal statement has been issued by Sugiono on this matter, President Prabowo has declared that Indonesia will send more aid to the Palestinians (SINDOnews 2024).

While it is beyond this chapter's scope to reach any firm conclusion as to how Indonesia might redefine "middle power" in the coming years, it will continue to be an interesting country case that will challenge IR scholars. We can expect each new administration to adapt its foreign policy to the shifting geopolitical order as it addresses domestic challenges. In closing, we might consider two possible scenarios for how Indonesia might evolve as a middle power.

The optimistic scenario sees Indonesia fully claiming its position as a middle power in not just the ideational and social but also material aspects. In Emmers and Teo's (2015) terms, this would mean Indonesia being "functional" in addition to being "behavioural," especially for critical security and geopolitical challenges. In this outcome, Indonesia not only acts as a bridge builder and honest broker to help resolve some of the region and the world's most pressing challenges, like food and energy security—which are key domestic priorities for President Prabowo. Indonesia being materially strong would mean it can fend off any potential attacks on its sovereignty or territory, without help from partners in the region. A prime example would be any serious conflict or dispute involving Indonesia's sovereignty over the Natunas Islands, which are already at threat from Chinese ambitions and actions.

A second scenario envisages Indonesia as remaining an aspiring middle power whose key tools remain rhetorical ones, where its role as convenor of summits rather than problem-solver or leader continues to predominate its conduct of foreign policy. This is a "business as usual" scenario, meaning that Indonesia continues its current trajectory and does not gain the material clout envisaged in the first scenario. This might be due to domestic challenges, especially if there is another global recession in 2024–2025, complicating Indonesia's efforts to modernize its military, including its maritime forces, and to augment its domestic infrastructure and improve its connectivity with the world. Alternatively, Indonesia's diplomats and foreign policy leaders might not seize the momentum to creatively fashion a distinct middle-power role for the country, preferring to stick to tried and tested ways of conducting foreign policy or reiterating broad aspirations to have many friends and no enemies, without specifying exactly how such friendships will evolve and take shape.

In essence, the Widodo presidency left a mixed legacy on foreign policy for the Prabowo administration. Despite the challenges, Indonesia's influence in the region and world can grow because both the US and China are looking to establish stronger footholds in Southeast Asia, in their quest to outdo each other. If predictions about a more active and assertive foreign policy under Prabowo are correct, Indonesia is likely to move the needle forward on defining and securing itself as a middle power in the coming decade.

References

Abbondanza, G. 2022. Whither the Indo-Pacific? Middle Power Strategies from Australia, South Korea and Indonesia. *International Affairs* 98 (2): 401–421. https://doi.org/10.1093/ia/iiab231.

Agastia, I. G. B. D. 2020. Understanding Indonesia's role in the 'ASEAN Outlook on the Indo-Pacific': A role theory approach. *Asia Pacific Policy Studies* 7: 293–305. https://doi.org/10.1002/app5.308.

Aridati, I. Z. and J. Lau. 2024, April 26. State of Southeast Asia survey 2024: Indonesia doubling down on ASEAN and Japan. *Fulcrum*. https://fulcrum.sg/state-of-southeast-asia-survey-2024-indonesia-doubling-down-on-asean-and-japan/.

Arifia, I. 2020, December 29. Understanding the cause of declining Maritime Fulcrum agenda performance. *UNAIR News*. https://unair.ac.id/understanding-the-cause-of-declining-maritime-fulcrum-agenda-performance/.

ASEAN. 2024, October 3. *ASEAN Leaders' Declaration on the ASEAN Outlook on the Indo-Pacific for the Future-Ready ASEAN and ASEAN-Centered Regional Architecture*. https://asean.org/wp-content/uploads/2024/10/3-Final_ALD-on-the-AOIP-for-the-Future-Ready-ASEAN-and-ASEAN-Centered-Regional-Architecture.pdf.

Asia Maritime Transparency Initiative. 2024, October 28. *Seismic Strife: China and Indonesia Clash Over Natuna Survey*. https://amti.csis.org/seismic-strife-china-and-indonesia-clash-over-natuna-survey/.

Channel News Asia (CNA). 2024, November 4. *Joint Russia sea drills signal start of Indonesian foreign policy shift*. https://www.channelnewsasia.com/asia/indonesia-russia-first-joint-naval-drills-foreign-policy-4722071.

Dzakwan, M. H. A. 2023, October 25. *President Widodo's October 2023 Visit to China: Interpreting the Mixed Signals*. CSIS Indonesia. Commentaries CSISCOM00323. https://www.csis.or.id/publication/president-widodos-october-2023-visit-to-china-interpreting-the-mixed-signals/.

Emmers, R., and S. Teo. 2015. Regional security strategies of middle powers in the Asia-Pacific. *International Relations of the Asia-Pacific* 15: 185–216. https://doi.org/10.1093/irap/lcu020.

Faridz, Devianti. 2024, October 26. *Indonesia presses China for more investment in high-speed railways*. VOA. https://www.voanews.com/a/indonesia-presses-china-for-more-investment-in-high-speed-railways/7840093.html.

Grehenson, G. 2024, June 4. *Indonesia foreign minister Retno Marsudi: We will always defend Palestine*. https://ugm.ac.id/en/news/indonesia-foreign-minister-retno-marsudi-we-will-always-defend-palestine/.

Intan, G. 2024, November 3. *Prabowo: Indonesia Tidak Boleh Jadi Kacung Negara Manapun.* VOA. https://www.voaindonesia.com/a/prabowo-indonesia-tidak-boleh-jadi-kacung-negara-manapun/7849666.html.

Jakarta Post, The. 2024, October 23. *New Foreign Minister Sugiono in Russia for BRICS Summit.* News Desk. https://www.thejakartapost.com/world/2024/10/23/new-foreign-minister-sugiono-in-russia-for-brics-summit.html

Jordaan, E. 2003. The concept of a middle power in international relations: Distinguishing between emerging and traditional middle powers. *Politikon: South African Journal of Political Studies* 30 (1): 165–181.

Jost, T., J. D. Kertzer, E. Min, and R. Schub. 2024. Advisers and Aggregation in Foreign Policy Decision Making. *International Organization.* https://doi.org/10.1017/S0020818323000280.

Karim, M. F. 2018. Middle power, status-seeking, and role conceptions: The cases of Indonesia and South Korea. *Australian Journal of International Affairs.* https://doi.org/10.1080/10357718.2018.1443428.

Lau, J. 2024a, November. Indonesia's Security Relationship with the EU: Prospects and Possibilities for Growth. In Y. Joshi, I. Nishida and D Chaturvedi eds. *The European Union as a Security Actor in the Indo-Pacific: Perceptions and Responses from the Region.* Springer.

Lau, J. 2024b, October 15. Outlook for the Prabowo Administration's Foreign Policy. *Fulcrum.* https://fulcrum.sg/outlook-for-the-prabowo-administrations-foreign-policy/.

Lau, J. 2023a, October 19. The implications of the Gaza conflict on Indonesia. *Fulcrum.* https://fulcrum.sg/the-implications-of-the-israel-gaza-conflict-on-indonesia/.

Lau, J. 2023b, July 10. Indonesia Looks East: Widodo's Engagement with Australia and Papua New Guinea. *Fulcrum.* https://fulcrum.sg/indonesia-looks-east-widodos-engagement-with-australia-and-papua-new-guinea/.

Lin, J. 2024, February 16. Advancing the ASEAN Outlook on the Indo-Pacific Beyond Indonesia's Chairmanship. *Fulcrum.* https://fulcrum.sg/advancing-the-asean-outlook-on-the-indo-pacific-beyond-indonesias-chairmanship/.

Lowy Institute. 2024. *Lowy Institute Global Diplomacy Index 2024.*

Pattiradjawane, R. L. 2017, October 21. Infantilism of Indonesian foreign policy. *The Jakarta Post.* https://www.thejakartapost.com/news/2017/10/21/infantilism-indonesian-foreign-policy.html.

Patton, S. and J. R. Sato. 2023, May 25. Decoding diplomatic dynamics in Asia: The numbers tell a captivating story | Lowy Institute. *The Interpreter.*

Roberts, C. B. and L. C. Sebastian. 2015. Ascending Indonesia: Significance and conceptual foundations. In C. B. Roberts, A. D. Habir, and L. C. Sebastian eds. *Indonesia's ascent: Power, leadership, and the regional order*. Palgrave Macmillan. https://doi.org/10.1057/9781137397416_1.

Sheany. 2018, February 12. Indonesia to Boost Efforts in Economic Diplomacy. *Jakarta Globe.* https://jakartaglobe.id/news/indonesia-boost-efforts-economic-diplomacy.

Shibata, Nana. 2024, September 2. Indonesia vows to 'defend interests of Global South' at Africa forum. *Nikkei Asia.* https://asia.nikkei.com/Politics/International-relations/Indonesia-vows-to-defend-interests-of-Global-South-at-Africa-forum.

SINDONews. 2024, October. Dukung Palestina, Prabowo Subianto Tegaskan Kirim Lebih Banyak Bantuan. https://www.youtube.com/watch?v=vPCwEWAzIOE.

Tempo. 2024, September 19. *Indonesia, European Union Strengthen Partnership through Security Policy Dialogue.* https://en.tempo.co/read/1918241/indonesia-european-union-strengthen-partnership-through-security-policy-dialogue#:~:text=Indonesia%20and%20the%20European%20Union,order%2C%20peace%2C%20and%20stability.

Teo, S. 2022a. Toward a differentiation-based framework for middle power behavior. *International Theory* 14 (1): 1–24. https://doi.org/10.1017/S1752971920000688.

Teo, S. (2022b). Middle power awkwardness? Indonesia's norm entrepreneurship in ASEAN. In G. Abbondanza and T. S. Wilkins eds. *Awkward Powers: Escaping Traditional Great and Middle Power Theory.*

Umar, A. R. M. 2023. The rise of the Asian middle powers: Indonesia's conceptions of international order. *International Affairs* 99 (4): 1459–1476. https://doi.org/10.1093/ia/iiad167.
Wendt, Alexander. 1992. Anarchy is what states make of it: The social construction of power politics. *International Organization* 46 (2): 391–425.
Widodo, J. 2014, October 20. *Di Bawah Kehendak Rakyat dan Konstitusi. Inaugural speech at the People's Consultative Assembly of the Republic of Indonesia (MPR-RI).*
World Bank. (2023). *Indonesia—Overview.* https://www.worldbank.org/en/country/indonesia/overview

Julia M. Lau is Editor, Fulcrum and Senior Fellow and Coordinator of the Indonesia Studies Programme at the ISEAS-Yusof Ishak Institute, Singapore. She was one of the co-editors for the multi-authored volume, *The Road to Nusantara: Process, Challenges and Opportunities* (ISEAS Publishing, 2023). Julia has lectured for various international relations and law, comparative politics, and research methods courses at Georgetown University, The Catholic University of America, and McDaniel College in the USA. She received her bachelor's (LLB) and master's (LLM) degrees in law from the National University of Singapore, and M.A. degrees in security studies and government from Georgetown. The views expressed in this chapter are the author's own.

Chapter 6
Indonesia-Africa Economic Relations: Leveraging Diplomatic Efforts, Bilateral Agreements, and Informal Networks

Ahmad Helmy Fuady, Bondan Widyatmoko, Muzzar Kresna, and Saiful Hakam

Abstract This study explores the strategic initiatives undertaken during President Joko Widodo's administration to enhance Indonesia's economic engagement with Africa. Key efforts include the establishment of new Indonesian embassies in major African countries, reciprocated by African nations such as Ethiopia, Kenya, and Tanzania opening embassies in Jakarta. These diplomatic expansions signify a strengthened commitment to bilateral relations. The creation of platforms such as the Indonesia-Africa Forum and the Indonesia-Africa Infrastructure Dialogue further supports dialogue and collaboration across various sectors. In response to challenges facing multilateralism, Indonesia has increasingly focused on bilateral agreements, including Preferential Trade Agreements (PTAs) and Free Trade Agreements (FTAs), to strengthen economic ties. President Widodo's official visit to Africa underscores the significance of these initiatives. Additionally, this study highlights the role of people-to-people diplomacy, specifically through the African trading community in Tanah Abang, Jakarta, which serves as a critical bridge between Indonesia and Africa. Despite its importance, this informal network remains underutilized and needs further enhancement to fully leverage its potential in fostering deeper diplomatic and economic ties. This study demonstrates Indonesia's proactive measures under President Widodo's leadership to expand and deepen its economic engagement with Africa, highlighting the need to enhance informal people-to-people connections as a complement to formal diplomatic efforts.

Keywords Economic engagement · Diplomatic expansion · Bilateral agreements · Indonesia-Africa forum · People-to-people diplomacy

---

A. H. Fuady (✉) · B. Widyatmoko · M. Kresna · S. Hakam
Research Center for Area Studies, The National Research and Innovation Agency, Indonesia (BRIN), Jakarta, Indonesia

I. Kartini and Kanisa (eds.), *Joko Widodo's Foreign Policy and Diplomatic Legacy*,
https://doi.org/10.1007/978-981-95-2085-5_6

6.1 Introduction

Before embarking on his trip to Africa, President Joko Widodo (Jokowi) addressed the media at Kualanamu International Airport in Deli Serdang Regency, North Sumatra Province, on August 20, 2023 (setkab.go.id 2023). He emphasized Indonesia's deep-rooted historical ties with Africa, recalling the nation's pivotal role in initiating and hosting the 1955 Asia-Africa Summit and its significant contribution to the establishment of the Non-Aligned Movement. During his visit, Jokowi aimed to underscore the 'Bandung Spirit' by promoting solidarity and cooperation among the Global South countries. He highlighted his visits to Kenya, Tanzania, Mozambique, and South Africa, noting that Kenya and Tanzania had recently opened embassies in Jakarta, signaling their commitment to strengthening ties with Indonesia. Mozambique was highlighted as the first African nation to establish a Preferential Trade Agreement (PTA) with Indonesia, while his visit to South Africa included participation in the Brazil, Russia, India, China and South Africa (BRICS) Summit and meetings with global leaders.

This brief statement by President Jokowi illustrates Indonesia's evolving perspective and approach toward Africa. The 1955 Asia-Africa Conference and its principles, encapsulated in the 'Bandung Spirit,' remain central themes in discussions about Indonesia-Africa relations. The conference established Indonesia as a leading voice in the Global South, promoting solidarity and cooperation among newly independent nations from Asia and Africa. This historical context continues to shape Indonesia's foreign policy and economic strategies, exemplified by initiatives such as the Indonesia-Africa Forum, which aims to enhance trade and investment between the regions.

For historical context, trade connections between Africa and Indonesia have deep historical roots, tracing back several centuries to the time when maritime trade routes traversed the Indian Ocean, linking Southeast Asia, South Asia, the Arabian Peninsula, and the eastern coast of Africa. These routes facilitated the exchange of goods, cultures, and ideas, establishing early commercial ties between the Indonesian archipelago and various African regions. Indonesian seafarers used double outrigger canoes to navigate the Cinnamon Route, creating early links between Nusantara (Indonesia) and Africa (Pramono 2017, pp. 14–15).

Between the seventh and fifteenth centuries, traders from the Indonesian archipelago, particularly from the Sriwijaya and Majapahit kingdoms in Sumatra and Java, played a crucial role in these trading networks. They were heavily involved in the spice trade, exchanging cloves, nutmeg, and other spices for textiles, gold, ivory, and various African products. The maritime links were further strengthened in the 1500 s through Islamic trade networks, as evidenced by artifacts such as Kenyan earthenware and Egyptian rifles found in the Sultanate of Cirebon (Pramono 2017, pp. 14–15).

During the colonial era, the eighteenth century saw the Dutch East India Company (VOC) develop modern shipping routes connecting Batavia (Jakarta) with Africa and Europe. This period also marked the introduction of the palm oil industry to

Indonesia, with oil palms imported from Mauritius and plantations established in Sumatra. This historical relationship has laid the groundwork for contemporary trade and economic interactions, which are becoming increasingly significant in today's global economy (Pramono 2017, pp. 14–15).

Indonesia's foreign relations landscape has experienced a notable transformation, with Africa emerging as a critical focus of its diplomatic and economic strategies. This shift is deeply rooted in the historical connections forged during the 1955 Bandung Conference, which established a foundation of solidarity and cooperation between newly independent nations from Asia and Africa. The Bandung Conference marked Indonesia as a prominent leader in the Global South, advocating for anti-colonialism and non-alignment—principles that continue to influence its foreign policy and economic endeavors. Scholars such as Walker (2005), Acharya (2014, 2016), Shimazu (2014), Lee (2010), and Ewing (2019) have explored historical and international dimensions, highlighting the enduring significance of the Bandung principles in shaping Indonesia's engagement with Africa. The conference's emphasis on cultural and political connections continues to inform Indonesia's approach to fostering collaboration with African nations. For the business realm, addressing trade and investment dynamics. Researches by Pramono (2012), Amalina et al. (2018), Sabaruddin (2017), and Sabaruddin and Sabariman (2023) underscore the importance of economic diplomacy to increase the role of Indonesia in Africa.

While these perspectives have provided valuable insights, they often remain entrenched in historical or economic narratives. To move beyond these frameworks, bold and proactive diplomatic efforts are necessary. President Joko Widodo's administration has demonstrated a keen recognition of Africa's potential, viewing it as a strategic market rich in natural resources, a growing population, and an expanding middle class. This view has spurred Indonesia to deepen economic ties, diversify trade partners, and contribute to Africa's development.

This study aims to examine the new initiatives undertaken during President Joko Widodo's administration to enhance Indonesia's economic and diplomatic engagement with Africa. This includes analyzing strategic measures such as establishing new embassies, creating platforms like the Indonesia-Africa Forum and the Indonesia-Africa Infrastructure Dialogue, and conducting high-profile state visits. Second, the research intends to evaluate the impact of these diplomatic expansions and bilateral agreements, including Preferential Trade Agreements (PTAs) and Free Trade Agreements (FTAs), on trade, investment, and overall bilateral relations between Indonesia and African nations. Finally, the study will explore the role of informal people-to-people diplomacy, particularly through grassroots networks like the African trading community in Tanah Abang, Jakarta, assessing how these informal connections contribute to strengthening diplomatic and economic ties and how they can be better utilized.

The importance of this study lies in its relevance to the current global economic dynamics. As Indonesia navigates a multipolar world with shifting economic conditions, understanding its proactive measures and strategic initiatives in engaging with

Africa provides valuable insights into the strategies of emerging economies. Additionally, by focusing on the role of informal networks in diplomacy, this research highlights the often overlooked but crucial role that grassroots connections play in complementing formal diplomatic efforts. This contributes to a broader understanding of how informal network contributes to the people-to-people diplomacy, enhancement of international relationships and foster deeper, more effective engagements.

6.2 Placing Non-State Network to the Indonesia-Africa Public Diplomacy

There is an increasing involvement of non-state actors in international affairs. The non-state actors in various forms such as non-governmental organization, academic institution, ethnic group organization, lobbyist group, labor union, private millitary organization including terrorist group organization contributes to peace truce but at the same time they could destabilize world power balance (Wijninga et al. 2014). In case of Ukraine–Rusia war, the religious group, the Russian Orthodox Church, and Ukraine Church both perpetuates the war through rationalization and direct assistance to the militia (Mulford 2016). BP, major multinational company in oil exploration is among the first corporation that recognized the risk of climate change and has been actively advocates green-house-gasses (GHG)s emmision on their business management (Lakhany 2006).

In the dicipline of International Relation, non-state actors involvement has been recognized as part of the public diplomacy. Eventhough public diplomacy itself has taken many different names such as multistakeholder diplomacy, paradiplomacy, track two diplomacy, citizen diplomacy and catalic diplomacy (van Doeveren 2011), the term highlight the importance of interaction among individual or groups outside the official negotiation processes (Kaye 2007). The involvement of non-state actors such as university, non governmental organization, business group in exercising soft power through dialogue, student exchange or commerce responsibility governance principal has recognized the variety of legal paradigm and social culture in international affairs (Raboin 2014; Westermann-Behaylo et al. 2015).

In a conflict situation, dispute settlement process or situation where formal diplomacy are unable to penetrate, building shared cultural meaning (van Doeveren 2011) through simultan negotiation (Chrismas 2012) are importance elements of public diplomacy. Exercising these element would transforming adversaries relation among involved parties. This perspective placed non-state actors in positive manner for their contribution in conflict resolution. In other words, non state actors who does not meet the state interest tend to be excluded from the scheme of public diplomacy, keeping the informality status.

There has been unresolved understanding in placing non state actors within the conception of public diplomacy (van Doeveren 2011). On one strand, the non-state actors is bounded within the diplomatic context. The non-state actors might advocates for various issues such as climate change, resourcess scarcity, sustainability, social injustice, human development, peace and security at different setting of national or international affairs. However, the advocation is serves the need and purposes of the state. The non-state actors are unable to escape state governance structure while the state has the liberty to incorporates the advocation into the diplomatic agenda. The other strand placed these advocations within the public context. The presence and absence of state diplomacy does not determine the outcome of mutual understanding build by the non-state actors. The process of advocation is independent against the state governance structure. Recognising these structures as (in)dependent is important in depicting the existence of non-state actors network.

The non-state actors network, by resisting formal incorporation into state structures, remain emancipatory and politically productive. Their autonomy fosters a more genuine form of relation and interaction that reflects the interest and identities of the people involved. In the case of Indonesia and Africa, the non state actors such as traders, business organization and students play a pivotal role in building public diplomacy. Leveraging cultural, academic and business exchages, these network build stronger ties, trust and foster mutual understanding between the people of Indonesia and Africa.

6.3 The Re-Emergence of Africa

Africa has re-emerged as a region brimming with potential for Indonesia's export expansion, transforming from its previous image as a continent primarily associated with poverty and conflict. Today, Africa is a dynamic and increasingly influential player in the global economy. According to the World Bank (2024), the continent is home to 18.3% of the world's population, representing a significant market for goods and services. Over the past decade, Africa has experienced remarkable economic growth, with several countries in the region undergoing rapid development and industrialization.

The World Bank's World Development Indicators further corroborate Africa's economic resurgence, revealing that many African nations have surpassed not only Indonesia but also numerous other emerging economies in terms of growth. This shift presents a substantial opportunity for Indonesia to enhance its trade relationships with African countries, leveraging the continent's expanding markets to bolster its export growth strategy.

In 2023, the economic landscape of Sub-Saharan Africa was particularly notable. An impressive 40% of the world's 50 fastest-growing economies were located in this region, underscoring Africa's burgeoning economic influence (World Bank 2024). Countries such as the Democratic Republic of the Congo and Rwanda demonstrated this trend with exceptional Gross Domestic Product (GDP) growth rates of 8.56% and 8.24%, respectively. These figures are strikingly higher compared to Indonesia's GDP growth rate of 5.05% in the same year, highlighting the narrative of Africa's rise.

Interestingly, the current trade figures between Indonesia and Africa do not fully reflect their historical ties. In 2023, the total value of trade (exports and imports) between Indonesia and 54 African countries reached USD 16.79 billion. While this is a notable figure, it represents only a small fraction of Indonesia's overall international trade activities.

Figure 6.1 illustrates Indonesia's trade dynamics with Africa over a ten-year period, from 2014 to 2023. According to UN Comtrade data, Indonesia's exports to Africa grew at an average rate of 4% per year during this period. Although this growth indicates increasing economic engagement, it is relatively modest compared to the potential that exists between the two regions. On average, total trade with Africa accounted for only 2.96% of Indonesia's total exports from 2014 to 2023 (Fig. 6.2). This underscores that, despite Africa's growing global importance, it remains a relatively small trading partner for Indonesia compared to regions like Asia, America, and Europe.

In 2023, Indonesia's exports to Africa were valued at USD 6.86 billion, which constituted just 2.65% of the country's total exports. In contrast, imports from Africa amounted to USD 9.87 billion, or 4.45% of Indonesia's total imports. This trade imbalance, with imports from Africa exceeding exports, points to the need for Indonesia to diversify and strengthen its export portfolio to Africa. Over the past two decades, Indonesia's exports to Africa have been dominated by primary commodities, with little diversification. Crude palm oil (CPO) (HS Code: 1511) has been the most significant export commodity, generating USD 5.75 billion in 2023 alone. The primary markets for Indonesian CPOs in Africa include countries like Egypt, South Africa, Djibouti, Kenya, Tanzania, Benin, and Senegal. This heavy reliance on a single commodity underscores the need for Indonesia to explore new export opportunities in Africa to reduce its vulnerability to market fluctuations in CPO. On

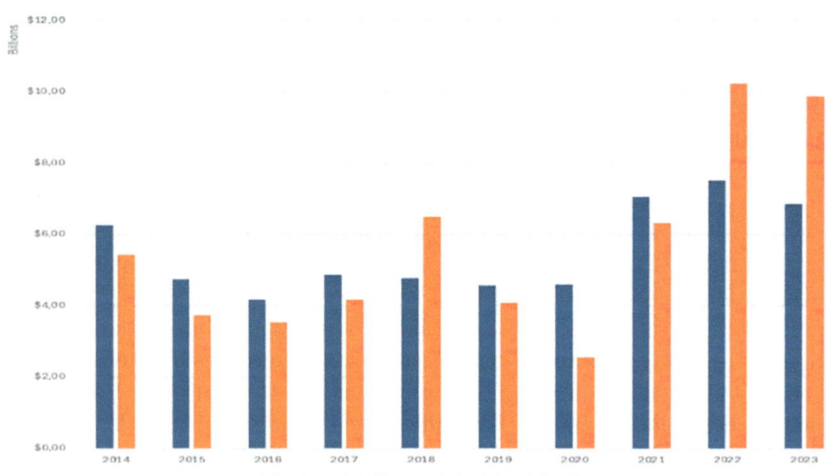

Fig. 6.1 Indonesia-Africa export–import (USD) *Source* UN Comtrade Database

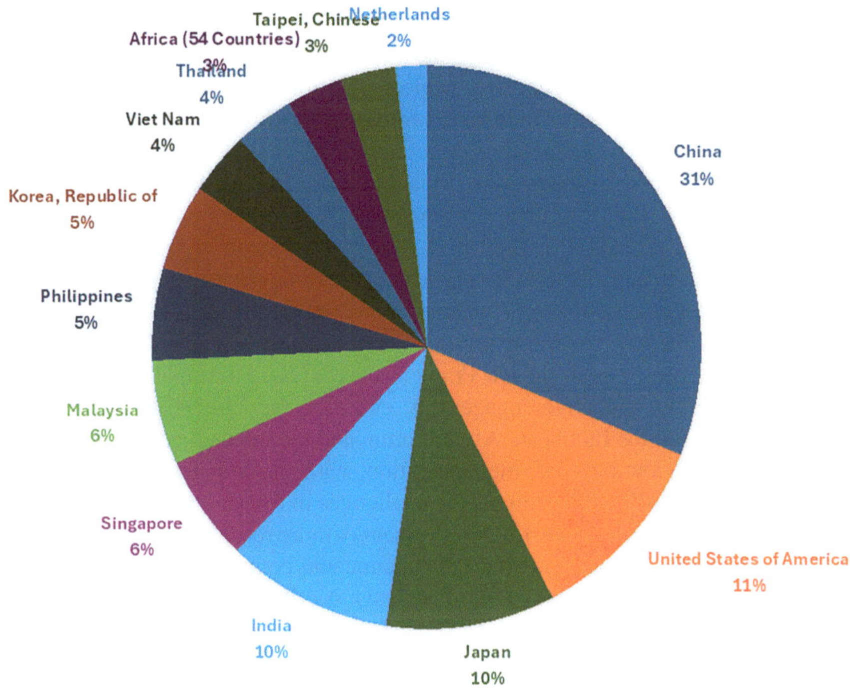

Fig. 6.2 Indonesia's export shares (%) *Source* UN Comtrade Database

the import side, Indonesia's largest import from Africa is petroleum. The reliance on African petroleum highlights a critical aspect of the trade relationship, where Indonesia depends on African resources to meet its energy needs.

While the trade relationship between Indonesia and Africa has seen some growth, it remains untapped relative to the historical ties and the potential that both regions offer. For Indonesia, there is a strategic opportunity to enhance trade with Africa by diversifying its export base, reducing reliance on primary commodities, and exploring new sectors such as manufacturing, technology, and services. Strengthening economic ties with Africa not only aligns with Indonesia's foreign policy goals but also positions the country to take advantage of Africa's growing economic influence on the global stage.

6.4 Africa as a Non-Traditional Market?

The conventional categorization of Africa as a "non-traditional market" within Indonesia's economic discourse is a stark reflection of the continent's often overlooked potential. This characterization, while prevalent, is increasingly incongruent with the burgeoning opportunities and geopolitical shifts that are reshaping the global economic landscape. Dorigné-Thomson's (2023) in-depth analysis of Indonesia's Africa engagement offers a valuable contribution but stands as a solitary beacon amidst a broader expanse of scholarly neglect. The Jokowi administration's efforts to elevate Africa's status within Indonesia's foreign policy architecture are commendable. However, the predominant focus on economic dimensions risks overshadowing the multifaceted nature of Indonesia-Africa relations. The persistence of the "non-traditional market" label in official documents such as the National Medium-Term Development Plan (RPJMN) is indicative of a deeper-seated underappreciation of Africa's potential to contribute to Indonesia's national development objectives.

This critique echoes broader concerns about the depth of Indonesia's Africa engagement. While the government's rhetoric is ambitious, concrete actions and measurable outcomes remain elusive. Pramono et al.'s (2012) seminal study provided a roadmap for more proactive engagement, outlining recommendations including the establishment of the Indonesia-Africa Forum and the intensification of technical cooperation. The slow pace of implementation, with the Indonesia-Africa Forum materializing only in 2018, highlights the challenges in translating policy into practice. The "Road to Africa" study (2018) represents a more recent attempt to assess the African market's potential. However, the dynamic and complex nature of the continent demands a continuous and nuanced analysis. A critical research gap persists in understanding the evolving dynamics of Indonesia-Africa economic relations.

In stark contrast, China and India have adopted far more assertive approaches to Africa (Cheru & Obi 2011; Wahlers 2015; Van Staden and Wu 2021). Driven by a combination of domestic economic imperatives and geopolitical ambitions, these Asian giants have invested heavily in the continent. China's strategy, often characterized as resource-driven and infrastructure-focused, has yielded significant dividends (Cheru & Obi 2011). India, on the other hand, has pursued a more multifaceted approach, combining economic interests with soft power diplomacy. Both countries have recognized Africa as a crucial arena for securing resources, expanding markets, and enhancing global influence.

Indonesia's aspirations to become a global player necessitate a more proactive and strategic approach to Africa. To compete effectively with China and India, the country must translate its political rhetoric into tangible economic outcomes. This requires a comprehensive understanding of Africa's diverse markets, the identification of strategic sectors for cooperation, and the development of innovative financing mechanisms to support Indonesian businesses. Moreover, Indonesia should leverage its historical ties with Africa, as exemplified by the Bandung Conference, to build upon existing goodwill and trust. By doing so, Indonesia can unlock the immense potential of the African continent and contribute to its own economic and geopolitical rise. Below are notable endeavors of Jokowi administration to transform historical connections and political rhetoric into tangible economic infrastructures.

6.5 Diplomatic Infrastructure Expansion

The African market is becoming increasingly competitive, with numerous countries striving to capitalize on its rapidly growing opportunities. Nations are not only engaging in bilateral forums with African states but also bolstering their diplomatic infrastructure in the region. According to *The Economist* (March 2019), between 2010 and 2016, a total of 320 new embassies were established across Africa. For instance, Turkey has significantly expanded its presence on the continent, increasing its number of embassies from 12 to 42 over the past 15 years.

In contrast, Indonesia, despite being the most active among ASEAN countries in Africa, maintains a relatively sparse embassy presence. Since opening its embassy in Pretoria, South Africa, in 1994, Indonesia has not established any new embassies in Africa. As of 2019, Indonesia has 16 embassies serving 50 accredited countries on the continent—5 in North Africa and 11 in Sub-Saharan Africa, including Madagascar. By comparison, there are 17 African diplomatic missions in Indonesia, comprising 11 embassies, 5 consulates, and 1 representative office.

Many Indonesian embassies in Africa face an overwhelming array of responsibilities. For example, the embassy in Nairobi, Kenya, is responsible for Uganda, the Democratic Republic of Congo, Somalia, Mauritius, Seychelles, and also serves as a liaison for UNEP and UN-Habitat. Similarly, the embassy in Dakar, Senegal, handles Gambia, Guinea, Guinea-Bissau, Mali, Ivory Coast, Cape Verde, and Sierra Leone, while the Abuja, Nigeria embassy covers Ghana, Liberia, Burkina Faso, Benin, Cameroon, Congo, Togo, Niger, São Tomé and Príncipe, Gabon, and ECOWAS. The Abuja embassy manages 11 countries with a combined population of approximately 324.82 million, representing 25.5% of Africa's total population. This extensive coverage presents a complex array of socio-economic, cultural, and political issues, placing a significant burden on these embassies to effectively manage Indonesia's economic diplomacy in Africa.

The extensive responsibilities of Indonesian embassies in Africa are not matched by adequate staffing levels. Moreover, Africa is not a preferred posting location for top Indonesian diplomats, given its relatively underdeveloped conditions compared to more advanced postings in Asia, Europe, and the Americas. There are also no special incentives for diplomats assigned to regions where educational, healthcare, and security facilities require substantial improvements.

To enhance diplomatic ties and cooperation with African nations, the Indonesian government has decided to establish a diplomatic mission in Cameroon. President Joko Widodo issued Presidential Regulation Number 69 of 2020 on June 8, 2020, authorizing the opening of an Indonesian Embassy in Cameroon (setkab.go.id 2020). This embassy will cover not only Cameroon but also additional accreditation for Chad, Equatorial Guinea, Gabon, Congo, and the Central African Republic. However, due to the COVID-19 pandemic and budget considerations, the establishment has not yet been realized.

Despite the modest overall progress in diplomatic infrastructure, the administration of President Joko Widodo has successfully facilitated the opening of new embassies by several African nations in Jakarta, including Ethiopia, Kenya, Tanzania, and Rwanda. In 2016, Ethiopia appointed its first Resident Ambassador to Indonesia, resuming operations in Jakarta in June 2023 after a temporary closure due to the pandemic. Ethiopian Airlines began operating flights to Jakarta in July 2018, marking the first direct link between Africa and Indonesia (Aviator 2018).

On March 18, 2022, Kenya inaugurated its embassy in Jakarta, a significant development in the long-standing diplomatic relations that began before Kenya's independence. This significant milestone represents a notable achievement in the long-standing diplomatic relations that were established even before Kenya gained independence. Indonesia had previously set the precedent by opening its diplomatic mission in Nairobi in 1982. The ceremony was attended by Amb Raychelle Omamo, Kenya's Cabinet Secretary for the Ministry of Foreign Affairs, Mrs. Retno Lestari Marsudi, Indonesia's Minister for Foreign Affairs, and numerous officials from both Kenya and Indonesia, including private and government representatives (Embassy of the Republic of Kenya -Indonesia 2022).

Similarly, Tanzania opened its embassy in Jakarta on June 22, 2023, emphasizing the importance of deepening economic cooperation between the two countries. On June 22, 2023, the Tanzanian government officially inaugurated its embassy in Jakarta, Indonesia. This move underscores the growing significance of the partnership and friendship between the two countries, which has been in place since 1964. At the launch ceremony, Dr. Stergomena Tax, Tanzania's Minister of Foreign Affairs and East African Cooperation, stated that the new embassy would serve as a crucial platform for deepening cooperation, particularly in the economic sector, between Tanzania and Indonesia. The establishment of the embassy after 59 years marks a new chapter in the enduring relationship between the two nations (jpnn.com 2023).

Rwanda formally inaugurated its embassy in Jakarta on June 6, 2023, with a ceremony attended by top officials from both nations, highlighting a commitment to strengthening bilateral relations. On June 6, the government of Rwanda formally inaugurated its embassy in Indonesia, located in Menteng Dalam, Tebet District, South Jakarta. The opening ceremony featured a ribbon-cutting event attended by Indonesia's Foreign Minister Retno LP Marsudi, Rwandan Foreign Minister Vincent Biruta, and Rwandan Ambassador to Indonesia Sheikh Abdul Karim Harelimana. Minister Biruta emphasized that the embassy's inauguration underscores the robust commitment of both Rwanda and Indonesia to strengthening their bilateral relations through enhanced cooperation. He described the new embassy as a vital "bridge" that will facilitate collaboration in mutual interests, trade, and investment between the two nations. Ambassador Harelimana characterized the opening as a significant milestone that reflects the deepening and strengthening of the friendship between Indonesia and Rwanda (The Jakarta Post 2024).

The establishment of these embassies serves various purposes. Primarily, it signifies the establishment of trust between two nations, as well as the acknowledgment of the future political, economic, and cultural benefits that can be derived from their partnership. It strengthens diplomatic and political relations by creating formal platforms for ongoing dialogue and collaboration between Indonesia and these African nations. This platform will efficiently address many issues that may arise during their cooperation. Moreover, it paves the way for expanded economic and trade connections, providing a solid framework for mutual investments and business opportunities. Furthermore, the presence of these embassies in Jakarta signifies a reciprocal commitment from African nations to bolster their interactions with Indonesia. This move aligns with Indonesia's broader strategy to enhance its influence and presence in Africa, promoting a more balanced and dynamic relationship between the two regions.

While progress in diplomatic infrastructure may seem incremental, the successful establishment of new embassies by African countries in Indonesia reflects the effectiveness of President Joko Widodo's diplomatic initiatives and suggests a promising trajectory for future cooperation and engagement. Following the reopening of the Ethiopian Embassy, Ethiopian Airlines is also launching flights to Indonesia. As of July 17, 2018, Ethiopian Airlines—Africa's largest aviation group and a SKYTRAX-certified Four Star Global Airline—began operating three weekly flights to Jakarta. The flights operate with Boeing 787–8 on the route, every Tuesdays, Thursdays, and Saturdays, with a stopover in Bangkok. This marks a historic milestone as it establishes the first direct link between Africa and Indonesia, and represents the first time an African airline has flown into Jakarta.

6.6 Platforms for Dialogue and Collaboration

Pramono et al. (2012) have suggested establishing the Indonesia-Africa Forum to create a unified policy approach towards Africa, bridge gaps in bilateral relations, and foster synergy among stakeholders. These strategies are designed to bolster Indonesia's economic diplomacy and build stronger relationships with African countries. The recommendations, however, took very long to be realized. The Indonesia-Africa Forum and Indonesia-Africa Infrastructure Dialogue, for instance, only happened at the end of 2018.

The Indonesia-Africa Forum (IAF), held in Bali from April 10–11, 2018, and the Indonesia-Africa Infrastructure Dialogue (IAID), conducted in Bali from August 20–21, 2019, are highlighted as significant milestones reflecting the resurgence of Indonesia's economic diplomacy with Africa. These forums build on past collaborations between Asian and African nations, such as the Non-Alignment Movement and the 2005 New Asia-African Strategic Partnership (NAASP). Additionally, Indonesia has aimed to bolster ties between Asian and African countries through events like the 60th Anniversary of the Asia-Africa Conference in 2015, where delegates committed to fostering stronger, more inclusive, and sustainable relationships.

The IAF seeks to identify and expand economic opportunities, enhance technical cooperation, and strengthen existing partnerships between Indonesia and Africa. According to its official website, approximately 500 participants attended the forum, including business leaders, high-level government officials, and stakeholders from 53 countries. The forum resulted in business agreements totaling USD 586.56 million, covering various sectors such as industry, infrastructure, finance, mining, textiles, aircraft maintenance, and commodity trading. Additionally, the IAF facilitated the signing of a Memorandum of Understanding (MoU) between the Ministry of Foreign Affairs and Indonesia's Exim Bank (*Lembaga Penjamin Ekspor* Indonesia, LPEI), aimed at improving economic diplomacy and encouraging business engagement in African markets.

The IAID, as a continuation of the IAF, focuses specifically on the infrastructure sector, building on discussions from the previous forum. Given the significant developments in Africa's infrastructure sector, the IAID provides a platform for the exchange of ideas and exploration of collaborative opportunities in infrastructure and related areas. The official IAID website reports that during the first two days of the forum, business agreements worth USD 822 million (approximately Rp. 12.3 trillion) were concluded.

6.7 Strategic Diplomatic Visits

President Joko Widodo's landmark trip to sub-Saharan Africa began on August 20, 2023, with his departure from Kualanamu International Airport in North Sumatra. His first destination was Nairobi, Kenya, where he arrived on August 21, 2023. In Nairobi, he met with President William Ruto to discuss a preferential trade agreement (PTA) and a bilateral investment treaty (BIT), and announced significant energy sector initiatives valued at $2.2 billion. On August 22, 2023, President Jokowi continued his journey to Dar es Salaam, Tanzania, where he held discussions with President Samia Suluhu Hassan. Their talks focused on PTAs, BITs, and new memorandums of understanding (MoUs) related to energy, mining, and health sectors.

On August 23, 2023, President Jokowi arrived in Maputo, Mozambique. Given Mozambique's existing PTA with Indonesia, the discussions aimed at deepening existing partnerships and exploring new trade opportunities. His tour concluded with participation in the 15th BRICS Summit, held in Johannesburg, South Africa, from August 24–25, 2023. This engagement sought to enhance Indonesia's role in global forums and strengthen South-South cooperation.

This visit was a historic milestone, marking President Jokowi's first official trip to sub-Saharan Africa and showcasing a notable advance in Indonesia's diplomatic engagement with the region. No previous Indonesian president had visited as many African countries as Jokowi. While President Sukarno played a crucial role in fostering Asia-Africa relations through the 1955 Bandung Conference, he did not personally visit Africa. President Suharto made a significant diplomatic gesture with his visit to South Africa in November 1997, following Nelson Mandela's visit to Indonesia. There are no records of President B.J. Habibie visited Africa during his short tenure from 1998 to 1999, which was marked by domestic challenges and the transition to democracy. President Abdurrahman Wahid visited South Africa in April 2000 as part of his trip to the G77 summit in Cuba, and President Megawati Soekarnoputri also traveled to South Africa in 2002, meeting notable figures including Nelson Mandela and President Thabo Mbeki. Additionally, President Susilo Bambang Yudhoyono made significant visits to Africa, including South Africa in March 2008 and Liberia in February 2013.

President Jokowi's visit aimed to revive the 'Bandung Spirit', emphasizing solidarity and cooperation among developing nations—a principle established during the 1955 Bandung Conference. Throughout the trip, Jokowi secured several bilateral agreements covering trade, investment, infrastructure, and health, expected to enhance economic ties and foster new collaborative opportunities. His efforts included pursuing a PTA and BIT with Kenya, underscoring Indonesia's commitment to strengthening trade relations with Africa. The visit's timing with the 15th BRICS Summit in South Africa also underscored Indonesia's strategic interest in engaging with major emerging economies. Overall, Jokowi's trip represented a pivotal moment in Indonesia's foreign policy, aiming to deepen diplomatic and economic connections with African nations and promote mutual development.

6.8 Bilateral Agreements and Trade Relations

The decline of multilateralism, rising protectionism, and the global economic crisis that began in 2008 have collectively reshaped international economic relations, leading many countries to increasingly pursue bilateral agreements to advance economic cooperation. Multilateralism involves multiple countries working collaboratively through international institutions to address global challenges and promote economic cooperation. However, in recent years, this approach has seen a decline. The effectiveness of multilateral institutions, such as the World Trade Organization (WTO) and various international summits, has been undermined by political and economic disagreements among member states. The slow progress and perceived ineffectiveness of these platforms have frustrated many countries, driving them to seek more manageable and potentially beneficial bilateral partnerships. Bilateral agreements offer a way to bypass the gridlock of multilateral negotiations and achieve more immediate and specific outcomes.

The trend towards protectionism has also contributed to the shift towards bilateral agreements. In response to economic uncertainties and domestic pressures, many countries have implemented protectionist measures such as higher tariffs and trade barriers to shield their local industries from foreign competition. This rise in protectionism has created a more fragmented global trade environment, where comprehensive multilateral agreements are less feasible. Instead, countries have turned to bilateral agreements that allow for more direct and tailored negotiations, providing a way to secure terms that are more favorable to their specific economic needs.

The global economic crisis that began in 2008, triggered by the collapse of financial markets and a severe recession, intensified economic instability and uncertainty worldwide. In the aftermath of the crisis, many countries faced slower economic growth and increased financial challenges. This economic hardship prompted nations to adopt more cautious and strategic approaches to international trade and investment. Bilateral agreements became an attractive option during this period, offering a means to negotiate trade relationships directly and secure targeted economic benefits. These agreements allow countries to navigate the complexities and uncertainties of the global economic landscape more effectively.

The convergence of declining multilateralism, increasing protectionism, and the fallout from the 2008 global economic crisis has driven many countries to favor bilateral agreements. These arrangements provide a more controlled and flexible approach to economic cooperation, helping nations to address their specific needs and adapt to a rapidly changing global economic environment. This approach has also been tried by Indonesia to penetrate the African markets.

Up to 2024, Indonesia has signed only one Preferential Trade Agreement (PTA) with African countries. The Indonesia-Mozambique PTA (IM-PTA) represents the first trade agreement between Indonesia and an African nation and is also Mozambique's first trade agreement with an Asian country. This PTA, which commenced on April 11, 2018, and was officially signed on August 27, 2019, in Maputo, Mozambique, encompasses trade in goods, including non-tariff measures, provisions on the origin of goods, and customs procedures (Marthini 2020). According to Marthini, the agreement grants Mozambique tariff preferences for 217 Indonesian products, such as palm oil, rubber, paper, furniture, fishery products, and processed foods. Conversely, Indonesia will extend tariff preferences to 242 Mozambican products, including cotton, nuts, sunflower seeds, aluminum ore, coffee, fishery products, vegetables, and fruits. The IM-PTA is anticipated to be a gateway for Indonesian products not only in Mozambique but across the African continent. However, despite its signing, the PTA is still awaiting ratification. In addition to Mozambique, Indonesia is negotiating PTAs with Tunisia, Morocco, and Mauritius and is exploring the possibility of agreements with Algeria, Djibouti, the South African Customs Union (SACU), the Economic Community of West African States (ECOWAS), and the East African Community (EAC).

Efforts by the Indonesian government to engage with African nations should provide benefits to both Indonesian businesses and the African governments and communities involved (Interview LI, May 5, 2020). The successful negotiation of the IM-PTA is partly attributed to the Indonesian Ministry of Trade's lobbying, which assured that this PTA would offer strategic advantages to the Mozambican government and its people (Interview LI, May 5, 2020). It is crucial to understand the principle of reciprocity and mutual benefit and to clearly define the model of economic diplomacy cooperation to avoid accusations of neo-colonialism (Interview PW, July 29, 2020).

Although there have been some advancements in formal diplomatic endeavors, unfortunately several indicators such as trade imbalance and low completion of trade agreements reflect a deficiency in the formal establishment of diplomatic relations between Indonesia and Africa. This issue may stem from an unbalanced perspective in diplomatic governance, where the historical and ongoing connections between African nations and Indonesia are evident in the daily life of urban Jakarta.

In urban Jakarta, the influence of African culture can be seen in various aspects such as cuisine, music, and fashion. This cultural exchange highlights the deep-rooted connections between Indonesia and Africa that have existed for centuries. By acknowledging and leveraging these historical ties, both regions can strengthen their diplomatic relations and work towards a more balanced and mutually beneficial partnership.

6.9 Penetrating Fixity of Symbolic Diplomacy to Africa Through Everyday Diplomacy

It has been widely acknowledged that Indonesia's ascension as a global economic powerhouse does not align with its available resources and influence that could bolster its foreign policy decisions (McRae 2014). This disparity significantly constrains Indonesia's capacity to navigate key global issues. A prime illustration of this is Indonesia's inability to counterbalance the escalating sway of the European Union (EU) in shaping international standards for primary commodity production and trade. To circumvent this challenge, a strategic shift towards less intricate regulations has been contemplated, with a focus on penetrating traditional markets such as those in African nations. However, diplomatic modalities such as historical resemblance, similarity in resource abundance, and the political symbolism of solidarity and resistance against colonialism (Shimazu 2014) through the establishment of diplomatic channels and market access are not enough to (re) build good will and mutual understanding (Cohen 2019). It does not appear to directly facilitate the enhancement of economic trade between the two regions.

For those modalities to be effective, it necessitates a substantial translation of will and a mutual understanding leading to recognition of equal footing between Indonesia and Africa, both of which are grappling with similar challenges. This recognition can be manifested through the enhancement of cooperation and communication between Indonesians and Africans. However, this endeavor is met with significant challenges, as there exists a prevalent perception in Indonesia viewing Africa as a region lagging behind and fraught with investment risks. In general, Indonesian business entities remain hesitant to engage in direct investments in Africa, as tales of investment failures due to fraudulent activities continue to circulate in both formal and informal settings. The negative portrayal of Africans is also evident in the daily lives of urban Indonesians, who often perceive Africans as being closely associated with criminal behaviors. The government has implemented stringent surveillance measures regarding the African presence within the country. Media outlets tend to sensationalize criminal acts committed by Africans rather than highlighting the growing social integration of Africans through sports and politics. This negative portrayal soon becomes a vicious cycle of distrust as Africans seeking general information about Indonesia discover how Indonesian government and society perceives them.

This fixity of negative perception potentially be dissolved by viewing Africans as global travelers whether as merchants or students. In this context, the presence of African global travelers is seen as global citizen who looks for better chances and opportunities and rises up its movement due to the burgeoning relationship between the continent and the rest of the world, especially China. The movement of Africans in China is not far from this context (Bertoncello et al. 2007; Siu & McGovern 2017). Rather than leveraging China's economic and political dynamics to strategically reshape African-Western relations (Edoho 2011), African traders navigate their movements from Chunking Mansions in Hongkong to the cosmopolitan "chocolate cities" of Guangzhou and Shenzhen (Bertoncello et al. 2007). They establish their operational headquarters for reverse trading, capitalizing on China's status as an international hub for electronics and textiles to meet the demand for such goods. These African traders are adeptly leveraging the opportunities presented by China's engagement for resource acquisition, market expansion, and human resourcing in Africa.

African presence in Southeast Asia seems to adhere to a similar rationale in response to the increasing bilateral economic cooperation with African nations. Commercial centers such as Chattaburi Province in Thailand and Tan Phu District, Ho Chi Minh Province in Vietnam are some trade hubs where Africans are navigating their south-south business operations (Table 6.1).

The narrative of Africans in Tanah Abang, Jakarta, mirrors that of other commercial hubs in Southeast Asia. Situated in Central Jakarta, the Tanah Abang commercial center serves as a bustling trading ground for African merchants hailing from various nations. They engage in commerce involving Indonesian goods like garments, footwear, bags, spices, and more recently, automotive components. In conducting these transactions, new African buyers can tap into the well-established African trade network, encompassing countries such as Nigeria, Mali, and Senegal, which already boast local suppliers, trading enterprises, and cargo services, along with a diverse array of sample products on display. Armed with comprehensive market insights, aspiring traders are not precluded from venturing into new business endeavors.

The Malians represent a prominent presence in the garment trading sector in Tanah Abang. Despite having an Association to support the Mali community, establishing trustworthy business relationships with the Malians can be challenging, even within their own community. The Malians prefer engaging in business independently, either by procuring directly from other traders in Tanah Abang or by visiting garment production hubs in Bandung, Cirebon, Pekalongan, and other cities. Indonesia is not only recognized as a haven for cost-effective goods by the Malians, but also the predominant Muslim population in the country creates a conducive environment for conducting business (Interview IC, 13 April 2022).

The interior of Tanah Abang prominently showcases the African influence in the city. In addition to the grand edifice of the African Trade Centre, where an array of products are exhibited, travel documents and visas are processed, and new African football athletes are registering their names, there are a multitude of African restaurants in the vicinity. At the African Food Center located on K.S. Tubun street, patrons can savor a delectable African delicacy consisting of Fufu paired with okra

Table 6.1 African Trade Hub in Thailand, Vietnam and Indonesia

Trade Hub	Historical Development	Trading Goods	Countries and Tribes involved
Chattaburi Province, Thailand	Historically renowned for gemstone and jewelry trade, particularly rubies and sapphires. Has been a center for trade and craftsmanship for centuries	Trade are primarily focused on inexpensive cloth, clothing and gems	South Africa, Tanzania, Somali Kenya, Nigeria (Zulu, Xhosa, Maasai, Igbo)
Tan Phu District, Ho chi Minh, Vietnam	Primarily an agricultural region, with rice cultivation as the main economic activity. Trade activities have intensified in recent decades due to Vietnam's economic liberalization that has attracted international workers including African	Although there are growing trade relations with Africa, particularly in agricultural products, Africa is specialized in clothing	Nigeria, Senegal, Madagascar (Zulu, Xhosa, Maasai, Igbo)
Tanah Abang District, Central Jakarta, Indonesia	A long-standing textile and fashion center, with a history dating back to the colonial era. Has evolved into a major wholesale market serving domestic and international buyers	Increasing trade relations with Africa, especially in textiles and apparel	Nigeria, Mali, Senegal (Igbo, Yoruba)

Sources Gozdziak and Chantavanich (2023), Modern Ghana (Monday, 16 February 2009); ThanhNien news (Tuesday, July 2022 2008); Online interview (2021)

vegetables and a choice of beef or chicken. The establishment is predominantly operated by individuals of Nigerian Igbo descent. These tribes not only oversee the intricate trade routes extending from Tanah Abang to various West African nations such as Nigeria, Ghana, and Cameroon, but they also exhibit remarkable acumen in the realm of culinary entrepreneurship (Online interview RD, July 2021).

Apart from engaging in trade, Africans also actively participate in socio-religious activities in Tanah Abang. For instance, traders from Mali are devout in their prayers at mosques throughout Tanah Abang, generously contributing to mosque restoration projects. Additionally, Africans demonstrate remarkable promptness in raising funds to support Indonesia during major disasters such as the Aceh Tsunami and volcanic eruptions in various parts of the country.

Tanah Abang is renowned as a prestigious hub for African merchants, with its name being highly esteemed and recognized by traders, particularly in Lagos, Nigeria. This exemplifies the enduring trade relations between Indonesia and Africa, which predate the Indonesian government's formal acknowledgment of the significance of economic diplomacy with Africa. The trade dynamics in Tanah Abang involving the Igbo tribes, the Malian, the Somali, and the Kikuyu, pave the way for the Indonesian government to consider these groups as crucial diplomatic entities in Africa. Organizations such as the Mali Association and the Indonesia Africa Trade Association (IATA) from the African diaspora should be nurtured to facilitate the integration of African life into the country. The involvement of cargo businesses like PT Balahap and PT Karya Baru Trans, as well as the African Food Centre and the philanthropic efforts of the African community in the country, illustrates the deep-rooted connection of African culture within Indonesian society. These dynamics of Africans in Tanah Abang shows the everyday live diplomacy between African and Indonesian.

Unfortunately, diaspora individuals or long-term residents of African descent in Indonesia, who act as intermediaries for African importers, engage in charitable activities, and participate in Indonesian socio-religious affairs, are seldom recognized as key players in economic diplomacy aimed. Taking them as part of the Indonesia economic diplomacy to Africa would portray the relations more as symbiotic rather than symbolic.

6.9.1 Conclusion

President Joko Widodo's foreign policy towards Africa represents a significant shift in Indonesia's diplomatic and economic approach. His administration views Africa as a promising new frontier for economic diplomacy, often referring to it as an untapped market with substantial potential for Indonesian products. This perspective marks a departure from traditional diplomatic practices, reflecting Widodo's background as a businessman and his pragmatic approach to international relations. The president's hands-on style is evident in his interactions with global leaders, entrepreneurs, and investors, emphasizing a results-oriented approach.

During Widodo's two terms in office, Indonesian embassies across Africa have become actively involved in promoting Indonesian products and tourism, particularly highlighting Bali. Ambassadors and diplomats have frequently organized business lobbying efforts and outreach to African entrepreneurs, aiming to position Indonesian goods as viable alternatives to the dominant, yet often lower-quality, Chinese products flooding the African market. This shift requires diplomats to adopt a more marketing-oriented role, aligning with the president's pragmatic and business-focused style.

Under Widodo's administration, bureaucrats and diplomats are required to meet specific performance targets, reflecting a new model of diplomatic work that involves active lobbying and business development. This approach contrasts sharply with the negative perceptions of Africa from the Suharto era, when the continent was seen as a region plagued by conflict, instability, and failure, and where posting as an ambassador was often viewed as a demotion.

Changing the perception of Africa from a negative to a more positive and realistic view is a challenging task, given Indonesia's limited knowledge of the continent. Although the Bandung Conference of 1955 laid the groundwork for Afro-Asian solidarity, it predominantly highlighted Asian leaders, with limited focus on African figures. Consequently, there is a lack of comprehensive understanding and awareness about Africa among Indonesians.

Interestingly, long before Widodo's diplomacy initiatives, Indonesian products like garments, food, footwear, and cooking oil were already present in African markets, particularly in West Africa. These goods were often imported by African traders who established commercial networks in Indonesia, especially around Tanah Abang market in Jakarta. This highlights the active role of African traders in facilitating trade and suggests a vibrant and proactive commercial exchange between Indonesia and Africa.

However, Indonesia's official economic diplomacy towards Africa has not adequately acknowledged the role of these African traders. This omission might stem from a lack of awareness or a deliberate choice to focus solely on governmental initiatives rather than recognizing existing informal trade networks. In summary, President Widodo's foreign policy toward Africa represents a pragmatic shift in diplomatic and economic strategies, emphasizing business opportunities and market expansion while seeking to overcome historical perceptions and build stronger bilateral relations.

References

Acharya, A. 2014. *Who Are the Norm Makers? The Asian-African Conference in Bandung and the Evolution of Norms.* 20(3). https://www.jstor.org/stable/24526222.

Acharya, A. 2016. Studying the Bandung conference from a Global IR perspective. *Australian Journal of International Affairs* 70 (4): 342–357. https://doi.org/10.1080/10357718.2016.1168359.

Amalina, A. A. F., T. Novianti, and A. Asmara. 2018. Analisis Kinerja Perdagangan Indonesia ke Negara Potensial Benua Afrika. *Jurnal Ekonomi Dan Kebijakan Pembangunan* 7 (1): 43–59.

Aviator. 2018. *Ethiopian to Start 1st Direct Flights from Africa to Jakarta.* Retrieved May 18, 2018, from https://newsroom.aviator.aero/ethiopian-to-start-1st-direct-flights-from-africa-to-jakarta/.

Bertoncello, B., S. Bredeloup, and N. Oates. 2007. The emergence of new African "trading posts" in Hong Kong and Guangzhou. *China Perspectives*, 1 (69), 94–105. http://www.jstor.org/stable/24053485

Cheru, F., C. Obi. 2011. Chinese and Indian Engagement In Africa: Competitive Or Mutually Reinforcing Strategies? *Journal of International Affairs*, 64(2), 91–110. http://www.jstor.org/stable/24385536.

Chrismas, R. 2012. Multi-Track Diplomacy and Canada's Indigenous Peoples. *Peace Research*, 44/45(2/1), 5–30. http://www.jstor.org/stable/24429459.

Cohen, M. I. 2019. Three Eras of Indonesian Arts Diplomacy. *Bijdragen Tot de Taal-, Land- En Volkenkunde*, 175(2/3), 253–283. https://www.jstor.org/stable/26699640.

Dorigné-Thomson, C. 2023. Indonesia's Engagement with Africa. The Palgrave Macmillan.

Edoho, F. M. 2011. Globalization and Marginalization of Africa: Contextualization of China—Africa Relations. *Africa Today* 58 (1): 103–124. https://doi.org/10.2979/africatoday.58.1.103.
Embassy of the Republic of Kenya -Indonesia. 2022. *Kenya Opens Embassy in Jakarta, Indonesia; Signs MoUs.* Retrieved from March 18, 2022, https://kenyaembassyjakarta.org/kenya-opens-embassy-in-jakarta-indonesia/.
Ewing, C. 2019. The Colombo Powers: Crafting Diplomacy in the Third World and Launching Afro-Asia at Bandung. *Cold War History* 19 (1): 1–19.
Goździak, E.M., and S. Chantavanich (Eds.). 202). *African Migration to Thailand: Race, Mobility, and Integration* (1st ed.). Routledge. https://doi.org/10.4324/9781003286554.
Hefele, P., Merkle, D., & Zhivkov, S. (2015). Taking Confucius to Africa: How Cultural Diplomacy Conveys China's Voice And Perspectives To The World. In G. Wahlers (Ed.), *Transatlantic Relations* (pp. 56–72). Konrad Adenauer Stiftung. http://www.jstor.org/stable/resrep10121.8.
https://web.archive.org/web/20080822082845/http://www.thanhniennews.com/features/?catid=10&newsid=40473.
https://www.modernghana.com/news/202691/1/tough-times-nigerians-flee-to-ghana-vietnam.html.
jpnn.com. 2023. *Setelah 59 Tahun, Negara Afrika InSetelah 59 Tahun, Negara Afrika Ini Akhirnya Buka Kedubes di Jakarta.* Retrieved June 22, 2023, from https://www.jpnn.com/news/setelah-59-tahun-negara-afrika-ini-akhirnya-buka-kedubes-di-jakarta
Kaye, D. D. 2007. Rethinking Track Two Diplomacy. In *Talking to the Enemy: Track Two Diplomacy in the Middle East and South Asia* (1st ed., pp. 1–30). RAND Corporation. http://www.jstor.org/stable/https://doi.org/10.7249/mg592nsrd.8.
Lakhany, F. 2006. How Important are Non-State Actors. *Pakistan Horizon*, 59(3), 37–46. http://www.jstor.org/stable/41394369.
Lee, C. J. 2010. *Making a world after empire: the Bandung moment and its political afterlives.* (Vol. 11). Ohio University Press.
Marthini, N. M. A. (2020). Manfaatkan Indonesia-Mozambique PTA: Dorong Perdagangan ke Benua Afrika. In Paparan pada Webinar Series 17 Juni 2020.
McRae, D. 2014. *More talk than walk: Indonesia as a foreign policy actor.* Lowy Institute for International Policy. http://www.jstor.org/stable/resrep10154.
Mulford, J. P. 2016. Non-State Actors in the Russo-Ukrainian War. *Connections*, 15(2), 89–107. http://www.jstor.org/stable/26326442.
Pramono, S. 2012. *Exploring Africa: Mainstreaming Indonesia's Economic Diplomacy in Non-traditional Markets* (Siswo Pramono, Ed.). P3K2 Aspasaf.
Pramono, S. 2017. *IORA Comprehensive Economic Partnership Agreement (CEPA).*
Raboin, B. 2014. The Emergence of Multi-Track Diplomacy In International Dispute Resolution: The Treaty Of Portsmouth And The Community That Made Peace Possible. *Willamette Journal of International Law and Dispute Resolution*, 22(1), 85–104. http://www.jstor.org/stable/26210499.
Sabaruddin, S. S. 2017. Grand Design Diplomasi Ekonomi Indonesia: Sebuah Pendekatan Indeks Diplomasi Ekonomi. *Jurnal Ilmiah Hubungan Internasional* 12 (1): 69. https://doi.org/10.26593/jihi.v12i1.2545.69-90.
Sabaruddin, S. S., and H. Sabariman. 2023. Analisis Potensi Pasar Non-Tradisional Dalam Meningkatkan Hubungan Ekonomi Indonesia dan Namibia. *Jurnal Lemhannas RI* 11 (1): 1–12.
setkab.go.id. 2020. *Indonesia Opens Embassy in Cameroon.* Retrieved June 22, 2020, from https://setkab.go.id/en/indonesia-opens-embassy-in-cameroon/.
setkab.go.id. 2023. *Presiden Jokowi Lakukan Kunjungan Kerja Ke Empat Negara di Afrika.* Retrieved August 20, 2023, from https://setkab.go.id/presiden-lakukan-kunjungan-kerja-ke-empat-negara-afrika/.
Shimazu, N. 2014. Diplomacy as theatre: Staging the Bandung Conference of 1955. *Modern Asian Studies* 48 (1): 225–252.

Siu, H. F., and M. McGovern. 2017. China–Africa Encounters: Historical Legacies and Contemporary Realities. *Annual Review of Anthropology*, *46*, 337–355. http://www.jstor.org/stable/44866094.

The Jakarta Post. 2024. *Rwanda Opens Embassy in Indonesia, Bilateral Cooperation Established.* June 7.

Van Staden, C., Y. S. Wu. 2021. Vaccine Diplomacy and Beyond: New Trends in Chinese Image-Building in Africa. *South African Institute of International Affairs.* http://www.jstor.org/stable/resrep34093.

van Doeveren, R. 2011. Moving beyond the Diplomat and the National Interest. In *Engaging the Arab World through Social Diplomacy* (pp. 16–24). Clingendael Institute. http://www.jstor.org/stable/resrep05438.5.

Walker, D. 2005. Nervous outsiders: Australia and the 1955 Asia-Africa conference in Bandung. *Historical Studies* 36 (125): 40–59.

Westermann-Behaylo, M. K., K. Rehbein, and T. Fort. 2015. Enhancing The Concept Of Corporate Diplomacy: Encompassing Political Corporate Social Responsibility, International Relations, and Peace Through Commerce. *Academy of Management Perspectives*, 29(4), 387–404. http://www.Jstor.Org/Stable/43822391

World Bank. (2024). *World Development Indicator.*

Wijninga, P., Oosterveld, W. T., Galdiga, J. H., Marten, P., Chivot, E., Gehem, M., Knowles, E., Maas, M., Schellekens, M., Silveira, J., Yang, M. Y., Zelinska, O., de Jong, S., Kogut, P., & van Luit, E. (2014). State And Non-State Actors: Beyond The Dichotomy. In J. van Esch, F. Bekkers, S. De Spiegeleire, & T. Sweijs (Eds.). *Strategic Monitor 2014: Four Strategic Challenges* (pp. 139–162). Hague Centre for Strategic Studies. http://www.jstor.org/stable/resrep12608.8.

Ahmad Helmy Fuady is a researcher at the Research Center for Area Studies (PRW), National Research and Innovation Agency (BRIN) Indonesia. He received his training in economics and development at the Gadjah Mada University, Yogyakarta, and the Australian National University, Canberra, where he earned his bachelor's and master's degrees, respectively. In 2012, he obtained his doctorate from the University of Amsterdam with a dissertation comparing development in sub-Saharan Africa and Southeast Asia. Fuady has worked on a wide range of development issues, including poverty alleviation, green economy, renewable energy, and local development. Currently, he coordinates the African Studies and Resource Nationalism Research Group at PRW-BRIN. Email ahmad.helmy.fuady@brin.go.id

Bondan Widyatmoko is a researcher affiliated with the Research Center for Area Studies, National Research and Innovation Agency (BRIN). Bondan obtained a doctoral degree from Kyoto University, Japan, in 2019. His study concentrates on European and African studies, with a particular emphasis on the dynamics of production, appropriation and circulation of the palm oil commodity. Bondan can be reached via email bond001@brin.go.id

Muzzar Kresna is an associate researcher currently affiliated with the Research Center for Area Studies at the National Research and Innovation Agency of the Republic of Indonesia (BRIN). He earned his Bachelor's degree from Airlangga University in Indonesia. Furthering his academic pursuits, he obtained a Master's degree in African Studies from Universiteit Leiden in the Netherlands. His research experience spans various geographic regions and topics. In 2014, he analyzed environmental policy in the Czech Republic, focusing on sustainable development in Eastern Europe. His work in 2015 and 2016 explored the impact of crises on social policies in Spain and social changes in Tanzania, respectively. Muzzar continued his research in Africa with studies on growth and change in Kenya (2017) and development in Ethiopia (2018). Between 2015 and

2017, he examined the role of young farmers in Indonesia's micro, small, and medium enterprises sector through knowledge transformation from Japanese internships. His most recent study in 2023 rethinks the structural transformation pathway in Tanzania, examining the transition from farmer to "Dalali" (a Swahili term often referring to brokers or middlemen). Muzzar Kresna's research shows a deep involvement in both the theoretical and practical dimensions of economic development, industrial policy, and social change, with a particular focus on Indonesia and Africa issues using multidisciplinary lenses. Email: muzz001@brin.go.id

Saiful Hakam completed his undergraduate degree in History at Gadjah Mada University (UGM) in 2005. The following year, he relocated to Jakarta to work as a researcher at the Research Center for Area Studies, the Indonesian Institute of Sciences (LIPI). At LIPI, he focused on research related to religion and culture, particularly within East Asia. In 2008, he moved back to Yogyakarta to pursue further studies at the Center for Religious and Cross-Cultural Studies (CRCS) at UGM. After earning his graduate degree from CRCS UGM in 2011, he rejoined LIPI as part of a team dedicated to the study of religion and culture. In 2017, he joined the African Studies Group at LIPI, which now known as BRIN. Email: saif003@brin.go.id

Chapter 7
Indonesian Gastrodiplomacy in Jokowi's Tenure: A Foreign Policy Analysis

Meilinda Sari Yayusman, Gregorius Andika Ariwibowo, Ahmad Nuril Huda, Prima Nurahmi Mulyasari, Rakhman Priyatmoko, and Agus Trihartono

Abstract During President Joko Widodo's (Jokowi) tenure, economic growth had been the main priority for Indonesia's development, including in its foreign policy. His "Nawa Cita" and "Trade, Tourism, and Investment" commitments encouraged Indonesia's foreign policy to strengthen economic diplomacy in every possible way, including utilizing the support of public diplomacy. An example includes the formulation of a gastrodiplomacy strategy, such as the Co-branding Wonderful Indonesia and the Indonesian Spice Up the World (ISUTW), which exercises the efficacy of culinary and cultural diplomacy to win over foreign people's hearts and minds through food. This study examines the formulation of Indonesian gastrodiplomacy during Jokowi's two terms in office, and its implementation in the case contexts of Australia and the Netherlands, two countries with a significant number of Indonesian diasporas and long-term historical relationships with Indonesia. Data for this study are collected through a mixed-method research approach, involving online surveys, observations, and in-depth interviews, and are analyzed with the concept of three

---

M. S. Yayusman (✉) · G. A. Ariwibowo · A. N. Huda · P. N. Mulyasari · R. Priyatmoko
Research Center for Area Studies, The National Research and Innovation Agency, Indonesia (BRIN), Jakarta, Indonesia
e-mail: meil002@brin.go.id

G. A. Ariwibowo
e-mail: greg002@brin.go.id

A. N. Huda
e-mail: ahmad.nuril.huda@brin.go.id

P. N. Mulyasari
e-mail: PRIM0003@e.ntu.edu.sg

R. Priyatmoko
e-mail: rakhman.priyatmoko@brin.go.id

P. N. Mulyasari
School of Humanities, Nanyang Technological University, Singapore, Singapore

A. Trihartono
Department of International Relations, University of Jember, Jember, Indonesia
e-mail: atrihartono@unej.ac.id

I. Kartini and Kanisa (eds.), *Joko Widodo's Foreign Policy and Diplomatic Legacy*,
https://doi.org/10.1007/978-981-95-2085-5_7

levels of foreign policy analysis. It found that the prominent foci of foreign policy to overview the implementation of Indonesian gastrodiplomacy during Jokowi's tenure involve individuals (leaders and groups of policymakers, including ministries and non-state actors), states (internal factors influencing a state's foreign policy behavior and action), and the international system (capabilities of states interacting with other states and outcomes of the policy). Indonesian gastrodiplomacy initiatives were formulated based on individual characteristics involving leaders and non-state actors, particularly Indonesian diasporas. In determining its goals, the government has considered internal factors such as the abundance of Indonesian spices, food history, and established Indonesian restaurants abroad, but overlooked its capabilities in the international context where regulations and consumer behaviors, such as healthy lifestyle and fresh food-oriented preferences, become concerns, depending on which country Indonesia will target for gastrodiplomacy.

Keywords Co-branding wonderful Indonesia · Foreign policy analysis · Gastrodiplomacy · Indonesia spice up the world · Jokowi administration

7.1 Introduction

It was during the Covid-19 pandemic, when social interactions mainly took place in digital platforms, that numerous online meetings and webinars to discuss Indonesian culinary practices in foreign countries became popular. Various stakeholders, including the government and non-state actors, virtually gathered to talk about the presence of Indonesian restaurants overseas, the prospects for culinary business in foreign countries, the export of Indonesian coffee beans and spices, and the endeavor of Indonesian diasporas to promote Indonesian culinary worldwide. These discussions were then followed by "The First Declaration of Indonesian Gastrodiplomacy," in December 2020, demonstrating a clear and unwavering commitment of the government to promote Indonesia's gastrodiplomacy.

In the second term of Jokowi's leadership, the Indonesian government collaborated with non-state actors such as Indonesian diasporas, culinary associations, chefs, and food experts, to develop a comprehensive gastrodiplomacy strategy, called "Indonesia Spice Up the World (ISUTW)," under the coordination of Indonesia's Coordinating Ministry of Maritime and Investment Affairs. Gastrodiplomacy is generally known as a public diplomacy strategy involving culinary and cultural diplomacy that is supported by financial investment to win the hearts of the foreign public and increase the nation's brand through food use (Rockower 2012). Officially launched on November 4, 2021, the ISUTW strategy is pushed to achieve two primary objectives by 2024, that is an increase of the export value of spices and seasonings from 1.2 billion to USD 2 billion and an establishment of up to 4,000 Indonesian restaurants abroad (Indonesia's Ministry of Tourism and Creative Economy 2021). Prior to this, or during the first term of Jokowi's administration, a program called "Co-branding Wonderful Indonesia" was initiated under the leadership of Arief Yahya at Indonesia's Ministry of Tourism in 2018 (Trihartono et al. 2020). The program invited Indonesian diaspora working in culinary business to join, and has successfully partnered with 100 restaurants abroad (V. Datau, pers. comm., June 10, 2020). However, due to the cabinet reshuffle in Jokowi's second-term administration, the program was halted and subsequently incorporated into the ISUTW.

Scholars have argued that the formulation of a gastrodiplomacy strategy demands the involvement of non-state actors, especially the diaspora groups, often viewed as agents of diplomacy and development (Kennedy 2020). This aligns with Jokowi's second-term leadership, which considers the Indonesian diaspora as an agent to the country's development. Likewise, while acknowledging the significance of human resources, Indonesia's short-term development plan saw that the potential of the Indonesian diaspora should be harnessed to face global competitiveness (Tempo 2019). In fact, long before Indonesia's gastrodiplomacy strategy was launched, Indonesian diasporas have played indefatigable roles in introducing Indonesian foods abroad by establishing restaurant openings, catering services, food festivals, and community gatherings (Yayusman et al. 2023a). While economic motives behind these actions may be present, Mannur (2010) suggests, however, that as such may also be driven by a sense of culinary nostalgia, or diaspora's intention to bring their nostalgic longing for the familiar and affective value of foods and smells from home.

This study aims to examine the formulation of Indonesia's foreign policy in gastrodiplomacy during Jokowi's presidency terms and its implementation, including among Indonesian diasporas in Australia and the Netherlands. Both countries are selected because of their long-standing historical relationships with Indonesia and their substantial numbers of Indonesian diasporas, with approximately 89,480 in Australia (Australia's Department of Home Affairs 2023) and 349,300 in the Netherlands (CBS 2022). We start this chapter by providing an extensive review on gastrodiplomacy strategies, particularly through the lens of foreign policy analysis. We then explore the historical account of how Indonesian foods found a foothold at the international levels. The last part is dedicated to examine the two main gastrodiplomacy strategies during Jokowi's administration Indonesia's Co-branding Wonderful Indonesia and Indonesia Spice Up the World, and the two instances of Indonesian diasporas' roles in Australia and the Netherlands. This chapter concludes with ways forward on the future trajectory of Indonesian gastrodiplomacy under the newly elected president of Indonesia, Prabowo Subianto.

7.2 Method

This study employed a mixed-methods approach, incorporating online surveys conducted in Australia and the Netherlands, in-depth interviews, field observations, and focus group discussions regarding Indonesian gastrodiplomacy. In Australia, two online surveys were administered from September 8 to October 8, 2021, serving as an exploratory tool to gather preliminary data on the Indonesian restaurant business in Australia, as well as to identify and map the existing Indonesian restaurants in the country. Using the Survey Monkey platform to distribute this questionnaire to 27 business owners and 100 members of the Indonesian diaspora in Australia, it was anticipated that the survey's outcomes would produce responses of similar nature and quality to those obtained through traditional paper-based surveys. In the Netherlands, an online survey was conducted from June 16 to July 16, 2022, using the similar platform to gather up to 82 respondents, including Indonesian diaspora and business owners.

Prior to distributing the online surveys, extensive measures were taken to create a well-organized questionnaire. Initially, the research team collected information from various sources, including academic papers, government documents, and regulations. This information encompasses the history of Indonesia's ethnic food and its potential for promotion as a culinary business and a diplomatic tool between the two countries. Subsequently, in the case of Australia, the research team consulted experts in gastrodiplomacy from both Australia and Indonesia, including a government official from the Indonesian Embassy in Canberra, professional chefs, and leaders of restaurant associations. The contributions of the proprietors of the Shalom and Ubud Restaurants, one of whom serves on the board of the Indonesian Restaurant Association (IRA) in Sydney, were also highly valuable. Additionally, the contributions of the owners of Rara de' Kitchen (Melbourne), Mammo Kitchen (Canberra), and professional chefs from "Dapur Bali" and "Shalom Indonesia Restaurant" in Brisbane were also significant. Furthermore, in the Netherlands, insights for the surveys were gained from government officials from the Indonesian Embassy in The Hague, the owner of "Iboe Tjilik" Indonesian Restaurant, and a co-founder of "Indofoodtours"—a tour of Indonesian food in the Netherlands.

Between 2020 and 2023, extensive research was conducted through in-depth interviews with various stakeholders, including government officials in Indonesia, such as the Directorate of Public Diplomacy in Indonesia's Ministry of Foreign Affairs, and the Deputy for Tourism and Creative Economy in Indonesia's Coordinating Ministry of Maritime and Investment Affairs. Interviews were also conducted with individuals from Indonesia's embassies in Canberra and The Hague, culinary experts in Indonesia, Australia, and the Netherlands, chefs, catering service owners, culinary business and restaurant owners, and representatives from culinary associations, such as Indonesia Gastronomy Networks (IGN), Indonesia Gastronomy Community (IGC), and Luv Indonesian Culinary. Field observations were conducted in 2023 to visit Indonesian restaurants in the Netherlands.

7.3 Gastrodiplomacy in Foreign Policy Analysis

In an attempt to introduce Indonesian foods to the global stage and economic growth, gastrodiplomacy has become a strategic instrument stipulated in Indonesia's foreign policy. Foreign policy may refer to the totality of policies of a country toward and in interactions with the environment beyond its borders, commonly utilizing instruments of statecraft such as diplomacy, military, or economic (Hudson 2012;

Hudson and Vore 1995). An examination about how such policy is formulated and implemented is crucial for assessing whether its objectives run accordingly, or further changes should be made. Thus, a foreign policy analysis of the Indonesian gastrodiplomacy strategy is critical. Previous research on Indonesian gastrodiplomacy mostly focus on identifying targeted countries that are potential for the implementation of the program, such as the Netherlands (Yayusman et al. 2023a; Huwaidaa 2022), Australia (Yayusman et al. 2023b;), and Africa (Yayusman and Mulyasari 2024), in addition to discussing Indonesian gastrodiplomacy as the most effective strategy (Muljabar 2024; Septyana and Saragih 2024; Trihartono 2023). However, an analysis of Indonesian gastrodiplomacy as an effective foreign policy to elevate Indonesia's nation brand and introduce its foods at a global stage remains lacking. Thus, this study aims to fill up this lacuna, by examining the foreign policy of Indonesian gastrodiplomacy during Jokowi's tenure.

To realize its economic growth, Indonesia chose gastrodiplomacy as one of the foreign policy instruments for "Trade, Tourism and Investment" aspiration. As a foreign policy, Indonesian gastrodiplomacy represents the acts of public diplomacy, where the national government attempts to winning the hearts and minds of foreign public, by employing foods as its main instruments; thus, it is also called culinary diplomacy. In such a diplomacy, Indonesian foods do not only serve as a material instrument, but also represent the richness of Indonesian culture, through which an array of food-making skills, food-presentation practices, and food's historical accounts are inherited across generations as the national culture; thus, it is known as cultural diplomacy. This understanding is aligned with Rockower's (2020, 2012) definition of gastrodiplomacy, that is, a national government's concerted public diplomacy campaign combining culinary and cultural diplomacy and backed up by monetary investment to raise its nation's brand status, not simply as an ad hoc or one-off culinary diplomacy.

Zhang (2015) also stated that gastrodiplomacy is intended to encompass a broad range of "users" and aims to deliver a specific message to the larger populations of other countries through food. However, according to Melissen (2005), key actors of public diplomacy also include non-state actors playing a pivotal role in undertaking relevant missions. Thus, the involvement of diaspora and other culinary practitioners as an "agent of diplomacy" cannot be overlooked. However, gastrodiplomacy is also defined as an attempt to achieve the state's diplomatic objectives through commensality, involving sitting together at the same table (Prud'homme 2023). This kind of gastrodiplomacy usually focuses on the utilization of food for gaining favor and support during a negotiation process with state officials, thus by definition it tends to exclude a foreign public at large. In this study, we will uphold the gastrodiplomacy contexts that aim to leveraging to the public at large.

Following the ten-year implementation of the Indonesian gastrodiplomacy, it is necessary to study it with foreign policy analysis, a subfield in international relations studies that seeks to explain foreign policy and its behavior with reference to human decision-makers as a theoretical background (Hudson; 2012; Brighi and Hill 2012). In our study, we determine our conceptual framework by harnessing the concept of three levels of analysis in foreign policy (Nye 2009; Breuning 2007; Rosenau 1971;

Singer 1961). According to Breuning (2007), the three levels of analysis correspond to the distinct foci of foreign policy analysis. This includes individuals who evaluate options and make decisions; the state that engages in foreign policy behaviors; and the interaction between states in the international system that produces outcomes (see Table 7.1). However, scholars have suggested that when conducting this analysis, it is important to select one level for an in-depth investigation and to consider the three levels as both indivisible and complementary (Snyder et al. 2002). Taking a cue from it, this study considers the three levels of analysis as a complementary approach to evaluate the future of Indonesian gastrodiplomacy.

At the individual level, the focus of the analysis is on leaders and decision-makers. This may encompass the president, his advisory board, and group interactions, including ministries and non-state actors such as the diasporas. The analysis of individuals can concentrate on their personalities, characters, or perceptions of how they interpret the world and respond to current global trends. The inclusion of group interactions at the individual level is crucial because it helps us to understand the dynamics of interpersonal interactions within the assigned group and to note that individuals do not make decision alone instead work together with others in a group or in a bureaucratic setting (Allison and Zelikow 1999). At the state level, the analysis is focused on internal factors that influence a state to engage in specific foreign policy behavior (Breuning 2007). This includes states' resources, cultures, and histories that can be utilized to determine which actions should be undertaken. Here, foreign policy behavior manifests as actions of the decisions made. Finally, at the international system level, the focus of the analysis is on interactions and comparisons between states. The international system is defined as a set of states whose interactions are guided by their relative capabilities, such as their power and wealth, which influence their possibilities for action and success on the global stage (Breuning 2007). The system-level analysis provides an assessment of states' capabilities and explanation to the outcomes, making it aligned with the system theory, where phenomena are seen as interrelated through the process of input, output, and feedback from the environment (Brighi and Hill 2012). State's interactions play a significant role in determining whether, in our case, the ambitious goals of Indonesian gastrodiplomacy are effectively implemented in other countries, particularly when they are entangled with different cultures and lifestyles in society, thus enabling us to examine the implementation of Indonesian gastrodiplomacy during Jokowi's tenure.

Table 7.1 Levels of analysis in the study of foreign policy

Level of analysis	Foreign policy focus
Individual	Decisions
State	Behaviors
System	Outcomes (including the process of input, output, and feedback from the environment)

Sources Brighi and Hill (2012), Nye (2009), Breuning (2007), Rosenau (1971), Singer (1961)

7.4 Indonesian Foods Gaining an International Acclaim

Indonesian foods have garnered significant international acclaim, with several dishes receiving recognition in global culinary rankings. In 2024, Taste Atlas ranked *empek-empek*, a traditional South Sumatran seafood dish, as the third-best seafood dish worldwide, surpassing even the renowned Japanese *sushi* (Jakarta Globe 2024). Previously, in a 2017 CNN reader survey, *rendang* (a slow-cooked meat dish) and *nasi goreng* (fried rice) secured the top two positions in the "World's 50 Best Foods" list (CNN 2017). Furthermore, *soto*, an Indonesian soup, was nominated by CNN Travel as one of the world's 20 best soups, noting its popularity not only in neighboring countries like Singapore and Malaysia but also in Suriname, South America, where it was introduced by Javanese immigrants (CNN 2024). These accolades highlight the diverse flavors and growing global appreciation for Indonesian culinary traditions.

Despite the international recognition of Indonesian cuisine, its global restaurant presence remains limited. As of 2021, there were only about 1,200 Indonesian restaurants worldwide (Indonesia's Ministry of Foreign Affairs 2021) significantly fewer than the 15,000 Thai restaurants. With the increasing awareness about the soft power potential of national cuisines, several Asian governments have actively promoted their culinary traditions at the international stage. Japan and Thailand stand out as successful examples of state-led culinary diplomacy, with their efforts forming part of broader nation branding and cultural outreach strategies (Rockower 2012, 2020; Farrer 2015). Recently, countries like Malaysia and Korea have followed their steps to enhance the global appeal of their national cuisines, demonstrating a growing awareness of gastrodiplomacy as a tool for cultural influence and economic opportunity.

Indonesian food is renowned for its vibrant array of flavors, ingredients, and cooking techniques found across its regional dishes. Such diversity represents a fascinating fusion of historical, traditional, and cultural traits that have evolved over centuries, shaped by the country's diverse geographical and climatic conditions (Rahman 2016). The country's strategic location along ancient trade routes, such as the famed Spice Route, attracted traders from China, India, the Middle East, and Europe, facilitating the exchange of spices, herbs, and cooking methods that eventually blended with indigenous ingredients and cooking techniques. Likewise, Indonesia's history of colonization by various European powers, including the Dutch and Portuguese, left a lasting impact on its cuisine, contributing to the unique and diverse culinary landscape that characterizes Indonesian food today.

While the equatorial climate in Indonesia ensures an abundance of tropical fruits and vegetables, fertile volcanic soil of its certain regions is a hotbed for rice cultivation, making it a staple in the Indonesian diet and the popularity of diverse rice-based dishes. Moreover, the country's tropical rainforests yield an array of spices, including cloves, nutmeg, and cinnamon, which have become fashionable commodities and pivotal to the development of Indonesian cuisines (Yayusman et al. 2023b). The remarkable diversity of Indonesian cuisine is evident in the extensive culinary offerings identified across the archipelago. Gardjito (2020) has recently identified an impressive 3,259 types of culinary offerings, almost a double of the number recorded by the cookbook Mustika Rasa, published in the 1960s (Hardjohutomo et al. 1967). Mustika Rasa is a government-involved project aimed to develop the national cuisine by localizing European-influenced recipes as an attempt to standardize cooking measurements and techniques, despite the regional variations (Hardjohutomo et al. 1967), indicating government's effort behind the development of Indonesian national cuisine (Rahman 2018).

Partly inspired by the success of gastrodiplomacy initiatives in other Asian countries, Indonesia officially launched its own program in 2021. However, long before this formal launch, there has been active promotion of Indonesian cuisine and spices through various channels. According to Yayusman and Mulyasari (2024), Indonesia's gastrodiplomacy journey can be traced back to 2008, with several key programs marking its evolution: Indonesia Inspired (2008), Indonesia–Spain Culinary Cooperation (2012), Spice it Up Programme at Frankfurt Book Fair (2015 and 2016), Wonderful Indonesia Co-Branding Programme (2018), and The First Congress of Indonesian Gastrodiplomacy (2020). These initiatives demonstrate Indonesia's long-standing commitment to leveraging its rich culinary traditions as a tool for cultural diplomacy and national branding on the global stage.

7.5 Gastrodiplomacy During Jokowi's Administration: Co-Branding Wonderful Indonesia and Indonesia Spice Up the World

This section is devoted to explore two gastrodiplomacy attempts designed and implemented during Jokowi's two-period administration, namely Co-Branding Wonderful Indonesia and Indonesia Spice Up the World, respectively coordinated by Indonesia's Ministry of Tourism and Creative Economy and Indonesia's Coordinating Ministry of Maritime and Investment Affairs. While having distinct objectives, both shared a common goal of promoting economic growth and enhancing the nation brand through the practice of gastrodiplomacy, as we will explore below.

7.5.1 The First Term: Co-Branding Wonderful Indonesia

Co-branding Wonderful Indonesia is a collaborative project between the state-initiated brand "Wonderful Indonesia" and its various partners from business, government, and community sectors. Popularly known as "Wonderful Indonesia" the project was introduced in 2018 with an aim to enhance the competitiveness of Indonesia's tourism industry at a global level. As such it is done by creating mutually beneficial synergies among the partners, one that allows them to strengthen their brand

and achieve outcomes greater than the sum of each party's individual contributions (Hendriyani 2023). While aiming to promote Indonesian tourism, this initiative fosters synergies between the government and private sectors to improve the competitiveness and global image of Indonesian tourism, including through foods.

The collaboration allows all partners involved to benefit from the branding image of Wonderful Indonesia, enhancing the visibility and sales of their products. For instance, the collaboration between the KitKat and Grab brands is reported to have boosted their brand awareness and sales (Indonesia Travel, n.d.). As a concept, Wonderful Indonesia was initiated by the Minister of Tourism in response to the country's increased ranking in the World Economic Forum (WEF). In 2019, Indonesia's brand rankings improved to the 40th position, surpassing "Truly Asia Malaysia" and "Amazing Thailand" brands, thus creating a brand value for public–private partnerships (PPP). It is important to note, however, that the cooperation is not necessarily in the form of financial transactions, as the government should not receive money from the private sector except for joint events.

Subsequently, the Ministry of Tourism formed a co-branching team to organize and facilitate this collaboration. Since its launch, the co-branding has involved 129 partners from a wide range of industries (Indonesia Travel, n.d.), with collaborative activities including design competitions, shopping programs, and discussion forums to plan more sustainable collaborations (Yusuf 2024). Not to mention its collaboration with 100 restaurant-owner Indonesian diasporas living in Europe, America, and Asia (V. Datau, pers. comm., June 10, 2020). Until the program ended, ten overseas Indonesian restaurants were successfully co-branded, including: (1) Ron Gastrobar in Amsterdam; (2) Djakarta Bali in Paris; (3) Yono's Restaurant in Albany, New York; (4) Kasih Restaurant in Los Angeles, California; (5) Fluffy Lamb in Perth; (6) Sendok Garpu in Brisbane; (7) Ubud Restaurant in Sydney; (8) Sari Ratu in Singapore; (9) Bumbu Desa in Kuala Lumpur; and (10) Indonesia Grill and Gastrobar in Houston, Texas (Trihartono et al. 2020a). While these restaurants served as promotional outlets for Wonderful Indonesia, known as Wonderful Indonesia Corner (Syana 2019; Dwi 2018), the initiative encouraged the restaurants to promote Indonesian foods in the host country (Yayusman and Mulyasari 2024), especially those regarded as "the national cuisine," such as *rendang* and *sate*, as well as to promote ten priority tourist destinations in Indonesia, wishing that it will increase the interest of international tourists in visiting Indonesia (Syana 2019; Dwi 2018; Baskoro 2022).

By becoming a co-branding partner, these restaurants can leverage the power of Wonderful Indonesia's brand to enhance its national brand value. They do so by disseminating branding material provided by the Ministry of Tourism, which includes information about Indonesian tourists and culinary destinations (Yayusman and Mulyasari 2024). The initiative was expected to increase the contribution of the culinary sector to the Indonesian creative economy, which is currently making a significant contribution to the Gross Domestic Product (GDP). The culinary sector is said to accounts for approximately 42% of the country's creative economies, making it a crucial sector to promote (Media Indonesia 2018). By adopting this strategy, successfully implemented in Thailand with its thousands of overseas restaurants,

Indonesia obviously sought to expand its culinary reach level and promote its rich cultural and natural heritage at the global level.

7.5.2 *The Second Term: Indonesia Spice up the World*

Inaugurated on November 4, 2021, at the Expo 2020 in Dubai, Indonesia Spice Up the World (ISUTW) is a government-led program aimed at boosting the economic value of tourism, trade, and investment through gastronomy industry. It does so by employing a set of public diplomacy efforts to attract foreign people and achieve its economic goals, thus it is part of gastrodiplomacy. Its primary objectives are three folds: to promote Indonesian cuisine abroad, increase the export of spices, and develop Indonesian restaurants overseas. To achieve these goals, it involves collaboration with multiple agencies, including the central government, districts, entrepreneurs, academics, chefs, culinary experts, diasporas, private sectors, culinary communities, and media.

Under the lead of Odo Manuhutu, the Coordinating Ministry of Maritime and Investment Affairs was responsible for coordinating the ISUTW, along with other ministries, including the Ministry of Foreign Affairs, the Ministry of Tourism and Creative Economy, the Ministry of Health, the Ministry of Agriculture, the Ministry of Education, Culture, Research, and Technology, PT Garuda Indonesia, and the National Research and Innovation Agency (BRIN). The ISUTW had two primary objectives: (a) increasing the value of spices and spice exports by USD 2 billion and (b) establishing 4,000 Indonesian restaurants in foreign countries by 2024 (Indonesia's Ministry of Tourism and Creative Economy 2021). Initially, this program aimed to enhance the country's gastro-destination places to attract foreign tourists. It also set Africa and Australia as its targeted gastrodiplomacy regions. However, over its implementation, ISUTW's targeted regions were expanded to all countries globally Sartin (2021a, b), and by the end of 2022, 12 pilot ISUTW project countries were established, including the United States, China, Japan, the Netherlands, Australia, the United Kingdom, Germany, France, the United Arab Emirates (UAE), Turkey, Qatar, and South Africa (Indonesia's Coordinating Ministry of Maritime and Investment Affairs 2022).

ISUTW also selected five "national" foods to be promoted in its campaigns. These are *nasi goreng, sate, soto, rendang*, and *gado-gado*. The idea is to make these dishes as a catalyst for popularizing other Indonesian dishes among international audiences. Furthermore, the program provides partnership facilities to support Indonesian restaurants, such as assistance with business redesign, spice and seasoning supply networks, and promotional materials for tourism products and creative economies (Indonesia's Ministry of Tourism and Creative Economy 2021). As of the beginning of 2024, this program was scheduled to establish a ministerial regulation for the creation of ISUTW's committees to ensure effective implementation in the future. Economic constraints have emerged as a significant factor impeding its realization and the objective of establishing 4,000 restaurants has not yet been achieved. Wahab

(2023, pers. comms.,) stated that the government cannot provide direct financial support, for instance, for the establishment of Indonesian restaurants abroad. Alternatively, national banks such as BNI, are now offering loans to selected potential restaurateurs for the development of restaurants. This program can serve as an alternative, albeit not a definitive solution, for the realization of 4,000 restaurants.

7.6 Diasporas in Indonesian Gastrodiplomacy: Case Studies in Australia and the Netherlands

As part of "agent of diplomacy," Indonesian diasporas play a prominent role in the implementation of Indonesian gastrodiplomacy. Throughout its formulation process, particularly in the creation of the ISUTW, the contribution of Indonesian diasporas is noteworthy. The organization of "the First Congress of Indonesian Gastrodiplomacy" during the COVID-19 pandemic was a result of the initiative of Indonesian diasporas, led by Robert Manan, the leader of Luv Indonesian Culinary and proprietor of Indonesian restaurants in Atlanta, United States. This event effectively convened various stakeholders, including government representatives, culinary experts, chefs, researchers, and many others to increase awareness and establish commitments for the global promotion of Indonesian cuisine. Subsequently, the formulation of Indonesian gastrodiplomacy has progressed with increased rigor, culminating in the enactment of ISUTW in 2021.

Our previous research (Yayusman et al. 2023a, 2023b; Yayusman and Mulyasari 2024) demonstrated that Indonesian diasporas in Australia and the Netherlands can serve a crucial function in the promotion of Indonesian cuisine and the execution of Indonesian gastrodiplomacy. The historical connections with these two countries constitute a primary factor in the substantial presence of Indonesian diasporas in both nations. In essence, it can be asserted that prior to the implementation of Co-branding and ISUTW initiatives, Indonesian diasporas have been actively promoting Indonesian cuisine through restaurants, catering services, and cultural festivals. The following presents our findings regarding the role of Indonesian diasporas in Indonesian gastrodiplomacy.

7.6.1 Indonesian Diasporas in Australia

The Indonesian diaspora in Australia has experienced significant growth over the past few decades. They have established robust connections between the two nations through various diplomatic channels, including gastrodiplomacy. As of 2023, official records indicate that more than 89,480 Indonesians reside in Australia, encompassing students, professionals, and other individuals residing temporarily or permanently (Australia's Department of Home Affairs 2023). They maintain Indonesian traditions, language, and culinary practices while actively introducing Indonesian culture to the Australian populace, frequently through culinary events and traditional dishes (Jati 2021; Yayusman et al. 2023b; Yayusman and Mulyasari 2024). The Indonesian Embassy in Canberra has identified approximately 140 Indonesian restaurants across eight states and two territories in Australia (G. Ismail, pers. comm. 2021). However, a recent survey conducted by the Public Diplomacy Directorate, Indonesia's Ministry of Foreign Affairs in July 2021 documented 151 Indonesian restaurants in Australia.

The spirit of Indonesian business owners in Australia has led to the establishment of associations such as Indonesian Restaurant Associations (IRA) based in Sydney. Additionally, other Indonesian diaspora groups, including students, have formed communities, such as the Indonesian Students Association (PPI) and various community organizations. These entities actively promote Indonesian foods throughout the nation through diverse channels, primarily take-away services (45.5%), catering services (36.4%), restaurants (22.7%), festivals (22.7%), and other means (36.4%) (Survey on Indonesian Gastrodiplomacy in Australia, BRIN 2021). Social media platforms, particularly WhatsApp groups and Facebook, have emerged as prominent promotional tools enabling restaurants and culinary business owners to reach both local and diaspora consumers. According to a survey conducted in 2021, the use of WhatsApp for Indonesian culinary promotion is the most prevalent, with approximately 79% of business owners employing this platform for promotional purposes. Other social media platforms, including Facebook (47%), Instagram (26%), Uberfood (21%), Menulog (5%), and Deliveroo (5%) were also used for culinary promotion.

Furthermore, Indonesian diasporas running culinary business, such as Shalom and Ubud Restaurants, have considered the significance of preserving the authentic taste of Indonesian cuisine while adapting their menus to accommodate local preferences. The survey indicated that 75% of local residents in Australia preferred savory flavors in their food selections. Concurrently, Indonesian students and cultural associations also made their own contribution, by organizing culinary events, cooking classes, and food festivals attracting diverse audiences (Yayusman et al. 2023b). Obviously, prior to the implementation of the Co-branding and ISUTW programs, Indonesian diasporas in Australia had actively promoted Indonesian foods for economic purposes and many other reasons.

Notwithstanding the establishment of the Co-branding and ISUTW programs, the Indonesian embassy in Canberra has implemented various support initiatives to enhance gastrodiplomacy in Australia, including the organization of annual culinary festivals, training sessions for Indonesian restaurant proprietors, and additional programs to introduce Indonesian cuisine to a wider audience. In Sydney, the Indonesian Consulate General collaborates with Indonesian food and beverage importers, such as the Sony Trading and Eastern Cross Trading, contributing to the availability of Indonesian products in the Australian market (Yayusman et al. 2023b). Despite these endeavors, limitations in logistical and financial infrastructure often affected the effectiveness of Indonesian gastrodiplomacy. Indonesian restaurants in Australia frequently encounter supply chain challenges, particularly in sourcing authentic spices and ingredients at reasonable prices, due to elevated import costs.

Furthermore, bilateral trade agreements such as the Indonesia–Australia Comprehensive Economic Partnership Agreement (IA-CEPA), effective per June 5, 2020, have facilitated the importation of Indonesian spices and food products through tariff elimination (Directorate General of National Export Development, Indonesia's Ministry of Trade, pers. comm. 2021). Although this policy has enhanced access to Indonesian ingredients, a comprehensive support system for restaurant promotion and marketing remains a significant impediment. Indonesian restaurants in Australia face high operational costs and competition, exacerbated by the absence of consistent government funding and a more structured distribution infrastructure. A more pronounced emphasis on establishing a reliable supply chain and developing trade routes for Indonesian culinary goods could augment ISUTW's impact, thereby facilitating the delivery of authentic flavors to local consumers by Indonesian restaurants in Australia.

The promotion of Indonesian food in Australia encompasses more than merely showcasing culinary diversity; it entails fostering a deeper cultural understanding and establishing enduring bonds of friendship. The efficacy of Indonesia's gastrodiplomacy is contingent upon the collaborative efforts of governmental entities, diaspora communities, and private sector stakeholders, all working toward the shared objective of elevating Indonesian cuisine on the global stage. Through these concerted efforts of Co-branding and ISUTW programs, Indonesian cuisine in Australia has the potential to emerge as a cultural staple, akin to Thai or Vietnamese foods, esteemed for its distinctive flavors and narratives of Indonesia's rich heritage and history.

7.6.2 Indonesian Diasporas in the Netherlands

The history of the Indonesian diasporas in the Netherlands is a long and rich narrative, encompassing various phases from the colonial period to the modern era. Indonesian diasporas there reflect a complex relationship between two countries with a deep historical connection. The relationship between Indonesia and the Netherlands began during the colonial period when the Dutch ruled the archipelago through the *Vereenigde Oost-Indische Compagnie* (VOC) and later the Dutch East Indies government. During this time, many Dutch people lived in Indonesia and were influenced by the local culture. The first wave of migration from Indonesia to the Netherlands began in the early twentieth century, involving students, domestic workers, and sailors (Ariwibowo 2024).

In the early 1900s, an "*Indisch*" eatery was established in The Hague by the *Vereeniging Oost en West* (East and West Association), aimed at popularizing Indies culture among the Dutch public. This place started serving its members and guests *rijsttafel* (a combination of rice and various dishes from Indonesian and *Indisch* cuisine adjusted to European tastes). This eatery eventually evolved into a large restaurant, and twenty years later, several other *Indisch* eateries emerged in the Netherlands (Ariwibowo 2024; Kuipers 2017). Following Indonesia's independence in 1945, a significant wave of migration occurred when around 300,000 people left

Indonesia for the Netherlands. Many were Dutch nationals who had previously lived in colonial Indonesia and Indonesians associated with KNIL (Royal Netherlands East Indies Army). This repatriation also brought Indonesian culture and culinary practices to the Netherlands, and many established small businesses, including restaurants, to make a living. They initially faced challenges in establishing these businesses due to a lack of necessary prerequisites, but the demand for ethnic goods and services gradually grew.

Culture played a crucial role in the social integration process for the Indonesian diaspora in the Netherlands, with culinary being one of the critical aspects. In the Netherlands, Indonesian diasporas began opening small restaurants to sustain themselves, introducing Indonesian food to the Dutch. For instance, in Amsterdam, the number of Chinese Indonesian restaurants increased from two in 1945 to 44 in 1960, while Indonesian restaurants grew from zero to six during the same period. Dishes like *bami*, *sate*, and *nasi goreng* emerged as the most favored Indonesian dishes in the Netherlands (Yayusman et al. 2023a). From the early 2000s to the 2010s, the culinary landscape in the Netherlands underwent significant changes. The number of Indonesian ethnic restaurants increased rapidly during this period, offering dishes from various regions of Indonesia, such as Minangkabau, Palembang, Java, Manado, Bali, and Maluku. Restaurants established by the Indonesian diaspora not only functioned as business ventures but also as efforts to strengthen cultural identity. They offered a variety of "authentic" and fusion dishes appealing to European tastes, and introduced Indonesian culinary heritage to a broader audience (Yayusman et al. 2023a).

Data from the Directorate of Public Diplomacy, Indonesia's Ministry of Foreign Affairs (2021), indicated that Indonesian restaurants in the Netherlands comprise approximately 295 Indonesian restaurants, serving as significant cultural intermediaries and showcasing authentic Indonesian flavors while embodying Indonesia's heritage and history. Indonesian diasporas have innovatively established restaurants representing their hometowns in Indonesia, such as *Lapek Jo*, with Padangnese specialties, and *Prabomoelih* with Palembangnese dishes. Both establishments serve Indonesian foods with flavors closely resembling those found in the country of origin. With its strong historical ties to Indonesia and existing resources and established restaurants, the Netherlands plays a significant role in Indonesian gastrodiplomacy. However, the Indonesian government has yet to fully leverage the presence of the diaspora and target the Netherlands as a strategic country in the context of strengthening Indonesian gastrodiplomacy. Indonesian restaurants in the Netherlands, established by the diaspora, may serve as "frontline messengers" in introducing Indonesian cuisine to the Dutch public and anchor gastrodiplomacy (Yayusman et al. 2023a).

According to available data, in 2022, approximately 349,300 Indonesians lived in the Netherlands (CBS 2022). This figure includes Indonesians who migrated legally and those who have acquired Dutch citizenship through mixed marriages or naturalization processes. This population comprises various social and economic backgrounds, reflecting the diversity of the Indonesian community in the Netherlands. The population of Indonesian nationals in the Netherlands based on 2022 data from CBS (Statistics Netherlands) is as follows (Table 7.2).

Table 7.2 Numbers of Indonesian diasporas in the Netherlands as per 2022

Categories	Population
Born abroad	106,300
Born in the Netherlands, one or both parents were born abroad	257,700
Born in the Netherlands, one parent was born abroad	190,400
Born in the Netherlands, both parents were born abroad	67,700

This data indicates that most Indonesian diaspora in the Netherlands were born in the Netherlands but maintain strong connections to Indonesia through their parents or grandparents.

As said earlier, in response to the growing influence of Indonesian cuisine, the Indonesian government launched the ISUTW initiative to increase spice exports and promote Indonesian restaurants globally. However, the ISUTW's full impact in Europe, particularly the Netherlands, remains limited due to logistical and regulatory challenges. In addition, in the Netherlands, potential Indonesian entrepreneurs face high opening costs and stringent regulations governing hygiene standards, food presentation and storage, tax requirements, and the relationships between restaurants and their surrounding communities. Although crucial to operating in the Dutch market, these regulations add to the difficulty of establishing Indonesian culinary businesses. They are arguably often handled independently by these Indonesian culinary entrepreneurs without extensive support from the Indonesian government.

7.7 Analyzing Indonesian Gastrodiplomacy: The Three Levels of Analysis

Jokowi administration has demonstrated innovative leadership and significant progress in leveraging the food sector for economic development. Seeing gastrodiplomacy as a form of soft power, the national government recognized that resources such as spices and authentic cuisines, and non-state actors, such as diaspora communities, are crucial to realize its gastrodiplomacy objectives. However, the decision to employ Indonesian gastrodiplomacy as a foreign policy option for enhancing economic growth and as an instrument of economic diplomacy was the result of a deliberate decision-making process in which the leadership of Jokowi and his staff played a crucial role. This section analyzes the dynamic formulation of Indonesian gastrodiplomacy, its strategies of implementation, and the outcomes of its execution, by employing three levels of analysis: individual, state, and international system.

At the individual level, the foreign policy focus is on the underlying factors that represent a leader's personality characteristics, perceptions, and interpersonal relations with their advisory board, ministries, and other relevant actors in policy formulation. During his two terms in office, President Jokowi pursued a robust and ambitious goal of holistic economic development throughout the nation, although his tenure was criticized to be predominantly oriented toward infrastructure development (Warbuton 2016; Salim and Nagara 2018). In his "Nawa Cita," or nine priority agenda items for Indonesia, an economic emphasis is obvious: enhancing the productivity and competitiveness of Indonesian citizens in international markets and realizing economic independence by empowering strategic sectors in the national economy (Wuryandani et al 2015). The spices and food sectors were among these considerations. In his foreign policy, a sentiment for Trade, Tourism, and Investment (TTI) has also become central mechanisms for economic growth. Jokowi did not work alone to achieve his ambitious economic goals. The implementation of its program, in this context, Indonesian gastrodiplomacy, was deliberated and formulated with relevant ministries, such as the Ministry of Tourism (2014–2019) and the Coordinating Ministry of Maritime and Investment Affairs (2019–2024). The promotion of Indonesian culinary heritage was strategically appealing because of the richness of the culture and history embodied in a variety of dishes. As a middle-power country, to achieve significant global prominence, the use of soft power crucial, wherein food can serve as a tool for nation branding and engagement with the global community (Rockower 2012). Gastrodiplomacy has emerged as a feasible policy option.

In the context of Indonesian gastrodiplomacy, the former Minister of Tourism, Arief Yahya, initiated the Co-branding Wonderful Indonesia. The program proved notably successful, as *rendang* was voted as one of CNN's 50 best foods in the world in 2017. Initial engagement with Indonesian diaspora communities was established by implementing the co-branding initiatives, involving 100 Indonesian restaurants overseas. Following a cabinet reshuffle in President Jokowi's second term, the initiatives were placed under the Coordinating Ministry of Maritime and Investment Affairs, and the promotion of Indonesian cuisine and nation branding was extended to actively include non-state actors. During the COVID-19 pandemic, Indonesian diasporas worldwide, both temporarily residing in Indonesia and actively working abroad, convened to discuss Indonesian cuisine and strategies for its global promotion. Several embassies also conducted a series of webinars of Indonesian gastrodiplomacy. For instance, in the Netherlands, the Indonesian embassy organized a webinar in March 2021, inviting Indonesian chefs, business consultants, and Indonesian restaurant owners to discuss opportunities in the Indonesian food industry and relevant regulations for establishing businesses there. Furthermore, Indonesian diaspora organizations, such as the Indonesian Diaspora Networks Global (IDN-G), incorporated gastrodiplomacy as a theme in their online discussions. Stemming from leadership initiatives for economic growth, various creative activities in realizing Indonesian foreign policy for the economic sector have been undertaken, including gastrodiplomacy. In the second term, government's implementation of the Indonesia Spice Up the World (ISUTW) strategy received greater support from diaspora communities, culinary experts, culinary associations, and business owners, particularly in formulating the goals and narratives of the ISUTW. This demonstrates that group interactions influence government decision-making. The individual layer of the decision-making process was not limited to the ultimate decisions of leaders or presidents but also encompassed intensive interactions with other relevant actors with underlying motivations and interests to achieve the same goal.

Now let's turn to the state level of the analysis, in which we focus on how the government executed its gastrodiplomacy strategy, and how its behavior and action about it reflects Indonesia's internal factors, such as natural resources, human resources, and other domestic aspects. Indonesian government was severely criticized for its lack of coordination when preparing and planning the project (Salim and Nagara 2018). However, in the first term of Jokowi administration, the objectives to be achieved from Indonesian gastrodiplomacy were pretty clear and the role of the Ministry of Tourism was well-defined. By employing "Co-branding Wonderful Indonesia," Indonesian gastrodiplomacy benefited from the compelling tagline of "Wonderful Indonesia," thus leveraging the beauty of Indonesian landscape and diversity to achieve its gastrodiplomacy-related goals. In this initiative, the Indonesian government engaged with Indonesian diaspora working as restaurateurs abroad. The objective is to promote Indonesian cuisine while simultaneously promoting the beauty of Indonesia, by incorporating images and other Indonesian creative arts into their restaurants. During this term, five traditional dishes were designated as "the national food," selection of which involved Indonesian culinary experts and academics. These dishes are: *nasi goreng, sate, soto, rendang,* and *gado-gado.* Despite many critiques about local variants of these dishes, the five national foods continue to be utilized for promotion in the second term of the gastrodiplomacy project.

The cabinet reshuffle in the second term of Jokowi's presidency transferred the responsibility of Indonesian gastrodiplomacy to the Coordinating Ministry. This transition indicates the country's level of seriousness to the project's success. However, it necessitates modifications to the goals set out by the previous ministry. With the notable involvement of non-state actors in the second term, ISUTW was enacted during the pandemic. The implementation strategy aimed to achieve two primary objectives: a) increasing the value of spices and spice exports by USD 2 billion; and b) establishing 4,000 Indonesian restaurants in foreign countries by 2024. Concurrently, non-state actors, including Indonesian Gastronomy Networks (IGN), Indonesian Gastronomy Community (IGC), Luv Indonesian Culinary, and various other culinary associations, participated in the development of the ISUTW narrative as an implementation guideline. ISUTW also created an implementation toolbox, enhancing its practicality. Strategies such as the re-activation of promotion and events related to ISUTW could be preliminarily implemented for branding and promotion purposes. Meanwhile, certain dynamics and uncertainties emerged in the stipulation of the goals. ISUTW initially had three objectives, including the improvement of Indonesian gastro-destinations in selected regions in Indonesia. However, this goal was omitted during the formulation process. Furthermore, ISUTW, which initially selected two target regions (Australia and Africa) as pilot projects for implementation, ultimately decided not to confine its implementation, thereby creating opportunities to achieve these ambitious goals globally. During its implementation in 2022, discussions led to the identification of 12 target countries for preliminary ISUTW implementation, considering factors such as inbound tourists, outbound tourists, and the value of spice exports. These evolving goals and actions arguably influence the overall implementation of Indonesian gastrodiplomacy.

At the international system level, the focus of the analysis is on interactions and comparisons between states and on the foreign policy outcomes. It is noteworthy that the capabilities of Indonesia as a country and its citizens are crucial considerations in implementing Indonesian gastrodiplomacy. While its objectives are sufficiently clear; implementation of the project is significantly influenced by commitment and capabilities in the interactions. Therefore, it is essential to comprehend the nature of other states and the capacity of Indonesia to influence the foreign public. The utilization of Indonesian diasporas and their restaurant or other food businesses is highly advantageous. In this regard, Australia and the Netherlands can be regarded as the two most promising countries, not only because of the number of Indonesian diasporas there, but also their enthusiasm to get involved in implementing Indonesian gastrodiplomacy, highlighting their potential as "agents of diplomacy" in realizing the goals of Co-branding and ISUTW. Moreover, existing Indonesian restaurants and the familiarity of foreign populations in both countries with Indonesian cuisine are key factors for success. Among them, some have knowledge of Indonesian foods, with *sate* (50%) being the most favored Indonesian food in Australia and *nasi goreng* (66.7%) being the most favored Indonesian food in the Netherlands (Survey of Indonesian Gastrodiplomacy in Australia and the Netherlands 2021, 2022).

However, it is not merely a matter of sufficient capabilities that the Indonesian diaspora possesses in interacting in the international arena, but also the consistency and concrete support of the Indonesian government in implementing this policy. The outcomes of this policy are arguably limited. The ambitious goal of establishing 4,000 restaurants by 2024 has not been well implemented, as the concrete mechanism remains in flux and incomplete. The fact that the Indonesian government cannot provide direct financial support to business owners abroad to open restaurants diminishes the expectations of Indonesian citizens in executing the program. If any support is available, mechanisms such as loans from national banks represent one of the most feasible approaches, as evidenced by their application to one Indonesian ethnic restaurant in the Netherlands, *Lapek Jo*.

Indonesia has proposed a specific mechanism through Indonesia Eximbank (LPEI) to provide financial support for this program. However, four years of implementation appear insufficient to realize this initiative. A recent critical assessment is that the diminishing role of the Coordinating Ministry of Maritime and Investment Affairs under the newly elected president, Prabowo Subianto, may affect further implementation of the Indonesian gastrodiplomacy program. Consequently, no governmental institution may assume responsibility for this program. A new scenario for Indonesian gastrodiplomacy is likely to emerge, either as a continuation of the previous program with modifications or as an entirely different approach. Despite these challenges, a key consideration for future Indonesian gastrodiplomacy is to ensure interaction with other states and assess the response, familiarity, and consumer behaviors of foreign environments to Indonesian cuisine before pursuing ambitious aspirations. It is crucial to identify the countries that align with Indonesian capabilities.

7.8 Conclusion

This chapter has examined the formulation of Indonesian gastrodiplomacy strategy and its implementation, including a consideration of prominent roles of Indonesian diasporas in Australia and the Netherlands, by using three-level foreign policy analysis: the individual, the state, and the international system. With the increasing awareness about the potential of gastrodiplomacy for cultural influence and economic growth, while at the same time Asian countries like Thailand and Japan have actively promoted their culinary traditions at the international stage, Indonesian government during Jokowi's two-term administration has attempted to establish its own gastronomy strategy, manifested in the form of Co-branding Wonderful Indonesia and Indonesia Spices Up the World programs. Given the richness of the culture and history of Indonesian cuisines, along with the power of food as a tool for nation branding and engagement with the global community, the use of culinary heritage to achieve significant global prominence is crucial for a middle-power country like Indonesia.

It goes without saying that Jokowi's economic-development-oriented "Nawa Cita" vision played a significant role in the initiation of Indonesian gastrodiplomacy strategy. However, the formulation and implementation of the strategy have involved relevant ministries such as the Ministry of Tourism (2014–2019) and the Coordinating Ministry of Maritime and Investment Affairs (2019–2024) and other state-owned institutions and government think-tanks such as PT Garuda Indonesia and the National Research and Innovation Agency, respectively. Non-state actors such as culinary-related professional associations and Indonesian diasporas working in culinary business, both living in and outside the country, were also actively involved as part of the actors succeeding the Indonesian gastrodiplomacy strategy. We noted that group interactions among these actors significantly influence government's decisions about its gastrodiplomacy strategies, to the point that at the level of the individual, the decision-making process was not limited to the state leaders, but also encompassed intensive interactions between the government and other relevant actors, all with underlaying motivations and interests to achieve the same goals. This means that the formulation of Indonesian gastrodiplomacy initiatives was based on the characteristics of individuals involving in the projects, including the state leaders and non-state actors, particularly Indonesian diasporas.

This study has also shown that, while the goals of Indonesian gastrodiplomacy are clearly defined, they are set as "changeable," that is, from "promoting Indonesian cuisines and touristic destinations to international audiences" during Jokowi's first-term presidency, to "increasing the number of overseas Indonesian restaurants and spice export values" in his second term. This change was also followed by the shift of coordinating leadership of the program, from the Ministry of Tourism to a coordinating ministry, indicating the increasing seriousness from the side of the government to succeed its gastrodiplomacy objectives. Five dishes were also selected as "the national food," which subsequently is used as a catalyst to promote Indonesian dishes globally. Not to mention government's effort to invite non-state actors to participate in developing Indonesian gastrodiplomacy narratives and implementation guidelines, and the selection of several countries as targeted destinations of the program. Obviously, the implementation of the program has considered the internal factors such as Indonesian spice abundance, its food histories and connections with other "international" communities, and the number of Indonesian diasporas and the presence of Indonesian restaurants in foreign countries. Still, our analysis of the international system level indicates that Indonesian gastrodiplomacy strategy has overlooked the international contexts where regulations and consumer behaviors, such as healthy lifestyle and fresh food-oriented preferences, become concerns. Moreover, while Australia and the Netherlands may serve as the most promising countries to become the pilot project of Indonesian gastrodiplomacy, due partly to their long-standing connections with Indonesia, the government seems to overlook their potentials.

Summing up this chapter, a new strategy is needed to improve Indonesian gastrodiplomacy. Rather than establishing fine-dining Indonesian restaurants, the implementation of food trucks and small food stalls introducing Indonesian cuisine as a grassroots movement of nation branding may serve as an effective initial approach to nation branding and achieve broader visibility across all societal strata overseas. Not to mention, with the rise of healthy lifestyle in certain countries abroad, Indonesian gastrodiplomacy needs to think of adding one of Indonesian plant-based dishes into its "national food" list.

References

Allison, G. T. and P. Zelikow. 1999. *Essence of decision: Explaining the Cuban Missile Crisis.* Longman.

Ariwibowo, G. A. 2024. Pendidikan selera dalam perkembangan restoran Hindia Belanda dan rijsttafel di Belanda pada periode kolonial. *Paradigma: Jurnal Kajian Budaya* 14 (1): 56–76. https://doi.org/10.17510/paradigma.v14i1.1382.

Baskoro, R. M. 2022. Kisah selera dari negeri rempah: Memahami gastrodiplomasi dari perspektif Indonesia. *Indonesian Perspective* 7 (2): 227–249. https://doi.org/10.14710/ip.v7i2.50780.

Breuning, M. 2007. *Foreign policy analysis: A Comparative Introduction.* London: Palgrave Macmillan.

Brighi, E., and C. Hill. 2012. Implementation and behaviour. In *Foreign policy: Theories, actors, cases*, ed. S. Smith, A. Hadfield, and T. Dunne, 147–167. Oxford University Press.

CBS (Statistics Netherlands). 2022. A new classification of population by origin. *CBS.* https://www.cbs.nl/en-gb/longread/statistische-trends/2022/new-classification-of-population-by-origin?onepage=true.

CNN. 2017. *Your pick: World's 50 best foods.* Retrieved July 23, 2024, from https://edition.cnn.com/travel/article/world-best-foods-readers-choice/index.html.

CNN. 2024. *20 of the world's best soups.* Retrieved July 23, 2024, from https://edition.cnn.com/travel/article/worlds-best-wellness-soups/index.html.

Dwi. 2018. Co-branding Wonderful Indonesia dan 100 restoran milik diaspora. *MIX.* https://mix.co.id/citybranding/cities-campaign/co-branding-wonderful-indonesia-dan-100-restoran-milik-diaspora.

Farrer, J. 2015. Introduction: Traveling cuisines in and out of Asia: Toward a framework for studying culinary globalization. In *The globalization of Asian cuisines: Transnational networks and culinary contact zones*, ed. J. Farrer, 1–19. Palgrave Macmillan.

Gardjito, M. 2020. *Memahami makanan Indonesia* [Understanding Indonesian food]. Yogyakarta.

Hardjohutomo, H. 1967. *Buku masakan Indonesia Mustika Rasa*. Departemen Pertanian.

Hendriyani, I. G. A. D. 2023. Siaran pers: Wonderful Indonesia Co-Branding Award 2023 jadi momentum tingkatkan penjualan produk dalam negeri. *Kemenparekraf*. https://www.kemenpar.go.id/kebijakan/siaran-pers-wonderful-indonesia-co-branding-award-2023-jadi-momentum-tingkatkan-penjualan-produk-dalam-negeri.

Hudson, V. M. 2012. The history and evolution of foreign policy analysis. In *Foreign policy: Theories, actors, cases*, ed. S. Smith, A. Hadfield, and T. Dunne, 13–35. Oxford: Oxford University Press.

Hudson, V. M., and C. S. Vore. 1995. Foreign policy analysis yesterday, today, and tomorrow. *Mershon International Studies Review* 39 (2): 209–238. https://doi.org/10.2307/222751.

Hussin, H. 2018. Gastronomy, tourism, and the soft power of Malaysia. *SAGE Open* 8 (4): 1. https://doi.org/10.1177/2158244018809 2.

Huwaidaa, M. 2022. The gastrodiplomacy: Indonesia's culinary publicity in the Netherlands. *Jurnal Sosial Humaniora Dan Pendidikan* 1 (1): 74–78. https://doi.org/10.56127/jushpen.v1i1.119.

Indonesia's Coordinating Ministry of Maritime and Investment Affairs. 2022. *Hasil rapat koordinasi program Indonesia Spice Up the World* [Minutes of meeting for Indonesia Spice Up the World program]. Unpublished document.

Indonesia's Ministry of Foreign Affairs. 2021. *Gastrodiplomasi Indonesia: Peran dan dukungan Kemlu dalam Indonesia Spice Up the World*. Focus Group Discussion conducted on July 27, 2021. Unpublished document.

Indonesia's Ministry of Tourism and Creative Economy of the Republic of Indonesia. 'Kemenparekraf Dukung Peluncuran "Indonesia Spice Up the World" di National Day Expo Dubai 2020 [Ministry of Tourism and Creative Economy Supports the Launch of "Indonesia Spice Up the Word" in the National Day Expo Dubai 2020]'. 2021, November 5. *Majalah Bandara*. https://www.majalahbandara.com/kemenparekraf-dukung-peluncuran-indonesia-spice-up-the-world-di-national-day-expo-dubai-2020/.

Jakarta Globe. 2024. *Taste Atlas ranks Pempek third best seafood dish worldwide, surpassing sushi*. Retrieved July 23, 2024, from https://jakartaglobe.id/lifestyle/taste-atlas-ranks-pempek-third-best-seafood-dish-worldwide-surpassing-sushi.

Jati, W. R. 2021. Being away from home in Australia: The Indonesian diaspora in Canberra. *Jurnal Humaniora* 33 (2): 93. https://doi.org/10.22146/jh.66455.

Kennedy, L. 2020. Diaspora and diplomacy. In *Routledge handbook of public diplomacy*, ed. N. Snow and N. J. Cull, 213–223. Routledge.

Mannur, A. 2007. Food in multi-ethnic literatures. *Melus* 32 (4): 11–31.

Mannur, A. 2010. *Culinary fictions: Food in South Asian diasporic culture*. Temple: Temple University Press.

Melissen, J. 2005. The new public diplomacy: Between theory and practice. In *The new public diplomacy: Soft power in international relations*, ed. J. Melissen, 3–27. Palgrave Macmillan.

Muljabar, H. 2024. Research of trend gastro diplomacy in Indonesia. *Journal of Islamic World and Politics* 8 (1): 85–93. https://doi.org/10.18196/jiwp.v8i1.72.

Nye, J. S. 2009. *Understanding international conflicts: An introduction to theory and history*, 7th ed. Pearson: Pearson Longman.

Nye, J. S. 2011. Power and foreign policy. *Journal of Political Power* 4 (1): 5–24. https://doi.org/10.1080/2158379X.2011.555960.

Prud'homme, A. 2023. *Dinner with the President*. Alfred A. Knopf.

Rahman, F. 2016. *Jejak rasa Nusantara: Sejarah makanan Indonesia*. Gramedia Pustaka Utama.

Rahman, F. 2018. Kuliner Sebagai Identitas Keindonesiaan. *Jurnal Sejarah* 2 (1): 43–63.

Rockower, P. 2012. Recipes for gastrodiplomacy. *Place Branding and Public Diplomacy* 8 (3): 235–246. https://doi.org/10.1057/pb.2012.17.

Rockower, P. 2020. A guide to gastrodiplomacy. In *Routledge handbook of public diplomacy*, ed. N. Snow and N. J. Cull, 205–212. New York: Routledge.
Rosenau, J. N. 1971. *The scientific study of foreign policy*. Free Press.
Salim, W., and S. D. Negara. 2018. Infrastructure development under the Jokowi administration: Progress, challenges, and policies. *Journal of Southeast Asian Economies* 35 (3): 386–401. https://doi.org/10.1355/ae35-3e.
Sartin. 2021, November 23. Prospects of Indonesian restaurants abroad in Indonesia Spice Up the World. Presented at the *National Seminar of Welcoming the Launch of Indonesia Spice Up the World* conducted by the Research Center for Area Studies, National Research and Innovation Agency.
Sartin. 2021, November 23. Prospects of Indonesian restaurants abroad in *Indonesia Spice Up The World*. Paper presented at the National Seminar of Welcoming the Launch of Indonesia Spice Up The World, Research Center for Area Studies, National Research and Innovation Agency.
Septyana, V., and R. P. Saragih. 2024. The branding narrative of "Indonesia Spice Up The World" through the participation of the Indonesian diaspora in South Korea as brand ambassadors. *Bricolage: Jurnal Magister Ilmu Komunikasi* 10 (1): 119–130.
Singer, J. D. 1961. The level-of-analysis problem in international relations. *World Politics* 14 (1): 77–92. https://doi.org/10.2307/2009557.
Snyder, R. C., H. W. Bruck, and B. Sapin. 2002. Decision-making as an approach to the study of international politics. In *Foreign policy decision-making (revisited)*, ed. R. C. Snyder, H. W. Bruck, and B. Sapin, 21–152. Palgrave Macmillan.
Tempo. 2019. *Enam poin pidato Visi Indonesia yang disampaikan Jokowi*. Retrieved November 1, 2024, from https://www.tempo.co/politik/enam-poin-pidato-visi-indonesia-yang-disampaikan-jokowi-727481
Trihartono, A., H. B. Patriadi, and A. E. Hara. 2023. *Gastrodiplomasi Indonesia*. Pandiva Buku.
Trihartono, A., F. Z. Purwowibowo, B. Pamungkas, and C. Nisya. 2020. The early stage of Indonesia's gastrodiplomacy: In the middle of nowhere? *IOP Conference Series: Earth and Environmental Science* 485 (1): 1–8. https://doi.org/10.1088/1755-1315/485/1/012008.
Warburton, E. 2016. Jokowi and the new developmentalism. *Bulletin of Indonesian Economic Studies* 52 (3): 297–320. https://doi.org/10.1080/00074918.2016.1249262.
Wuryandani, D., L. Adam, P.S.H. Rasbin, N. W. Sayekti, T. M. Lisnawati, S. Izzaty, and S. Wurini. 2015. *Mewujudkan agenda prioritas nawacita*. Pusat Pengkajian, Pengolahan Data dan Informasi (P3DI). Retrieved November 18, 2024, from https://berkas.dpr.go.id/pusaka/files/buku_lintas_tim/buku-lintas-tim-public-14.pdf.
Yayusman, M. S., and P. N. Mulyasari. 2024. Indonesia's spice-based gastrodiplomacy: Australia and Africa continents as the potential markets. *Journal of ASEAN Studies* 12 (1): 51–77. https://doi.org/10.21512/jas.v12i1.8004.
Yayusman, M. S., P. N. Mulyasari, A. Trihartono, and Gusnelly. 2023a. Culinary diaspora: An anchor for Indonesian gastrodiplomacy in the Netherlands. *The Hague Journal of Diplomacy* 11 (1): 558–583. https://doi.org/10.1163/1871191X-bja10167.
Yayusman, M. S., U. K. Yaumidin, and P. N. Mulyasari. 2023b. On considering Australia: Exploring Indonesian restaurants in promoting ethnic foods as an instrument of Indonesian gastrodiplomacy. *Journal of Ethnic Foods* 10 (43). https://doi.org/10.1186/s42779-023-00207-1.
Zhang, J. 2015. The foods of the worlds: Mapping and comparing contemporary gastrodiplomacy campaigns. *International Journal of Communication* 9: 568–591.

Meilinda Sari Yayusman is a researcher at the Research Center for Area Studies, National Research and Innovation Agency (BRIN), Jakarta, Indonesia. Her research focuses on European governance and EU studies as well as Indonesian foreign policy, particularly Indonesian public diplomacy and gastrodiplomacy. She has recently published several articles including Culinary Diaspora: An Anchor for Indonesian Gastrodiplomacy in the Netherlands (The Hague Journal of Diplomacy, 2023), On Considering Australia: Exploring Indonesian Restaurants in Promoting

Ethnic Foods as an Instrument of Indonesian Gastrodiplomacy (Journal of Ethnic Foods, 2023), and Indonesia's Spice based Gastrodiplomacy: Australia and Africa Continents as the Potential Markets (Journal of ASEAN Studies, 2024).

Gregorius Andika Ariwibowo is a researcher at the Research Center for Area Studies, National Research and Innovation Agency, Jakarta. His research spans cultural heritage, gastronomic history, and urban studies. He has co-authored several academic papers, including "From Panchuran to Waterleiding: Clean Water Solutions in Colonial Bandung, West Java, Dutch East Indies (1898–1934)" (History of Science and Technology, 2023) and "Heritage in Motion: Safeguarding the Cultural Legacy of Wayang Kulit Kedu, Indonesia" (TRAMES, 2024).

Ahmad Nuril Huda is a researcher at BRIN's Research Center for Area Studies. He obtained his PhD from Leiden University's Institute of Cultural Anthropology and Development Sociology. His academic and research interests focus on issues related to religion in everyday life, foodways culture, gender and sexuality, new media/digital technologies, and the politics of (in)visibilities in the field contexts of Southeast Asia. His paper "Making Film as an Ethical Practice: the case of the traditionalist Islamic boarding school students in Indonesia" was recently published with Springer in Contemporary Islam.

Prima Nurahmi Mulyasari is a PhD student at the School of Humanities, Nanyang Technological University, Singapore. Her research focuses on culinary history, transnational history, and urban history. She has co-authored several academic papers, including "Culinary Diaspora: An Anchor for Indonesian Gastrodiplomacy in the Netherlands" (The Hague Journal of Diplomacy, 2023), "Considering Australia: Exploring Indonesian Restaurants in Promoting Ethnic Foods as an Instrument of Indonesian Gastrodiplomacy" (Journal of Ethnic Food, 2023), "A Global Pela: The Moluccan Diaspora in the Sister City of Vlissingen-Ambon" (Diaspora Studies, 2024), and "Indonesia's Spice-Based Gastrodiplomacy: Australia and Africa Continents as Potential Markets" (Journal of ASEAN Studies, 2024).

Rakhman Priyatmoko is a researcher at the Research Center for Area Studies, Social Sciences and Humanities Research Organization, National Research and Innovation Agency (BRIN). He completed his Masters in Tourism Studies, Postgraduate School, Universitas Gadjah Mada, Yogyakarta. The author's research interests revolve around issues related to Islam-related tourism, such as halal tourism, Islamic tourism, Muslim-Friendly tourism and its supporting industries.

Agus Trihartono is a Professor at the Department of International Relations and is also a co-founder of Centre of Gastrodiplomacy Studies (CGS), at the University of Jember, Jember, Indonesia. He works on soft power diplomacy and comparative politics.

Chapter 8
Indonesia's Foreign Policy in Pursuing Human Rights Norms to Combat Trafficking in Persons

Tri Nuke Pudjiastuti, R. A. Rizka Fiani Prabaningtyas, Athiqah Nur Alami, Faudzan Farhana, and Irin Oktafiani

Abstract This chapter encapsulates Indonesian foreign policy, regulations, and policy implementation to advance the protection of Indonesian migrant workers (IMW) from trafficking in persons (TIPs). Since his election in 2014, President Joko Widodo has underscored a pro-people foreign policy aimed at safeguarding Indonesians both domestically and abroad. This chapter will examine Indonesia's foreign policy in addressing TIPs, encompassing land, maritime, and cyber trafficking activities. It employs norm entrepreneurship as a conceptual framework to investigate how Indonesia frames, institutionalizes, and keeps promoting human rights norms in addressing trafficking issues through the conduct of human rights diplomacy in combating TIPs. We argue that there has been progress in adopting a human rights perspective to combat TIPs and protect IMW as TIPs victims, indicating Indonesia's proactive role as a norm entrepreneur in advocating TIPs as human rights issues. This progress is seen at a bilateral, regional, and multilateral level. However, despite the existing efforts and advancements, Jokowi's policy on TIPs continues to encounter substantial challenges due to the long-established perception of TIPs as a more political-security matter at the national and regional level, as well as the unequal interests at the multilateral efforts. Therefore, Jokowi legates problems of inclusive TIPs to the next presidency, including promoting a more balanced paradigm that takes into account the socio-cultural and economic aspects of TIPs.

Keywords Trafficking in persons · Indonesian foreign policy · Indonesian migrant workers · Human rights · Norm entrepreneur

T. N. Pudjiastuti (✉) · R. A. R. F. Prabaningtyas · A. N. Alami · F. Farhana
Research Center for Politics, The National Research and Innovation Agency (BRIN), Jakarta, Indonesia

I. Oktafiani
Research Center for Population, The National Research and Innovation Agency (BRIN), Jakarta, Indonesia

I. Kartini and Kanisa (eds.), *Joko Widodo's Foreign Policy and Diplomatic Legacy*,
https://doi.org/10.1007/978-981-95-2085-5_8

Table 8.1 Indonesia handled the victims of TIP during 2015–2019

Across Regions	Total cases
East and Southeast Asia	802
Middle East	858
Africa	235
Oceania	35
Europe	33
South and Central Asia	12
Victims of TIP	**1975**

Source 2015–2019 Laporan Kinerja Gugus Tugas PP-TPPO, 2020

8.1 Introduction

Trafficking in persons (TIPs) is a complex crime that is a form of modern slavery and a violation of human rights for a long time. For Indonesia, the main cause of TIP correlates with the number of Indonesian migrant workers overseas. There were 141 TIP cases in 2014, and in 2017, IOM released the number of TIP victims, which is 6.904, including 5.907 women and 970 girls. Most of them were undocumented migrant workers (Nupus 2017), but the actual cases are probably bigger. Therefore, the number of cases and victims of TIP leads to an urgent situation to solve TIP's problem because trafficking is not only a case that needs the government's responsibility for its citizens but also a situation against humanity that should be eradicated around the world. It is estimated that 90% of victims are Indonesian migrant workers (IMW).[1] This situation presents a significant challenge for the governance of TIP eradication in Indonesia, both domestically and internationally, as it aims to safeguard Indonesian citizens who work as IMWs from falling into TIPs. Indonesia was a major source country and, to a much lesser extent, a destination and transit country for women, children, and men subjected to sex trafficking and forced labor, which is a part of trafficking in persons (TIPs). As one of the main senders of migrant workers, the number of Indonesians who want to work overseas remains high every year. Unfortunately, some people use this situation to lure the candidate workers into their trafficking tricks. In some cases, people get into the trap because TIP's perpetrators sometimes cover their "dirty job" as legal migration administrators who can help people work overseas.

At the beginning of his administration, Jokowi made one of Indonesia's foreign policy priorities, which is to protect all Indonesians abroad, including migrant workers, regardless of their status and situations. The protection focuses on the efforts to prevent Indonesians from becoming involved in crime or becoming the victims of organized crime. According to the Table 8.1 below, the Ministry of Foreign Affairs has succeeded in helping TIP victims spread across various regions during 2015–2019.

[1] Interview with Director-General for Protocol and Consular Affairs/Chief of State Protocol, Ministry of Foreign Affairs, Republic of Indonesia, 2024.

The data indicates that the major problems in TIPs are primarily found in the East and Southeast Asia and Middle East regions. According to the Indonesian Task Force on Combating TIPs, the countries with the largest number of TIP cases are the People's Republic of China (283), Malaysia (275), Syria (178), the United Arab Emirates (176), and Saudi Arabia (83) (Indonesia's Ministry for Women, 2020).

However, the complexity of the TIP problem makes it hard for Indonesia to move from the Tier 2 category, even in 2022, it dropped to Tier 2 Watchlist[2] as the criteria of the Palermo Convention or General Assembly resolution 55/25 of the 15 November 2000 United Nations Convention against Transnational Organised Crime criteria. However, it does not mean there have been no achievements over the past nine years in efforts to prioritize human rights as a goal or tool in foreign policy related to eradicating TIPs. The results of our study and data collection demonstrate that Indonesian diplomacy has made numerous achievements based on human rights norms, starting from the policy level, dialogue, cooperation, and strengthening the capacity of officers in bilateral, regional, and multilateral relations.

This chapter uses norm entrepreneurship to analyze the role of Indonesia as a norm entrepreneur and utilize the human rights approach to TIP handling during Jokowi's tenure. This chapter answers two main questions: First, how does the practice of Jokowi's foreign policy play Indonesia's role as a norm entrepreneur in promoting human rights norms on TIPs handling? Second, what are the challenges to realizing human rights norms in the TIPs issue and advancing protection for IMW?

8.2 Human Rights Diplomacy as the Strategy to Perform Norms Entrepreneurship

In the ever-evolving realm of international relations and the diplomatic sphere, the concept of human rights diplomacy has risen globally, signifying the importance of human rights as a goal and tool of diplomacy. According to Mullerson (2014), human rights diplomacy is defined as the use of foreign policy instruments to promote human rights and the use of human rights issues to further other foreign policy aims. It differs from traditional diplomacy in that it employs diplomatic negotiation and persuasion to advance and safeguard human rights rather than governments' national interests (O'Flaherty et al. 2011). It encompasses efforts by states and other entities to address human rights concerns, either through multilateral forums or bilateral diplomacy.

Several studies investigating human rights diplomacy have been carried out on the roles of various actors, including regional and international organizations and non-governmental organizations (NGOs) in the practices of such diplomacy (King 2011; Salama 2011; O'Flaherty et al. 2011; Ramcharan 2011; Weschler 2011; Finnemore and Sikkink 1998; Hicks 2011). Other scholars have paid attention to how

[2] From 2014 to 2023, Indonesia was at Tier 2 level, only in 2022 it dropped to Tier 2 Watchlist, see: https://www.state.gov/reports/2022-trafficking-in-persons-report/indonesia/ and https://www.state.gov/reports/ 2024-trafficking-in-persons-report/indonesia/. Accessed on 1 June 2024.

human rights diplomacy has been channeled, including through conference diplomacy (Kedzia 2011) as well as through summit diplomacy and network diplomacy (Pease 2016). Moreover, a study by Benedek (2011) focused on a multistakeholder approach in conducting human rights diplomacy, underscoring how this type of diplomacy has been approached. Collectively, these studies outline notable characteristics of human rights diplomacy based on its actors, channels, and approaches, differing from a traditional conception of diplomacy that is mainly state-centric.

Given the extensive previous studies on the characteristics of human rights diplomacy, there are still few studies that address factors that influence the practice of human rights diplomacy, including norms and values that may shape its practices. Norms play a critical role in human rights diplomacy as they establish the standards and expectations for behavior regarding human rights on an international level (Mullerson 2014). They play roles as guiding tools for the objectives of human rights diplomacy, which involves promoting and protecting human rights through various foreign policy instruments, even if it sometimes risks deteriorating interstate relations (Mullerson 2014, p. 12). Thus, norms not only define the goals of human rights diplomacy but also as tools that shape its strategies and interactions among states, international bodies, and non-state actors. Moreover, the position of human rights in the foreign policy of a country is influenced not only by international structure but also by different perspectives on which foreign policy is based, including norms that various actors believe (Baehr and Castermans-Holleman 2004). These arguments underline the close link between norm entrepreneurship and human rights diplomacy as part of a country's foreign policy strategies.

Human rights diplomacy serves as a tool to enable states to perform their role as norm entrepreneurs. The realm of global politics has witnessed a growing interest in the concept of norm entrepreneurship, which refers to individual and collective actors who eagerly strive to promote the establishment, diffusion, and institutionalization of norms (Finnemore and Sikkink 1998; Bjorkdahl 2007). As part of the constructivist turn in International Relations (IR), the concept underlines the constitutive impact of ideational factors such as ideas, norms, and values in shaping the behavior of actors (states and non-states), including their interests and preferences. Therefore, the notion of "norm entrepreneurship" has been significant in elucidating the effective establishment and dissemination of international norms.

Some scholars, such as Richter (2018), Fry (2020), and Hariyadi (2021), drew from Finnemore and Sikkink's (1998) approach to the norm life-cycle model to understand the process and stage of norm evolution from its emergence to its internalization in particular case studies. Other works on norm entrepreneurship have focused on agency-based analysis that investigates the role of various actors as norm entrepreneurs. For example, Finnemore and Sikkink (1998) and Khagram et al. (2002) took the case of transnational advocacy networks of non-government organizations that play roles in bringing about normative change. Besides focusing on non-state actors, other research on international norms still believes in the important role of the states in shaping norms (Carr and Baldino 2015; Stuenkel 2016; Bjorkdahl 2007; Kurusu 2018; Teo 2022; Wunderlich 2020).

Given the extensive previous discussion on the role played by states and non-state actors as norm entrepreneurs, this paper aims to enrich the discussion by bringing the experiences of Indonesia as a norm entrepreneur in trafficking issues. It argues that Indonesia has played a role in advocating human rights norms to combat trafficking in persons that involve Indonesian migrant workers. The trafficked persons, including women and girls, children, persons with disabilities, and migrant workers, have experienced human rights violations. Hence, they are entitled to protection, regardless of their nationality or citizenship, highlighting a clear link between trafficking and human rights. By arguing that trafficking is a human rights issue, the response to trafficking should apply a human rights-based approach. It is an approach that is "normatively based on international human rights standards and that is operationally directed to promoting and protecting human rights" (United Nations 2014, p. 8).

This paper builds upon Teo's arguments (2022), taken from Finnemore and Sikkink (1998) and Carr (2015), on the three key tasks of norm entrepreneur: framing the new norms, establishing or relying on an organizational platform to promote the norm, and sustaining their advocacy efforts in the face of criticism. A norm entrepreneur frames the new forms through "language that names, interprets and dramatizes them" (Finnemore and Sikkink 1998, p. 897). Moreover, in order to promote norms, there should be a process of institutionalization in specific sets of international rules and organizations. Finally, since introducing a new norm is uneasy, a norm entrepreneur should be willing to keep promoting the new norms to make them sustainable in the midst of resistance and criticism.

By applying the task of norm entrepreneur as a conceptual framework, this paper will investigate Indonesia's role in shaping human rights norms on combating trafficking involving Indonesian migrant workers through how Indonesia frames, institutionalizes, and keeps promoting the human rights norms in addressing trafficking issues. It aims to answer the questions by taking into account three different levels of Indonesia's engagement on trafficking issues, namely at the bilateral, regional (in the context of ASEAN), and multilateral levels (Bali Process). Through Indonesia's norm entrepreneurship at these levels, the Jokowi administration has demonstrated a notable legacy on the significance of the human rights approach to address the issue of trafficking that involves Indonesian migrant workers.

8.3 Practices of Human Rights as Norms in Jokowi's Foreign Policy to Combat TIPs

The incorporation of human rights into Indonesia's diplomatic practices is not a recent development; it has evolved significantly since the country's 1998 reformation (Mohsin 2019). Human rights diplomacy has been instrumental in Indonesia's efforts to achieve national interests, particularly in addressing TIPs issues. This approach is evident in bilateral and multilateral engagements, particularly under Jokowi's administration from 2014 to 2024.

8.3.1 *Framing New Norms in Combating TIPs on Migrant Workers: Continued Efforts*

Historically, protecting citizens abroad has been a priority for Indonesia's MOFA. Under Hassan Wirayuda's leadership (2001–2009), two directorates focused on human rights and citizen protection were established. Therefore, Jokowi's efforts can be seen as a continuation of these initiatives, aiming to enhance data collection on Indonesians abroad and promote safe migration practices during his first and second term (Penyampaian Prioritas Politik Luar Negeri Republik Indonesia 2019–2024, n.d). Following up on this foreign affairs priority, Indonesia adopted two patterns of diplomatic strategies: an indirect strategy that employs a human rights framework for addressing migrant worker protections and a direct approach that fosters constructive dialogue with nations sharing interests in combating TIPs.

In the past decade, Indonesia's foreign policy has emphasized the protection of its citizens abroad, especially Indonesian migrant workers (IMW). Foreign Minister Retno Marsudi highlighted this commitment in her 2015 statement, ensuring that migration occurs only under specific conditions: either the destination country has regulations protecting foreign workers, or there is a bilateral agreement prioritizing IMW protections (Humas 2015). These strategies manifest at bilateral, regional, and multilateral levels, though progress varies according to the nature of cooperation involved.

a. Bilateral Level

Understanding the strengths and weaknesses of each level, Indonesia engaged in constructive dialogue to combat TIPsat various levels. At the bilateral level, Indonesia has engaged in negotiations to secure agreements ensuring IMW protections. As a result, Indonesia has managed to reach an understanding with six countries, as set out in the Table 8.2:

The political climate and priority levels on the issue of migrant workers' protection vary in each country. Consequently, the negotiation process may have distinct ups and downs in achieving agreement. For instance, in 2016, an MoU was signed between Indonesia and Malaysia. As Malaysia failed to adhere to the terms of the initial MoU, a new MoU was drawn up and agreed on April 1, 2022, in Jakarta. The new MoU establishes a one-way mechanism for all processes of placement, monitoring, and repatriation of Indonesian workers in Malaysia.

Meanwhile, although in 2015, Indonesia put a moratorium on the placement of Indonesian domestic migrant workers in the Middle East countries, an agreement between Indonesia and Saudi Arabia in 2018 bypassed this ban. Initially, Indonesia prohibited work for individual employers in Middle East countries due to frequent abuses experienced by IMW there. Middle Eastern countries have been criticized for their poor treatment of migrant workers under the kafala sponsorship system, which requires migrant workers to have a domestic sponsor, usually their employer, who is responsible for their visa and legal status. However, Indonesia and Saudi Arabia then agreed to use the One Channel System (SPSK) in 2018, which stipulates

Table 8.2 MoU between Indonesia and other countries in protection of Indonesian migrant workers

No.	Year	Countries	Concerning
1.	23 February 2016	China—Macau	On The Facilitation and Protection of Indonesian Overseas Workers in Macau Special Administrative Region of The People's Republic of China
2.	19 February 2019	Singapore	On The Placement and Protection of Indonesian Overseas Workers In Singapore
3.	25 June 2019	Japan	On a Basic Framework for Proper Operation of The System on Foreign Human Resources with The Status of Residence of Specified Skilled Worker
4.	25 June 2019	Japan	On The Technical Intern Training Program
5.	26 Juli 2021	The Federal Republic of Germany	On the Placement and Protection of Indonesia Health Professionals in the Federal Republic of Germany
6.	18 Maret 2022	The United States of America	Concerning The Facilitation of Labor Market Information for Potential Placement and Protection of The Indonesian Migrant Workers In The United States of America
7.	1 April 2022	Malaysia	On The Employment and Protection of Indonesian Domestic Migrant Workers in Malaysia

Source BP2MI, 2023

that Indonesians seeking work in the country must go through an approved *syarikah* (incorporated company), not individual employers. This system is expected to ensure the safety and decent work of IMWs in Saudi Arabia.

Furthermore, Indonesia uses constructive dialogue to focus on combating online scams related to TIPs. By employing a victim-centered approach, Indonesia navigates complex judicial systems to advocate for its citizens effectively. Notably, Indonesia's human rights diplomacy has led to the rescue of 360 citizens from death row and the resolution of over 200,000 legal cases in 9 years (Retno Marsudi, 2023). The collaboration with Australia on TIP issues is significant, dating back to the initiation of the AAPTIP program in 2007. This partnership has fostered capacity-building among law enforcement agencies in Indonesia (Australian Mission to ASEAN, n.d).

b. Regional (ASEAN) Level

ASEAN has gradually broadened its human rights framework to encompass standards for vulnerable groups, including women, children, persons with disabilities, and migrant fishers. Indonesia has taken the lead in advocating for human rights protections related to migrant workers and TIP issues. Despite the slow consensus-based decision-making in ASEAN, Indonesia has made strides in normalizing human rights dialogue, including the ASEAN Intergovernmental Commission on Human Rights (AICHR).

AICHR was established in 2009, and Indonesia continues to conduct constructive dialogue to normalize the human rights approach by formulating six strategies ensuring the implementation of the AICHR mandate and emphasizing the link between migrant worker rights and TIP prevention.[3] Recent initiatives, such as the AICHR—SOMTC Joint Workshop on Human Rights-based Approach to Combat Trafficking in Persons, Especially Women, and Children held on 5–6 November 2015 in Yogyakarta highlights Indonesia's influence in fostering collaboration on human rights (AICHR 2015). During the 27th ASEAN Political-Security Community Council Meeting as part of the 43rd ASEAN Summit in September 2023, Indonesia encouraged member states to commit to protecting victims of TIPs, using regional declarations as negotiation tools (ASEAN Secretariat, 2024).

c. Multilateral (Bali Process) Level

Jokowi made many efforts to place Indonesia on a multilateral platform, such as IORA, G20, APEC, OIC, and many more. Particularly on TIPs, there is only one forum that specifically discusses smuggling, TIPs, and other transnational crimes, namely the Bali Process,[4] which addresses TIPs and transnational crimes. Despite differing priorities with Australia—where Australia emphasizes people smuggling over TIPs—Indonesia has successfully established a Working Group on Human Trafficking. However, challenges remain in fully utilizing the forum's potential for sharing best practices and information (Arisandy 2017).

8.3.1.1 Establishing and Relying on Organizational Platform: National, ASEAN, and Bali Process

a. Institutionalization at the National Level

Since Jokowi's leadership began, human rights diplomacy has become embedded in Indonesia's foreign policy. The creation of a dedicated human rights directorate (Regulation of the Minister of Foreign Affairs of the Republic of Indonesia No. 6 of 2021) and *Peduli WNI* site demonstrates a commitment to protecting citizens abroad,

[3] Interview with the first and second Indonesian Commissioner of AICHR (2009–2012 and 2013–2016).

[4] The Bali Process comprises 45 Member States and 4 Member Organizations—the International Organization for Migration, the Office of the United Nations High Commissioner for Refugees, the United Nations Office on Drugs and Crime, and the International Labour Organization—as well as 18 Observer States and nine Observer Organizations.

particularly TIPs victims. The Indonesian National Human Rights Commission has also prioritized TIP awareness, collaborating with agencies responsible for worker protection, namely the Indonesian Migrant Workers Protection Agency (BP2MI).

b. Institutionalization at the Bilateral Level

One example of institutionalization at the bilateral level is between Indonesia and Australia, where intensive dialogue facilitated the implementation of the Indonesia–Australia Comprehensive Strategic Partnership (CSP). Key arrangements include cooperative efforts between law enforcement agencies and immigration departments, as follows:

(1) The arrangement between the Indonesian National Police and Australian Federal Police on Cooperation in Preventing and Combating Transnational Crime, signed on 12 August 2019
(2) Arrangement for Cooperation on Migration and Border Management between the Australian Department of Home Affairs and the Indonesian Directorate General of Immigration signed on 3 December 2019;
(3) The arrangement between the Maritime Security Agency of the Republic of Indonesia (Badan Keamanan Laut Republik Indonesia/Bakamla RI) and the Department of Immigration and Border Protection as represented by the Australian Border Force on Maritime Security. Cooperation was signed on 17 November 2017.
(4) Carry out the Plan of Action for the implementation of the Joint Declaration on Maritime Cooperation, signed on 16 March 2018, to achieve mutual economic and security benefits in the maritime domain in the areas of maritime security and safety, connectivity, and sustainable management of marine resources

c. Institutionalization at the ASEAN Level

Indonesia's advocacy for human rights within ASEAN led to the signing of the ASEAN Convention on Combating TIPs. The ASEAN Leaders signed the ASEAN Convention against Trafficking in Persons, Especially Women and Children (ACTIP) in 2015. Furthermore, the establishment of the Bohol TIP Work Plan and its successor, Bohol TIP Work Plan 2.0, illustrates ongoing efforts to combat TIP collaboratively among member states (ASEAN Secretariat 2022).

d. Institutionalization at the Bali Process

Indonesia's role in the Bali Process includes pushing for joint actions against TIPs, fostering bilateral agreements and initiatives that incorporate the private sector in prevention efforts in the High-Level Meeting on the Appraisals of the Global Plan of Action on Trafficking in Persons on the sidelines of the 72nd UN General Assembly (General Assembly 2010). These recommendations were the result of the Bali Process Government and Business Forum in Perth, Australia, on 24–25 August 2017, where both co-chairs pioneered the involvement of the private sector to prevent TIP, including forced labor and slavery.

e. Institutionalization with Relevant International Organizations

Indonesia also works with UN agencies operating in Indonesia. Despite its limitations, Indonesia provides space for the International Organization of Migration (IOM), UNICEF, The International Labour Organization (ILO), and the United Nations Office on Drugs and Crime (UNODC). UN agencies in Indonesia have continuously supported the efforts of the Government of Indonesia in preventing and handling TIP cases in Indonesia under its "3P Approach"—Prevention, Protection, and Prosecution with their own characteristics. They work both at the national and sub-national levels in close partnership with government and non-governmental organizations. They have built various collaborations, starting from anti-TIPs campaigns, developing anti-TIPs policies, helping to recover TIPs survivors, and building the capacity of officers in the field, including community organizations in efforts to help handle TIPs survivors.

8.4 Sustaining Advocacy Efforts: Lesson Learned

Intra-regional migration within ASEAN is influenced by various factors, including economic disparities and conflicts (Sharom et al. 2018). This context complicates the protection of migrant workers as the majority of the AMS view them on a security-based stance, often leaving them vulnerable to exploitation. Critically, ASEAN's traditional focus on security-related migration issues overlooks essential protection for irregular migrants in ASEAN (ILO 2015; Khasru 2018).

This condition makes the bargaining position of workers the lowest in the labor market, and it can lead to exploitation and even human trafficking (Basir 2020). While some AMS rely heavily on low-wage labor, which has been fulfilled by migrant workers (Bal and Gerard 2017; ASEAN Secretariat 2022), the absence of a regional mechanism that guarantees their protection creates legal loopholes that are not only exploited by migrant workers themselves but also by traffickers. Consequently, this situation has drawn much criticism from both civil society and non-governmental organizations, which play a crucial role in advocating for migrant workers' rights.

Despite ASEAN being known as an elitist and government-oriented organization, it is evident that ASEAN allows the involvement of Civil Society Organizations (CSOs) and NGOs in addressing the issue of migrant workers. Their involvement has been instrumental in the drafting of the ASEAN Consensus on the Protection and Promotion of the Rights of Migrant Workers. The Consensus was agreed upon at the ASEAN Summit in November 2017, accommodating several recommendations from the CSOs proposal entitled "Civil Society Proposal: ASEAN Framework Instrument on the Protection and Promotion of the Rights of Migrant Workers" in May 2009 and submitted to the ASEAN mechanism (Septiyana 2019). Four key provisions make up the core of this consensus's protection of migrant workers: the definition of migrant workers and undocumented migrant workers, basic rights for migrant workers and their families, and specific rights for migrant workers.

The involvement of CSOs and NGOs continues in the discussion of the issues of migrant workers, and it results in a more comprehensive normative framework in

ASEAN. ASEAN successively agreed on the ASEAN Declaration on the Rights of Children in the Context of Migration (2019), the ASEAN Guidelines on Effective Return and Reintegration of Migrant Workers (2020), the ASEAN Declaration on the Portability of Social Security Benefits for Migrant Workers in ASEAN (2022), the ASEAN Declaration on the Protection of Migrant Workers and Family Members in Crisis Situations (2023), and the ASEAN Declaration on the Placement and Protection of Migrant Fishers (2023) were all agreed upon successively after the ASEAN Consensus on the protection and promotion of migrant workers. These instruments enrich the protection aspect by expanding the scope of protection to include children in migration activities, return and reintegration stages, social security, crises, and fisheries workers.

Two things can be learned to encourage the use of human rights-based norms in the protection of migrant workers. First, the involvement of CSOs and NGOs as human rights defenders can ensure that the development of the norm framework is not limited to institutional formalities but does include the substance of human rights that need to be protected. Secondly, this involvement also allows criticism from the community to be constructively used to strengthen the norms that apply in the region. However, the path forward requires continued collaboration, advocacy, and institutional support to address the multifaceted challenges faced by migrant workers in ASEAN.

8.5 Challenges in Realizing Human Rights Norms in the TIPs Issue and Advancing Protection for IMW

Despite the above explanation of Indonesia's efforts to perform as a norm entrepreneur in TIPs, there are fundamental obstacles to the continuous promotion of human rights. For Indonesia, IMW protection has long become a main national interest in dealing with TIPs issues. However, national progress, regional commitment, and multilateral diverged interests have shaped the dynamics of Indonesia's response to combatting TIPs during the Jokowi administration and potentially complicate Indonesia's diplomacy on TIPs and IMW protection.

8.5.1 National Shifting to Political-Security Approach

Before Jokowi's presidency, The Government of Indonesia affirmed its commitment to end the practice of TIPs, including forced labor through Law No. 21/2007 on the Eradication of Trafficking in Persons and other related legal instruments. To support this commitment, the then President (SBY) issued Presidential Regulation no. 69/2008 concerning the Task Force to Prevent and Address the Criminal Act of Trafficking in Persons that assigned the Coordinating Ministry for Human Development and Cultural Affairs as The Coordinator of the National Taskforce and Ministry for Women's Empowerment and Child Protection as the steering chair. This assignment was crucial in the attempt to form a comprehensive victim-based approach to handling TIPs with five main strategies: Prevention, Protection, Prosecution, Policy, and Partnership. It is important to acknowledge that it was under Jokowi's presidency that the operational guidelines for the task force were quite well-developed, and the only publicly available national report from the task force was from 2015 to 2019.

However, in 2023, President Jokowi issued Presidential Regulation No. 49/2023, which regulates the structural change of the national task force for TIPs. In this new regulation, the main leading actor for the task force is the Coordinating Ministry for Political, Legal, and Security Affairs as the first chair and the Chief of the Indonesian National Police as the daily chairman. This change indicates that the priority of the Indonesian government has shifted to the prosecution aspect. It was then confirmed by the Coordinating Minister for Human Development and Cultural Affairs Indonesia (who has now become the second chair) and Coordinating Minister for Political, Legal, and Security Affairs that this change was due to their evaluation of the previous performance of the task force, the nature of TIPs that the weight of TIPs problems is more in law enforcement and criminal matters, and the demand of President Jokowi to accelerate efforts against TIPs (Tempo 2023). The appointment of the police as the daily chairman is related to the pattern developed in ASEAN, which is that the Indonesian delegation to SOMTC has so far been from the Indonesian national police, which is aligned with the mandate for police at the regional level. To some extent, this could be seen as a necessary approach for a more coordinated response for law enforcement, but there is a risk of ignoring the prevention and protection aspect of TIPs' handling.

8.6 The Challenges of Implementing ASEAN Commitment in the Protection of IMW

One year before the end of Jokowi's leadership at the ASEAN Summit in Labuan Bajo, when Indonesia was in the position of Chair of ASEAN, Indonesia was able to push for three declarations related to TIP. This achievement is a milestone in the legacy of his 10 years of leadership since achieving a declaration at the ASEAN level is not easy. While there have been attempts to constitute more institutionalized human rights norms in responding to TIPs, there remains limited progress in adopting the declaration at the ASEAN level into a more operational and in-line response in respected AMS policies on TIPs.

The first challenge relates to the regional view on TIPs among Southeast Asian countries. In ASEAN, the permanent mechanism for ASEAN member states to discuss and monitor the implementation of the ASEAN Work Plan on TIPs is the Senior Official Meetings on Transnational Crime (SOMTC) Working Group on Trafficking in Persons under the political-security pillar. While the Term of Reference of SOMTC declares that SOMTC "delegation could include members representing different areas of the TIP response (law enforcement, prosecution, victim support, repatriation, and reintegration)" (ASEAN 2021), the SOMTC meeting was typically attended by security sector actors from respective member states. It implies the established view within ASEAN about the nature of TIPs as a criminal act and therefore, requires a more security approach in the process of realizing the action plans on TIPs.

The second challenge comes from the nature of a declaration as a non-binding norm, and there are no sanctions for countries that violate it. Consequently, achievement in resulting regional declarations becomes ineffective in practice since there has been no operationalization agreed upon among ASEAN countries, making it difficult to implement at the regional level. Each country still uses its national standards when dealing with TIPs since there is no standard policy among ASEAN member countries. This ASEAN commitment is also not reflected in bilateral relations.

Furthermore, despite the interconnectedness between TIPs and migrant workers issues, ASEAN has a different set of law instruments and separate organizational mechanisms to address these issues. While TIPs are under the political-security community pillar, ASEAN cooperation on migrant worker issues is led by the ASEAN Labour Ministers Meeting (ALMM) under the socio-cultural community pillar. As a result, there is a blurred line within ASEAN when member states discuss the process of law enforcement and protection for IMWs who fall into TIPs. Human rights norms become the potential bridge to enable cross-pillar cooperation. As previously explained, AICHR, as the leading sectoral body on human rights issues in ASEAN, had led the way for a more cross-pilar collaboration with other sectoral bodies under different community pillars such as ACWC, and ASEAN Committee on the Implementation of the ASEAN Declaration on the Protection and Promotion of the Rights of Migrant Workers (ACMW). However, AICHR does not have the authority to carry out executions but only to promote human rights in ASEAN member countries. To gain more support on the advocacy of human rights norms in addressing TIPs–IMW issue linkage, Indonesia should continue to ensure that the works of AICHR in promoting a right-based approach in TIPs are consistently echoed by other sectoral bodies that have concerns on the prevention of TIPs and protection of the victims.

8.7 Ineffectiveness of the Bali Process to Indonesia's Foreign Policy to Protect IMW as TIPs Victims

Although it has been long-established, the Bali process has fluctuating dynamics that influence its effectiveness in combating human trafficking and smuggling. The early Jokowi period was a time of efforts to revive the forum after a vacuum of many years because the interests of the relationship between Australia and Indonesia influenced it. Many parties have high hopes for the Bali Process, considering that there are not many initiatives at regional and multilateral levels that attempt to bring countries affected by transnational crime together.

However, the establishment of the Bali Process was heavily influenced by Australia's interests, which are more focused on the issue of smuggling. The Bali Process cannot be separated from Australia's high national interest after the Tampa Crisis occurred in Australian waters in 2001 (Douglas and Schloendhart 2012). At that time, Indonesia shared concerns about border protection from the Middle East asylum seekers and terrorism issues, which then led to an agreement to co-lead a multilateral forum to combat illegal migration and enforce border protection. Consequently, discussions on TIPs in the Bali Process tend to focus on people smuggling, particularly concerning refugee and asylum seeker issues.

Moreover, despite hopes for the potential of the Bali Process as an avenue for multilateral diplomacy on TIPs, Indonesia has not employed serious efforts to utilize this forum to advance its national interest in TIPs and the protection of migrant workers. The potential is seen in the various levels of meetings held by the Bali Process, ranging from Working Groups, AD Hoc Groups, and Steering Committees to Ministerial Meetings. However, these ranges of institutional mechanisms within the Bali Process do not serve as a consultative forum that can prevent situations from escalating and better manage the diverged interests of involved countries collectively. Rather, the forum's role remains limited to information sharing, which limited its role in promoting improvements in national and regional contingency planning that would allow for more predictable and effective responses to forced migration. It could also provide greater support to key member states in developing policies, legal systems, and implementation capacities in a way that also provides for an orderly system of protection against smuggling and trafficking.

8.8 Conclusion

Given Indonesia's position as the major sending country of migrant workers worldwide, there is an urgent need to apply human rights perspectives in its foreign policies on citizen protection. In the two terms of President Joko Widodo's presidency, Indonesian foreign policy has actively promoted and made substantial progress in promoting human rights standards in addressing TIPs, particularly concerning Indonesian migrant workers. Indonesia's normative entrepreneurship, which was positioning itself as a leader in advocating for these norms at various levels, has demonstrated its contribution to addressing the issue of human trafficking affecting Indonesian migrant workers, emphasizing the importance of a human rights approach.

This improvement has been accomplished by establishing the new standard within bilateral, regional, and international coordinating mechanisms. From a bilateral level, endeavors are undertaken through productive dialogue and interstate collaboration, addressing TIPs as components of transnational matters and incorporating TIPs with other transnational concerns, such as those of marine workers. Secondly, there are ongoing initiatives to promote productive discussions among all member countries in the region who have a shared interest in combating TIPs. The initiatives involve strengthening the involvement of several stakeholders in regional organizations like ASEAN and offering assistance to United Nations agencies in Indonesia to fulfill the commitment to address Trafficking in Persons (TIPs) at the international level. Thirdly, there are ongoing endeavors to develop or depend on an organizational framework to advance the norm. Furthermore, they persist in their advocacy endeavors despite encountering criticism.

In addition, there are certain obstacles in implementing human rights standards in the context of the TIPs issue and improving internal protection. As the country is transitioning toward a political-security focus, it exhibits an inadequate strategy. This change represents a transition in the Indonesian government's focus toward prosecution, as the TIPs issue includes not only the political-security agenda but also socio-cultural issues such as the endeavors to support the victims. Besides that, the difficulties faced at the ASEAN level arise from the conventional approach of ASEAN Member States toward TIPs issues which categorizes the efforts within the political-security framework. Therefore, the process of translating regional dedication to accepted national standards on TIPs is a difficult one. Moreover, the lack of efficiency of the Bali process, combined with Indonesia's foreign policy to safeguard individuals who have been trafficked and exploited as victims of TIPs, has led to imbalanced ties between Australia and Indonesia, which seem to arise from different political objectives.

Our recommendation for Indonesian foreign policy on handling TIPs bilaterally, regionally, and multilaterally needs to continue to prioritize human rights diplomacy as a tool and goal. Entrepreneurial diplomacy is important to be implemented by strengthening collaboration with non-state actors to achieve the goal. In addition, it is important to be in line with Indonesia's domestic policy. National interests are important to balance the interests of security politics and the interests of mainstreaming human rights for the victims.

References

AICHR. 2020, July 20. Press release: Aichr—SOMTC joint workshop on human rights-based approach to combat trafficking in persons, especially women and children, 5–6 November 2015, Yogyakarta, Indonesia. AICHR. Retrieved July 15, 2024, from https://aichr.org/news/press-release-aichr-somtc-joint-workshop-on-human-rights-based-approach-to-combat-trafficking-in-persons-especially-women-and-children-5-6-november-2015-yogyakarta-indonesia/.

Arisandy, Y. 2017. September 29. Indonesia dorong aksi melawan kejahatan perdagangan orang. Antara News. Retrieved June 25, 2024, from https://www.antaranews.com/berita/655479/indonesia-dorong-aksi-melawan-kejahatan-perdagangan-orang.

ASEAN. 2021. ASEAN Senior Officials Meeting On Transnational Crime Working Group On Trafficking In Persons (WG on TIP) Terms Of Reference. Retrieved August 15, 2024, from https://asean.org/wp-content/uploads/2021/01/DOC-9-ADOPTED-TOR-for-WG-on-TIP-1.pdf.

ASEAN Secretariat. 2022. *ASEAN Migration Outlook*. Jakarta: ASEAN Secretariat.

ASEAN Secretariat. 2024. Overview on External Relations. ASEAN.ORG, n.d. https://asean.org/our-communities/asean-political-security-community/outward-looking-community/external-relations/.

Baehr, Peter R., and M. Castermans-Holleman. 2004. *M. The Role of Human Rights in Foreign Policy*. New York: Palgrave Macmillan.

Bal, C. S., and K. Gerard. 2017. ASEAN's governance of migrant worker rights. *Third World Quarterly*, 1–21.

Basir, S. M. 2020. Irregular Migrations in Southeast Asia: Challenges for Protection. Indonesian *Journal of International Law* 145–180.

Benedek, W. 2011. "The Relevance of the Multi-Stakeholder Approach and Multi-Track Diplomacy for Human Rights Diplomacy". In *Human Rights Diplomacy: Contemporary Perspectives*, ed. Michael O'Flaherty, et al. Martinus Nijhoff Publishers.

Bjorkdahl, A. 2007. Swedish Norm Entrepreneurship in the UN. *International Peacekeeping* 14 (4): 538–552.

Carr, A. 2015. *Winning the Peace: Australia's Campaign to Change the Asia Pacific*. Melbourne: Melbourne University Press.

Carr, A., and D. Baldino. 2015. An Indo-Pacific Norm Entrepreneur? Australia and Defence Diplomacy. *Journal of the Indian Ocean Region* 11 (1): 30–47.

Douglas, J. H., and A. Schloenhardt. 2012. *Combatting Migrant Smuggling with Regional Diplomacy: An Examination of the Bali Process*. [Research Paper]. The University of Queensland Migrant Smuggling Working Group. Retrieved from https://law.uq.edu.au/files/6723/Douglas-Schloenhardt-BaliProcess.pdf.

Finnemore, M., and K. Sikkink. 1998. International Norm Dynamics and Political Change. *International Organization* 52 (4): 887–917.

Fry, J. E. 2020. "Networks in the norm life cycle and the diffusion of environmental norms" (Order No. 28156601). Available from ProQuest One Academic. (2471451439). Retrieved from https://www.proquest.com/dissertations-theses/networks-norm-life-cycle-diffusion-environmental/docview/2471451439/se-2.

General Assembly. 2010, August 12. United Nations Global Plan of Action to Combat Trafficking in Persons (64/293). UNODC. Retrieved from https://www.unodc.org/documents/commissions/CCPCJ/Crime_Resolutions/2000-2009/2009/General_Assembly/A-RES-64-293.pdf.

Hariyadi, R. P. 2021. FLEGT License in Indonesia-European Union Cooperation: Norm Life Cycle Analysis. *Global Strategies* 15 (2): 218–236. https://doi.org/10.20473/jgs.15.2.2021.218-236.

Hicks, P. 2011. "Human Rights Diplomacy: The NGO Role". In *Human Rights Diplomacy: Contemporary Perspectives*, ed. Michael O'Flaherty, et al. Martinus Nijhoff Publishers.

Humas. 2015. January 8. Inilah Prioritas Politik Luar Negeri Indonesia 5 Tahun Ke Depan. Sekretariat Kabinet Republik Indonesia. https://setkab.go.id/inilah-prioritas-politik-luar-negeri-indonesia-5-tahun-ke-depan/.

Indonesia and China Discuss Strengthening Comprehensive Strategic Partnership. Kemlu.go.id. Retrieved July 1, 2024, from https://kemlu.go.id/portal/en/read/5736/berita/indonesia-and-china-discuss-strengthening-comprehensive-strategic-partnership06/February/2024.

Indonesia's Ministry for Women's Empowerment and Child Protection. 2020. 2015–2019 Laporan Kinerja Gugus Tugas PP-TPPO. Retrieved from https://aseanactpartnershiphub.com/wp-content/uploads/2022/03/IN-NPA-Report-2015-2019-Bahasa.pdf.

International Labour Organization (ILO). 2015. Migration, irregular status, and deportation: An outline of international standards, regional trends. International Labour Organization (ILO).

Khasru, S. M. 2018. November 20. Opinion. Retrieved from Nikkei Asia: https://asia.nikkei.com/Opinion/Migration-the-forgotten-part-of-ASEAN-integration.

Kedzia, Z. 2011. "Conference Diplomacy and Human Rights". In *Human Rights Diplomacy: Contemporary Perspectives*, ed. Michael O'Flaherty, et al. Martinus Nijhoff Publishers.

Ketua Harian Gugus Tugas Penanganan TPPO Dialihkan ke Kapolri, Ini Alasannya. Tempo. 4 July 2023. Retrieved July 13, 2024 from https://nasional.tempo.co/read/1744381/ketua-harian-gugus-tugas-penanganan-tppo-dialihkan-ke-kapolri-ini-alasannya.

Khagram, S., J. V. Riker, and K. Sikkink, eds. 2002. *Restructuring World Politics: Transnational Social Movements, Networks, and Norms* (NED-New edition, vol. 14). University of Minnesota Press. http://www.jstor.org/stable/ https://doi.org/10.5749/j.ctttsc5q.

King, T. 2011. The European Union as a Human Rights Actor. In *Human Rights Diplomacy: Contemporary Perspectives*, ed. Michael O'Flaherty, et al. Martinus Nijhoff Publishers.

Kurusu, K. 2018. Japan as a norm entrepreneur for human security. In M. McCarthy (Ed.), *The handbook of Japanese foreign policy* (pp. 321–336). Routledge.

Mohsin, A. 2019. Politik Luar Negeri Republik Indonesia Masa Lampau, Kini, dan Masa Depan: Suatu Tinjauan dan Saran ke Depan. *Jurnal Ilmu dan Budaya* 41 (62): 7225–7244.

Mullerson, R. 2014. *Human Rights Diplomacy*. London: Routledge.

Nupus, H. 2017. Jumlah korban perdagangan manusia di Indonesia menurun. *aa.com*. Retrieved July 20, 2024, from https://www.aa.com.tr/id/headline-hari/jumlah-korban-perdagangan-manusia-di-indonesia-menurun/1005679.

O'Flaherty, M., Kędzia, Z., Müller, A., & Ulrich, G. Eds. 2011. *Human Rights Diplomacy: Contemporary Perspectives*. Leiden, The Netherlands: Martinus Nijhoff Publisher. https://doi.org/10.1163/ej.9789004195165.i-301

Opening Remark Australian Ambassador to ASEAN in Signing of Memorandum of Subsidiary Arrangement (MSA)—Indonesia. Australia-Asia Program to Combat Trafficking In Persons. https://asean.mission.gov.au/aesn/HOMSpeech15_07.html.

Pease, K. 2016. *Human rights and humanitarian diplomacy*. Manchester: Manchester University Press.

Penyampaian Prioritas Politik Luar Negeri Republik Indonesia 2019–2024. Portal Kementerian Luar Negeri Republik Indonesia. n.d. https://kemlu.go.id/portal/id/read/725/pidato/penyampaian-prioritas-politik-luar-negeri-republik-indonesia-2019-2024.

Perkuat Konsistensi dan Keseriusan Perang Terhadap TPPO, Kepala BP2MI Hadiri Rapat Koordinasi Komnas HAM. BP2MI. 2023, March 6. Retrieved August 1, 2024, from https://www.bp2mi.go.id/berita-detail/perkuat-konsistensi-dan-keseriusan-perang-terhadap-tppo-kepala-bp2mi-hadiri-rapat-koordinasi-komnas-ham.

Protocol to Prevent, Suppress, and Punish Trafficking in Persons Especially Women and Children, supplementing the United Nations Convention against Transnational Organized Crime (Trafficking Protocol). 2000. General Assembly resolution 55/25. United Nations.

Ramcharan, Bertrand G. 2011. The Human Rights Diplomacy of the UN Secretary-General". In *Human Rights Diplomacy: Contemporary Perspectives. Martinus*, ed. Michael O'Flaherty, et al. Nijhoff Publishers;

Richter. A. 2018. The Norm Life Cycle of UN Reform: "Delivering as One and UN System Wide Coherence." *Journal of International Organizations Studies* 8 (2): 25–46. final-03440908f.

Safe and fair migration through Migrant Workers Resource Center. ILO. Retrieved July 1, 2024 from https://www.ilo.org/resource/safe-and-fair-migration-through-migrant-workers-resource-center.

Salama, I. 2011. Human Rights Diplomacy from a UN Perspective: A Complement to Advocacy. In *Human Rights Diplomacy: Contemporary Perspectives*, ed. Michael O'Flaherty, et al. Martinus Nijhoff Publishers.

Septiyana, I. 2019. The Involvement of Indonesian Civil Society Organizations in the Policy-Making Process of Migrant Workers Protection in ASEAN. *Journal of Indonesian Social Sciences and Humanities (JISSH)* 1–19.

Sharom, A., H. R. Purnama, M. Mullen, M. Asuncion, and M. Hayes. 2018. *An Introduction to Human Rights in Southeast Asia First Edition*. Southeast Asian Human Rights Studies Network (SEAHRN).

Stuenkel, O. 2016. Brazil and R2P: A Case of Agency and Norm Entrepreneurship in the Global South. *International Relations* 1–16.

Tempo. 2023. Ketua Harian Gugus Tugas Penanganan TPPO Dialihkan ke Kapolri, Ini Alasannya. Retrieved August 16, 2024, from https://nasional.tempo.co/read/1744381/ketua-harian-gugus-tugas-penanganan-tppo-dialihkan-ke-kapolri-ini-alasannya.

The Annual Press Statement of the Minister for Foreign Affairs of The Republic of Indonesia. 2024. Retrieved January, 9, 2024 from https://kemlu.go.id/portal/id/read/5658/pidato/pidato-menlu-pada-pernyataan-pers-tahunan-menteri-luar-negeri-ri-tahun-2024.

The CoE kicks off Activities to Produce High-Quality Data on Trafficking in Persons in Indonesia. United Nations UNODC-KOSTAT. Retrieved July 16, 2024, from https://coekostat.unodc.org/coekostat/en/news/the-coe-kicks-off-activities-to-produce-high-quality-data-on-trafficking-in-persons-in-indonesia.html.

The Member States of The Bali Process. Retrieved July 18, 2024, from https://www.baliprocess.net/about-the-bali-process/#:~:text=The%20Bali%20Process%20brings%20Member,Asia%2DPacific%20region%20and%20beyond.

Teo, S. 2022. Middle Power Awkwardness? Indonesia's Norm Entrepreneurship in ASEAN. In *Awkward powers: Escaping traditional great and middle power theory*, ed. G. Abbondanza and T. S. Wilkins. New York: Springer Nature.

United Nations Human Rights Office of the High Commissioner. 2014. Human Rights and Human Trafficking. Fact Sheet No. 36. United Nations.

UNHCR. n.d. Trafficking in Persons. Retrieved July 1, 2024, from https://www.unhcr.org/what-we-do/protect-human-rights/asylum-and-migration/trafficking-persons.

Weschler, J. 2011. Human Rights Diplomacy of the United Nations Security Council. In *Human Rights Diplomacy: Contemporary Perspectives*, ed. Michael O'Flaherty, et al. Martinus Nijhoff Publishers.

Wunderlich, C. 2020. *Rogue States as Norm Entrepreneurs: Black Sheep or Sheep in Wolves' Clothing?* Cham: Springer.

Tri Nuke Pudjiastuti is a Research Professor at the Research Center for Politics, National Research and Innovation Agency (BRIN), Indonesia, focusing on forced migration, ASEAN, and border studies in Southeast Asia. She graduated with a master's degree in international migration from the Department of Human Geography and Environment at Adelaide University, South Australia. She obtained her Doctor of Forced Migration in the Department of Criminology at the University of Indonesia in 2014. Since 2015, she has been the co-convenor of the Asian Dialogue on Forced Migration (ADFM), which is a Track II Meeting of the Bali Process.

R. A. Rizka Fiani Prabaningtyas is a researcher at the Research Centre for Politics, National Research and Innovation Agency (Badan Riset dan Inovasi Nasional/BRIN), Indonesia. She graduated from the Gadjah Mada University, Indonesia, with a bachelor's degree in international relations. She obtained her master's in International Relations (Advanced) from the Australian National University in 2018. In BRIN, she is a member of two research teams, namely Indonesia's foreign policy and forced migration. Her research interests include international relations, particularly Indonesia's foreign policy, forced migration, and gender in IR. Currently, her research discusses issues on refugees in Indonesia, education for displaced children, gender in Indonesia's foreign policy, and Indonesia's foreign policy on the Indo-Pacific.

Athiqah Nur Alami is a researcher at the Research Center for Politics, the National Research and Innovation Agency (Badan Riset dan Inovasi Nasional-BRIN), Indonesia. She completed her Ph.D. from the Department of Southeast Asian Studies at the National University of Singapore. Her main interests are Indonesia's foreign policy, labor migration, and gender in international relations, particularly to apply a gender lens in analyzing both migration and foreign policy and diplomacy issues. She has been part of the research team on Indonesia's foreign policy for the last ten years. Her team has produced several publications, including a book on "Foreign Policy and Energy Security Issues in Indonesia" in Springer in 2017. Currently, Athiqah and her team have been conducting research on Indonesia in the Indo-Pacific, gender in Indonesia's foreign policy, and education for displaced children.

Faudzan Farhana is a researcher at the Research Centre for Politics, National Research and Innovation Agency (Badan Riset dan Inovasi Nasional/BRIN), Indonesia. She obtained her LL.M. in international maritime law from Swansea University, UK, in 2019. Her research interests are strategic issues in ASEAN, international maritime law, and forced migration. Her latest publications with the ASEAN Research Group and Forced Migration Studies, BRIN include ASEAN Maritime Security: The Global Maritime Fulcrum in the Indo-Pacific (Springers, 2022), Access to Education for Refugee Children in Indonesia During the COVID-19 Pandemic: Challenges and Adaptation Strategies (ASEAS Vol 16 No.1, 2023), and Kerja Sama ASEAN dalam Penanganan Kejahatan Transnasional di Wilayah Laut Indonesia–Malaysia (ASEAN Cooperation in Handling Transnational Crime in Indonesia–Malaysia Maritime Area) (BRIN Press, 2023).

Irin Oktafiani is a researcher at the Research Center for Population, the National Research and Innovation Agency (Badan Riset dan Inovasi Nasional-BRIN), Indonesia. Her main interests are anthropology and Indonesia's labor mobility. She just completed her Master of Anthropology from the School of Archeology and Anthropology at the Australian National University.

Chapter 9
Assessing Indonesia's Health Diplomacy During Covid-19

Yessi Olivia, Yusnarida Eka Nizmi, Umi Oktyari Retnaningsih, Purwasandi, and Rahmi Yulia

Abstract This article analyzes Indonesia's health diplomacy when Indonesia and other countries in the world faced the COVID-19 pandemic. Health diplomacy here is understood as the ideas, practices, and goals state and non-state actors wish to achieve in formulating global and universal health policies. The writers argue that Indonesia's fragile health system greatly influenced Indonesia's health diplomacy. Compared with other countries in the Asian region whose governments could control the pandemic at a fast pace, the Indonesian government appeared unable to adopt effective and efficient policies for handling COVID-19. As a result, the fatality rate due to the coronavirus in Indonesia was considerably high and took the lives of not only patients but also medical personnel. Therefore, Indonesia actively approached many parties to seek assistance, whether in vaccines or other medical equipment. Intriguingly, due to its status as one of the largest economies in the world, Indonesia inevitably had to help other countries anticipate COVID-19 even though the amount of assistance provided by Indonesia was lower than what Indonesia had received. By looking at the politics of public health in Indonesia and how they influence Indonesia's health diplomacy, the writers hope to contribute to the growth of the literature on health diplomacy in general and Indonesian foreign policy in particular.

Keywords Health diplomacy · COVID-19 · Pandemic · Politics of public health · Foreign policy

9.1 Introduction

In 2020, the world was caught off guard by the spread of an infectious and deadly SARS-CoV-2 virus, or COVID-19 in short. The first case of COVID-19 was reported from Wuhan province, China, in December 2019 by the Wuhan Municipal Health Commission. The WHO Representative Office in China caught up on the news. It informed the WHO Western Pacific Regional Office about the "viral pneumonia"

Y. Olivia (✉) · Y. E. Nizmi · U. O. Retnaningsih · Purwasandi · R. Yulia
Department of International Relations, Universitas Riau, Pekanbaru, Indonesia

I. Kartini and Kanisa (eds.), *Joko Widodo's Foreign Policy and Diplomatic Legacy*,
https://doi.org/10.1007/978-981-95-2085-5_9

cases where the patients were unresponsive to standard treatments (WHO 2020c). As the number of cases rapidly grew, the WHO's International Health Regulations (IHR) Emergency Committee recommended that the WHO's Director-General declare to the world the status of the outbreak as the Public Health Emergency of International Concern (PHEIC) in January 2020. By February 2020, COVID-19 had reached 28 countries; as of June 2020, almost all countries worldwide reported confirmed Covid-19 cases (WHO 2020c). From January 1, 2020 to December 31, 2021, the reported number of deaths directly or indirectly caused by the virus was nearly 14.9 million (WHO 2022). The death toll is very shocking; it makes the COVID-19 pandemic parallel to other great pandemics that have hit the earth.

Not only has the COVID-19 pandemic caused a staggering number of casualties, it has also caused a significant shock to the global economy. Despite WHO's recommendation against implementing strict travel and trade policies (WHO 2020a, 2020b), countries pulled their emergency brakes and scrambled to contain the virus. Various measures, such as border closures, quarantines, curfews, and lockdowns, were taken in some countries (Wang and Mao 2021). Undeniably, these approaches affected the economy. A preliminary assessment by economists from the World Bank in 2020 predicted that the pandemic cost the fall of the total gross domestic product (GDP) to 4%. The COVID-19 pandemic hit developing countries the hardest, with the fall of their GDP to 2.5% compared to 1.8% for developed countries (Maliszewska et al. 2020). The coronavirus also caused the impoverishment of so many people, as the United Nations estimated around 34.3 million people suffered under the extreme poverty line in 2020 (United Nations n.d.).

The COVID-19 case shows us some important lessons. First, humans' vulnerability to infectious diseases. It becomes apparent that there is no fortress strong enough to stop the spread of disease, especially in a world becoming ever more interconnected due to advancements in transportation technology. Nevertheless, coherent and decisive preventive action is needed to minimize the impact of the spread of the virus. The COVID-19 pandemic demonstrates how indecisiveness and poor emergency preparedness can be fatal. The virus originated in Wuhan, an area known as a significant technology and automobile hub where millions of people move in and out of the region. Human mobilization is domestic and international, as people travel from and to Wuhan from around the world. While the first COVID-19 cluster was reported in 2019, only on January 21, 2020 did the Chinese acknowledge "the risk of human-to-human transmission" and decided to close the area (Wu et al. 2020). The decision, however, was overly late as the virus had spread to other cities in China and outside the country as people traveled to celebrate Chinese New Year.

Secondly, the rise of nationalism mode of survival (see Bieber 2022; Çetin 2020; Fisher et al. 2020; Jenne 2022). It could be argued that fear triggered some countries' responses to the pandemic. Many governments readjusted their border policies to stop the virus transmission from outside their countries and made traveling less convenient for international visitors. In addition to restricting people's movements, the world witnessed governments exploiting the COVID-19 pandemic to curb civil liberties. As reported by Freedom House, a human rights group, since the beginning of the pandemic, there has been an increasing report of governments engaged in "abuses of power, silencing their critics, and weakening or shuttering important institutions" (Repucci and Slipowitz 2020, p. 1). What is also a concerning phenomenon due to the rise of nationalism is how developed countries stockpiled COVID-19 vaccines (Al Jazeera 2021; BBC 2020). Unsurprisingly, developing countries, including Indonesia, protested this move. In the seventy-sixth session of the United Nations General Assembly in September 2021, developing countries demanded vaccine inequity and criticized rich countries "for sharing surplus doses" and helping low-income countries (United Nations 2021).

This paper analyzes how the Indonesian government navigated the country amid the COVID-19 pandemic. The research focuses on Indonesia's health diplomacy under the Joko Widodo (Jokowi) administration. The writers argue that Indonesia's foreign policy during the pandemic reflected Indonesia's status as a middle-income country where, on one side, Indonesia, like other developing countries, struggled to secure COVID-19 vaccines. As the pandemic hit Indonesia's economy, Indonesia struggled to recover financially. Thus, Indonesia employed multiple approaches to get the help that it needed. On the other side, Indonesia, which has been building a path as a middle-power country, has used the pandemic as a way to leverage its status as an upper-middle economic country by assisting its partner countries.

The organization of this paper is as follows. In the first section, this paper discusses health diplomacy as a framework of analysis. Health diplomacy is a subset of diplomacy that has been rising, especially amid COVID-19. An analysis of Indonesia's health system is brought to the discussion as it is crucial to explain Indonesia's policy in responding to the pandemic. The rest of this paper assesses Indonesia's health diplomacy to secure COVID-19 vaccines and Indonesia's foreign aid to help other countries handle the virus.

9.2 Health Diplomacy and Global Governance on Health

The academic discourse on health diplomacy underscores its collective nature, a significant aspect of the importance and impact of collective efforts. It is not just a concept but a practice and a goal that state and non-state actors collectively strive to achieve in formulating global and universal health policies. This collective nature is reflected in the various definitions of health diplomacy. Kickbusch et al. (2007), for example, emphasize the goal of health diplomacy, where multiple actors are involved in negotiations aimed at "shaping and managing the global order on health." Ruckert et al. (2016) highlight that health diplomacy is a practice taken by states and non-state actors in formulating global health policies. Some emphasize the process of doing health diplomacy. Adams et al. (2008) regarded it as a political activity, and Bond et al. (2001) saw it as a confidence-building and negotiation activity. Meanwhile, Smith and Irwin (2016, p. 1) see health diplomacy as a series of processes, including "finalizing agreements between multilateral or bilateral aid donors and recipient countries, to the processes of making binding and non-binding international agreements in health or related to health." Despite the different emphases, a common thread emerges in health diplomacy: the collective efforts to enhance the global health order.

The study of health diplomacy is particularly intriguing due to its rich and complex origins. As the name suggests, the concepts of diplomacy and health stem from two distinct fields of study: international relations and health. In International Relations, discussions about diplomacy are closely intertwined with foreign policy, which embodies the aims and strategies states wish to achieve, or as International Relations scholars often phrase it, national interest. To quote Morgenthau, a renowned international relations academic, diplomacy is a step policymakers take to achieve national interests. He argued that a country's diplomacy quality is a crucial determinant of its ability to maintain national strength and state morale (Morgenthau 1993). Similarly, Kissinger, former United States Secretary of State, viewed diplomacy as a tool in the context of power politics and efforts to achieve national interests (Kissinger 2014).

But how can health issues be included in a country's foreign policy priorities, giving rise to the concept of health diplomacy? The answer lies in understanding how international cooperation on health has developed over the years in response to public health issues. Kickbusch et al. (2007) are leading academics in this subject. According to them, the basis for global cooperation that focuses on health began with multilateral meetings such as the First International Sanitary Conference in 1851 and the First International Sanitary Convention in 1892. The establishment of international institutions such as the Office International d'Hygiene Publique and the League of Nations Health Organization, which merged under the World Health Organization (WHO) after World War II, played a crucial role in this development. With their robust systems and dedicated professionals, these institutions have been instrumental in shaping global health diplomacy. The WHO, in particular, is essential because it introduced health into high politics, a significant milestone in the history of global health diplomacy. As written in the WHO Constitution, health is considered one of the prerequisites for security and peace. The WHO's role in setting global health standards, coordinating responses to health emergencies, and advocating for health as a fundamental human right further underscores its significance in health diplomacy, keeping you informed and aware of the organization's influence. In the early years since its founding, WHO faced complex challenges, starting from coordinating the eradication of disease, becoming an arena for high-level meetings during the Cold War, and facing newly independent countries that did not yet have a national health system. Dynamic world developments mean that WHO will continue to face new challenges, ranging from infectious diseases and global inequality to global health politics, which involve various actors and institutions at various levels.

The global political landscape has significantly influenced the development of international cooperation in the health sector. Before the First World War, regulations on international affairs were limited. The global economic crisis in the 1930s and the devastating World War II also profoundly impacted international relations. Things changed after World War II when there was an urge to initiate international cooperation to avoid more significant conflict. The United Nations, a system of international collaboration coordinated by the UN through specialized organizations and partner organizations, played a pivotal role in driving global change, providing a sense of reassurance about the direction of global health efforts. The UN, with its specialized agencies like the World Health Organization (WHO), Food and Agriculture Organization (FAO), and United Nations Children's Fund (UNICEF), has been instrumental in setting global health standards, coordinating responses to health emergencies, and advocating for health as a fundamental human right. The international financial and trade system, regulated through institutions such as the IMF, World Bank, and WTO, initiated economic globalization, crucial for world stability. The 1970s–1980s showed how economic agendas under neoliberalism strengthened and spread despite the Cold War between the United States and the Soviet Union.

The attention to global governance on health began to increase after the Cold War ended. The shift in the way international actors view security is one of the things that has driven increased attention to health issues. Initially, security was a concept inherent only to the state. A country is said to be safe if it can eliminate things that are considered to threaten its territorial integrity. Toward the end of the Cold War, the traditional security paradigm was challenged by people who used a bottom-up approach to viewing security. According to them, security should start from individuals (humans). The sense of security obtained from fulfilling all basic needs such as food, work, shelter, and health will accumulate into the security of society and the state. This idea of human security, which emphasizes the protection of individuals and communities from direct and indirect threats, was realized when the United Nations Development Program (UNDP) issued a special report highlighting its importance. In the report, UNDP argues that to respond to global challenges, countries must prioritize human development over security in a strategic (military) context. Health security must be guaranteed by the state, in addition to food, economic, environmental, personal, community, and political security, making you feel secure and reassured about global health efforts. The UNDP report identified several problems related to health security, namely the issue of poverty in both rich and developing countries. Poor groups, especially women, are a group that is notoriously vulnerable to disease and do not have access to health services. UNDP, as a leading advocate for human security, is not the only international organization that includes health indicators in its development reports. With increasing focus on sustainable development after the Cold War ended, organizations such as the World Bank and WHO also analyzed health and its relationship to development.

The end of the Cold War has made global politics increasingly complex. The world is no longer hostage to the competition between the Soviet Union and the United States. However, this situation does not mean that the United States rose as the only dominant player. After the Cold War, countries like China and India emerged as competitors to other powerful countries, adding complexity to global politics. These emerging powers are reshaping the global political landscape and influencing the direction of global health efforts. China, for instance, has been increasingly involved in global health initiatives, providing aid and expertise to countries in need. The world order also experienced drastic changes due to information technology innovation and a free economic system. One of the prominent trends was the increasing growth of global governance to solve global problems. As suggested by Almeida et al. (Almeida 2020), there are at least three trends in global governance. Firstly, the relationship between domestic and international politics is strengthening, marked by the various actors involved in governance to solve global problems. Secondly, the involvement of non-state actors in international institutions and global governance is increasing, a significant trend that is reshaping the dynamics of global politics. This shift is significant as it broadens the range of perspectives and expertise brought to bear on global issues, making the global governance system more inclusive and effective. Thirdly, the emergence of various forms of global management has resulted in multiple rules and mechanisms to ensure provider compliance.

In the twenty-first century, the following actors influence the global health governance. The leading players are in the G20 countries as a collection of countries that control 80% of world trade and 2/3 of the world's population. These are a group of countries that provide foreign aid or are the most prominent donors in world health programs. Apart from the G20 group, the world is increasingly seeing the existence of countries such as Brazil, China, India, and South Africa playing an active role in world health programs. Regional organizations, such as ASEAN, the European Union, and the European Union, and state groupings, such as South-South Cooperation, must also be addressed because they increasingly have a good bargaining position to encourage change in the global health system. The twenty-first century period is also marked by the acceptance of countries in the world toward integrating health issues into foreign policy. In 2006, the governments of France and Norway initiated Foreign Policy and Global Health as an informal consultation forum for countries to discuss health issues and bring them to the UN General Assembly meetings. A year later, foreign ministers from Brazil, France, Indonesia, Norway, Senegal, South Africa, and Thailand issued statements on the need to include health within the scope of their countries' foreign policies. According to them, this is important because "health is fundamental to economic growth and development" and "health threats may compromise a country's stability and security" (Pibulsonggram et al. 2007, p. 1). On November 26, 2008, the UN General Assembly adopted Resolution 63/33 on global health and foreign policy as a continuation of the initiative of countries involved in the Foreign Policy and Global Health initiative (United Nations 2008). Apart from encouraging countries worldwide to consider health issues in their foreign policies, the resolution also emphasizes the role of the UN system in assisting countries in implementing cooperation in the health sector.

Apart from state actors, present-day global governance on health also shows the involvement of various non-state actors. WHO remains the leading organization tasked with coordinating cooperation on health issues. WHO has created many initiatives to respond to emerging challenges, such as increasing awareness of noncommunicable diseases, eradicating HIV/AIDS, managing pandemics, and others. However, the WHO is no longer a single player. Experts in the field of public health are also in international organizations such as the UN Children's Fund (UNICEF), Food and Agriculture Organization (FAO), World Trade Organization (WTO), International Labor Organization (ILO), UN Environment Program (UNEP), and World Bank. Apart from these international organizations, hundreds of NGOs, faith-based organizations, and foundations such as the Bill & Melinda Gates Foundation have partnerships with WHO to discuss and create health programs.

The existence of increasingly complex global governance of health shows us several things. The first is related to the integration of health into foreign policy. Fidler (2006) argues that the relationship between health and foreign policy changed dramatically after the Cold War ended. Redefining security is one driving factor. Security is related to threats coming from enemy countries and the safe position humans feel (human security). Second, according to Fidler, globalization has forced national governments to rethink their domestic and foreign policies because of increasing interconnectedness and interdependence. Globalization, in this case, has a significant impact on many things, including health, through the emergence of issues such as the pollution, infectious diseases, etc. The third point relates to how groups of non-state actors increasingly gained political space and material to participate in international politics after the Cold War ended. Fidler claimed that the final structural factor is the increasing complexity of global health problems, for example, the spread of disease, which can affect the main functions of the state, starting from maintaining national security, securing economic interests, providing development assistance, and supporting human dignity. Fidler emphasized that this transformation process is not unidirectional. People who work in the field of health policy also inevitably have to participate in the practice of foreign policy and diplomacy because globalization has blurred the boundaries between "domestic health activities and international health efforts."

9.3 Indonesia's Health System

For context, COVID-19 spread to all corners of Indonesia during the second term of President Joko Widodo, popularly known as Jokowi. He won the 2019 presidential election and resumed office two months before COVID-19 struck the world. Just like the first period of his administration, Jokowi prioritized economic development, a crucial aspect that was sometimes implemented at the expense of civil and political rights. For his second term from 2019–2024, Jokowi set five work priorities that continued highlighting the importance of economic development (panrb n.d.). The priorities were accelerating and continuing infrastructure development, followed by human resource development, investment regulations to create jobs, bureaucratic reform, and spending budgets for economically profitable things. The COVID-19 pandemic, a significant challenge, potentially damaging the Jokowi's second-term agenda, further underscores the importance of economic development. The government faced a difficult choice: executing the plans or postponing them. The options then taken by the Jokowi government, as explained below, indicated that the government tended to prioritize the economy over the welfare of many people. This choice had significant implications for public health, social welfare, and the economy, which will be discussed in the following sections.

Before delving into the discussion, it is crucial to understand Indonesia's demography and health system. Indonesia is an archipelagic country in the form of a republic. The country's 270 million citizens are spread across thousands of islands, with five being the largest, stretching from the east to the west of the archipelago. This population statistic ranks Indonesia fourth in the world's largest population. In a sociodemographic context, Indonesian society is heterogeneous, consisting of various ethnicities and cultures. Javanese are the largest ethnic group (40%), followed by Sundanese, Malays, and Bataks. Around 55% (154 million people) of Indonesia's population live on the island of Java, followed by the island of Sumatra. These statistics make Java the island with the highest population density in Indonesia (1,171 people per km^2) (Mahendradhata et al. 2017). This figure is a significant contrast compared to the population density level on the island of Papua (8–9 people per km^2), even though the island is larger than Java (Ananta et al. 2016). Based on data from Indonesia's Statistic Bureau (BPS 2024), the total birth rate in Indonesia in 2020 was 2.18, a decrease of 0.23 from data in 2010. Indonesia's population growth rate is declining, with 1.17 in 2022, 1.13 in 2023, and 1.11 in 2024. The median age of the Indonesian population in 2023 is 29.8, slightly below the median of the world population (30.4) and Asia (31.7). Another critical piece of information is that most of its population (69.25%) is in the productive age category (15–64 years).

Since Indonesia's political transformation from an authoritarian country to a democracy in 1998, its economy has improved significantly. The economic crisis that hit Asian countries, including Indonesia, in 1997 brought Indonesia's Gross Domestic Product (GDP) down to—13.1% (World Bank n.d.). After executing several economic reforms, the B.J. Habibie administration stabilized the economy, including reducing inflation and boosting economic growth. Through prudent economic policies, economic growth rose significantly to 4.9% in 2000 and has continued upward. Indonesia is not immune to shocks from the outside. Indonesia's economic growth rate during the COVID-19 pandemic 2020 was recorded at—2.1%, down drastically from 5% in 2019. However, with a relatively good macroeconomic profile, Indonesia achieved status as the tenth-largest economy in the world; it joined the G20 group that comprises the countries with the best economies in the world. However, several development problems, especially poverty, continue to linger. Even though the poverty rate in Indonesia shows a downward trend, the government cannot overlook the problem. As of 2023, around 9.36% or around 25.90 million of Indonesia's population are living under the poverty line. Once again, Java has the largest population of poor people, followed by those living on the island of Sumatra (BPS 2023).

The political changes in 1998 had an impact not only on Indonesia's economic system but also on its health system. During the New Order era, Indonesia was an authoritarian country operating in a very centralized system. It can be said that the government budget for health sector during the New Order era was small in scale, especially for health services that had a preventive function. As the regime vigorously pursued economic development, it focused only on the welfare of a small part of society who were deemed crucial to support government goals. The government mandated social health insurance to civil servants through *Asuransi Kesehatan—Askes*; the police and military through *Asuransi Sosial Angkatan Bersenjata Republik Indonesia—Asabri*; and the formal private sector through *Jaminan Sosial Tenaga Kerja—Jamsostek*. Meanwhile, basic health services were only began with the establishment of Community Health Centers (*Puskesmas*) (Mahendradhata et al. 2017).

These community health centers are spread across all sub-districts in Indonesia and have been a crucial element to Indonesia's present health system. State-funded hospitals, which were available in the district and provincial levels were largely underfunded. There were not many private hospitals, and they were run by civil society groups or doctors who taught at universities and opened small clinics. However, in the 1980s, as the need for health services increased, the government finally left it to the market. With Minister of Health Regulation 920/1986 and later Minister of Health Regulation 84/1990, the government allowed the establishment of private hospitals owned by individuals, groups, and corporations. This shift toward a market-driven health system had both positive and negative implications. On one hand, it increased the availability of health services, particularly in urban areas. On the other hand, it also led to a significant disparity in the quality and affordability of health services, with many rural areas still lacking adequate healthcare.

The 1998 reforms changed everything. The enormous demand for regions to manage their government made Indonesia implement a decentralization policy in 1999. The division of roles between the central and regional governments is also seen in the management of the health system in Indonesia. Basically, decentralization means that regional governments take over some of the functions of the Ministry of Health. The Ministry of Health carries out functions such as making regulations, ensuring the availability of resources, managing several leading and specialist hospitals, and carrying out supervisory functions. Regional governments, divided into provincial and district governments, carry out health service functions. The provincial government, for example, owns hospitals and manages health services through the provincial health office. The government at the district level also carries out the same function, owning hospitals and managing health services through district health offices, which reach down to First Level Health Facilities in the form of Community Health Centers (*Puskesmas*).

The relationship between the central and regional governments regarding health services is complex. Health services at the district level are not under the provincial government structure, nor are hospitals at the district level under the authority of provincial hospitals. Hospitals at the provincial level are also directly responsible to the provincial government, not the Ministry of Health. Only a few programs managed directly by the Ministry are passed on to the provincial and district levels, such as immunization. After the 1998 Reformation, the Ministry of Health no longer dominated the health sector. There are many ministries and other bodies involved, for example, the Ministry of Home Affairs, which coordinates with regional governments; the Social Security Administration, which manages National Health Insurance; and the Food and Drug Monitoring Agency/BPOM, which is tasked with supervising food and medicine (Mahendradhata et al. 2017).

Aside from the structural changes in the health system due to decentralization and the formation of new institutions, another significant post-Reformasi change is the government's expansion of social security coverage. The 1998 reforms, triggered by the financial crisis in 1997, significantly impacted the increasing number of people living in poverty. In response, the government implemented several policies to protect the lower class, demonstrating its commitment to public health. These policies included the Health Sector Social Safety Network (*Jaringan Pengaman Sosial Bidang Kesehatan—JPS BK*), the first health insurance for the non-formal sector covered by the government. The government then introduced various new schemes, such as Health Insurance for the Poor (*Asuransi Kesehatan Masyarakat Miskin—Askeskin*), later changed to Community Health Insurance (*Jaminan Kesehatan Masyarakat—Jamkesmas*). In addition to central government social security, several regional governments have developed similar programs providing subsidies to people experiencing poverty to access health services. A significant milestone was reached in 2004 when Indonesia adopted the Law on the National Social Security System (*Undang-undang tentang Sistem Jaminan Sosial Nasional*), paving the way for establishing universal health coverage.

9.4 Indonesia in the Time of Covid-19

The Indonesian government reported the first case of the COVID-19 virus in March 2020, a moment that marked the beginning of a critical period for the nation (Gorbiano 2020). President Jokowi's announcement that two adults in Depok, West Java, had contracted the virus after coming into contact with a Japanese citizen who visited Indonesia in February 2020 was a stark wake-up call. At that time, Jokowi said that the government had prepared more than 100 hospitals with isolation rooms and internationally standardized equipment. However, this declaration was met with skepticism, as some questioned the discovery of the cases in March 2020, when neighboring countries, such as Malaysia and Singapore, reported their first COVID-19 cases in February 2020. The urgency of the situation was further underscored by the readiness of Indonesia's health infrastructure to detect and test the virus, which fueled the public's doubt. That same month, several cities in Indonesia reported cases similar to the coronavirus symptoms, but the authorities dismissed them as not COVID-19. Even expatriates in Jakarta were not immune to these concerns. A confidential report circulating among diplomats in Jakarta warned them about the Indonesian government's perceived weak handling of the coronavirus. In a press release, the WHO Representative for Indonesia acknowledged that the reports of COVID-19 cases were unsurprising, and they anticipated the number of cases would increase. The WHO representative stressed that early detection and laboratory readiness were "critical," underlining the importance of these measures in controlling the spread of the virus (Fadli and Arbi 2020). This underscores the necessity of these measures in the fight against the pandemic, emphasizing the urgency of the situation.

As predicted by many people, the spread of COVID-19 was increasing rapidly in Indonesia. In August 2020, the total number of confirmed COVID-19 cases increased to 125,396. Almost all regions in Indonesia (34 provinces and 480 districts/cities) reported COVID-19 cases. The five provinces with the highest COVID-19 cases are DKI Jakarta, East Java, Central Java, South Sulawesi, and West Java (Mukaromah and Akbar 2020). In November 2020, the number of COVID-19 cases exceeded 500,000 (Djalante et al. 2020). The situation became even more complicated when, at the beginning of the outbreak of COVID-19 in Indonesia, there was a severe shortage of masks, medical oxygen, and personal protective equipment needed by medical personnel. This shortage significantly hampered the ability of healthcare workers to provide adequate care, leading to increased mortality rates. Even though the government continued to reassure the public that they appropriately mitigated and handled the spread of the virus, some people cast doubt. A mathematical modeling expert from Indonesia's top university said that coronavirus transmission became increasingly undetectable, and the actual number of cases could be close to millions. With the emergence of the Delta coronavirus variant in 2021, some even called Indonesia the epicenter of the spread of COVID-19 in Asia (Dyer 2021).

The most alarming aspect of COVID-19 cases in Indonesia is the very high fatality rate. Table 9.1 shows data on confirmed coronavirus cases and death rates for Southeast Asian countries. It can be seen from the data that Indonesia ranks at the top for the number of patients who died due to COVID-19, even though the number of confirmed coronavirus cases in Indonesia is still far below Vietnam, whose confirmed cases are above 11 million. Researchers from the Eijkman-Oxford Clinical Research Unit (EOCRU), which examined COVID-19 cases in several hospitals in Jakarta, said that patients in the area died at a faster rate than in other countries. They revealed that the high death rate was due to delays in detecting and treating COVID-19 patients, as well as the lack of adequate healthcare infrastructure and resources. The high mortality rate was not only among patients but also among health workers. Research by Ekawati et al. (2022) shows that the number of health workers in Indonesia who have contracted COVID-19 is very high. The number of health workers who died as a result of COVID-19 in the 18 months of Ekawati et al.'s research was 1545. They came from various health fields, from doctors, nurses, midwives, and health laboratory workers. The results of this research imply the unpreparedness of the Indonesian health sector, both in terms of infrastructure and human resources, in facing the pandemic. Health facilities such as hospitals were overwhelmed by the increasing number of patients. At the pandemic's peak, the government received many reports from hospitals that lack the oxygen supply patients need. *Lapor Covid*, an online platform a civil society group initiated to disseminate information about COVID-19, said they had difficulty helping the public find health facilities to treat COVID-19 patients, highlighting the severe strain on the healthcare system (BBC 2021).

Due to the phenomenal characteristics of the COVID-19 pandemic, many governments were under the spotlight for how they anticipated the virus. Countries such as Taiwan, South Korea, and Vietnam have received praise because their governments quickly controlled conditions and reduced deaths due to the coronavirus (Barron 2020; Dabla-Norris et al. 2020). Vietnam, which shares borders with China, quickly took preventive measures such as temperature checks for people who had just arrived from Wuhan, China, to prevent the spread of the virus. Taiwan, which had experience dealing with SARS cases in 1990a, quickly developed a tracing system to detect COVID-19 cases.

Table 9.1 Fatality rate in Southeast Asian Countries

Country	Confirmed	Deaths	Case-fatality (%)	Deaths/100k population
Brunei Darussalam	279,661	225	0.1	51.43
Cambodia	138,719	3,056	2.2	18.28
Indonesia	6,738,225	160,941	2.4	58.84
Laos	218,023	758	0.3	10.42
Malaysia	5,044,718	36,967	0.7	114.22
Myanmar	633,950	19,490	3.1	35.82
Philippines	4,077,452	66,188	1.6	60.40
Singapore	2,235,294	1,722	0.1	29.43
Thailand	4,728,182	33,918	0.7	48.59
Vietnam	11,526,994	43,186	0.4	44.37

Source John Hopkins Coronavirus Resource Center https://coronavirus.jhu.edu/data/mortality (data retrieved on 18 July 2024)

How about Indonesia? Looking back at the outbreak of COVID-19 in 2020, it is evident that the government played down the threat posed by the spread of the coronavirus and failed to take preventive actions. Instead of providing explicit instructions on preventive measures to mitigate the virus, Indonesia's Minister of Health boasted of how the virus had not hit Indonesia and encouraged the people to pray (Rina 2020). Even when the government finally established a special task force to handle the virus, the government seemed unable to come in unison to convey the message. Several ministries devised questionable ideas, such as special herbal drinks and anti-corona necklaces, for combatting the coronavirus (see Kemenko PMK 2020; Yuniar 2020). Competition between political elites was also apparent, as shown when the central government and the Jakarta governor bickered on the issue of lockdown (Asmara 2020). However, the most telling part about how the government handled the coronavirus was who it chose to sit on the special task force. In March 2020, the Jokowi administration established a task force for interagency coordination. This task force was disbanded and replaced by the Committee for Handling COVID-19 and National Economic Recovery in July 2020 (Kemenko Bidang Perekonomian 2020). Even though the Health Minister was part of the task force, most committee members were mainly in charge of economic affairs. The decision to form this task force was undoubtedly a big question, considering that, at that time, COVID-19 cases were still very high.

As previously mentioned, the COVID-19 pandemic occurred during Jokowi's second leadership term. The government, determined to push for economic development, saw the coronavirus pandemic as a disruption. The policies taken by the Indonesian government in response to the coronavirus implied that the government preferred the economy to continue running. This preference for economic development led to decisions prioritizing economic stability over public health, such as implementing a partial lockdown through a Large-Scale Social Restrictions (PSBB) policy rather than a total lockdown on March 31, 2020. Jokowi expressed his rejection of the lockdown in a limited meeting on handling COVID-19 on March 24, 2024. President Jokowi said each country has its own culture and characteristics, and the total lockdown was incompatible with Indonesia. He finally admitted that the lockdown would disrupt the Indonesian economy in April 2024 (Ihsanuddin and Erdianto 2020). While aiming to protect the economy, this approach also had significant implications for public health and the healthcare system.

The impact of this lack of leadership and its lenient approaches almost paralyzed Indonesia's health facilities, which were overwhelmed by the drastically increasing number of patients. Apart from dealing with an overflow of patients, the health infrastructure also suffered from a shortage of medical equipment, part of which was caused by panic buying. This shortage of medical equipment, including masks, personal protective equipment, and medical oxygen, significantly hampered the ability of healthcare workers to provide adequate care and put patients at risk. It is not an understatement to say that strong leadership is essential when serious problems such as a pandemic strike, and Indonesia was seriously lacking that during the COVID-19 period.

9.5 Indonesia's Health Diplomacy During Covid-19: Balancing Needs and Responsibilities

The previous discussion of this paper has clarified the concept of health diplomacy. In essence, health diplomacy encompasses all diplomatic endeavors undertaken by state and non-state actors to shape the global health landscape. This paper has also discussed the evolution of Indonesia's health system, which has shown remarkable resilience and adaptability as Indonesia transformed into a democratic country. The section on what happened when the coronavirus hit Indonesia implied that while democratization has reformed the political and economic system and affected Indonesia's health system, it is still fragile. This section underscores the delicate balance of needs and responsibilities in Indonesia's health diplomacy during the COVID-19 crisis. On one hand, Indonesia, as a developing nation, desperately needs assistance to combat COVID-19. On the other hand, its status as one of the world's economic powerhouses bestows upon it the moral responsibility to aid other nations, thereby influencing the global health order significantly.

It is worth noting how influential *Reformasi* was to Indonesia's status in global politics (Nabbs-Keller 2013; Rüland 2017; Umar 2023; Xue and He 2022). Firstly, pressures to democratize the country derived not only from within but also from outside Indonesia. Cases such as the kidnapping of democracy activists, ethnic-based riots, and demands for the separation of East Timor gained a lot of traction from the outside, which then forced Indonesia's interim government to make dramatic changes. B.J. Habibie, who was appointed as the president, made a series of strategic decisions to liberalize the country, from reestablishing freedom of the press, instigating the first free general election, ratifying several international conventions on human rights to allowing the people of East Timor to determine their future through referendum. Since the Reformasi period, democracy and human rights have become part of Indonesia's foreign policy rhetorics, especially in its efforts to rebrand the country as the largest democracy with the most populous Muslim in the world. This position sharply contrasts the New Order era, where the government suppressed political and civil liberties, and discussions about democracy or human rights were seen as an attempt to interfere in Indonesia's internal affairs. With the significant improvement of the Indonesian economy after comprehensive economic reforms in 1998, which included measures to stabilize the currency, restructure the banking system, and attract foreign investment, and Indonesia occupying a position as one of the largest economies in the world, Indonesia has become a country that is confident and active in discussing global issues in international forums. Indonesia, especially during the leadership of Susilo Bambang Yudhoyono (SBY), initiated the Bali Democracy Forum (BDF) to maintain Indonesia's branding as a democratic country that upholds equality and Indonesian human rights. By using the middle-power diplomacy approach, Indonesia has also succeeded in convincing other ASEAN countries to incorporate democratic values and human rights into institutions that have been known as the cornerstone of Indonesia's foreign policy.

Regarding health diplomacy, two events put Indonesia on the map. The first event was Indonesia's breakthrough in issuing what is known as the virus sovereignty doctrine in 2007 (see Hameiri 2014; Lowe 2019). This doctrine, a view that says viruses circulating within a country's territory belong to that country, was a significant development in global health diplomacy. It was considered part of state sovereignty, giving the state the right to decide whether or not to share the virus sample with the rest of the world. The Indonesian Minister of Health at that time, Siti Supari, argued before the World Health Assembly (WHA) that there had been fraud in virus sharing, so Indonesia refused to share virus samples with outside parties. She claimed some parties distributed the virus to pharmaceutical companies who patented them. Supari said that although the ingredients for the vaccine came from developing countries, rich developed countries had priority in getting the vaccine first. Supari's provocative statement received a great response from other countries. Indonesia was considered to have violated the norms of scientific cooperation in fighting the pandemic, and people accused Supari of inciting anti-western sentiment. This virus sovereignty discourse ultimately forced the WHA to formulate a new agreement, the Pandemic Influenza Preparedness (PIP) Framework, which guarantees transparency, justice, and an effective system for virus sharing.

The second event was a joint statement by foreign ministers from Brazil, France, Indonesia, Norway, Senegal, South Africa, and Thailand on March 20, 2007 (Pibulsonggram et al. 2007). These countries are members of the Foreign Policy and Global Health Initiative (FPGH), a forum between countries that meets to discuss health issues and has played a crucial role in Indonesia's global health diplomacy efforts. In a joint statement known as the Oslo Ministerial Declaration, these countries said that health needs great attention. These countries are calling for the scope of foreign policy to be expanded, one of which is to include health issues. Foreign ministers from these countries said they would "increase awareness of our common vulnerabilities in the face of health threats by bringing health issues more strongly into the arenas of foreign policy discussions and decisions" (page 1). This inter-country forum has also become an essential medium for Indonesia and member countries during the COVID-19 era. Indonesia was chairman of Foreign Policy and Global Health in 2020, coinciding with the spread of coronavirus worldwide. At Indonesia's initiative, the FGPH countries issued a joint statement to respond to the COVID-19 pandemic. In the joint statement, the FGPH countries reaffirmed the need for international cooperation to anticipate COVID-10 and the importance of improving national health systems, building trust between authorities and the public, and improving the quality of health (Foreign Policy and Global Health 2020). When the forum countries met online on September 3, 2020, the Indonesian foreign minister, Retno Marsudi, raised several urgent issues (Kementerian Luar Negeri 2020). First is equal and fair access to vaccines, second is the importance of global health system governance, and third is the urgent need to improve national health systems. These positions emphasize the urgency of this issue and the need for immediate action.

What happened through the sovereignty virus debate and membership in FGPH shows several things. Firstly, Indonesia's desire to become an influential country in global politics. The Reformasi period has transformed Indonesia as one of the rising economic powers. This new status positively affects Indonesia's self-confidence in conducting international relations. Second, Indonesia has become more aware that it needs support and cooperation from other countries to overcome problems such as the pandemic. The COVID-19 experience shows that the rising trend of nationalism is the biggest challenge in global health diplomacy. As mentioned at the previous section, the coronavirus pandemic has significantly impacted the global economic slowdown. The world's poverty rate is increasing because many economic activities have stopped or been disrupted. Of course, lower- and middle-income countries are at high risk of falling into a financial crisis.

The next problem that Indonesia faced was vaccine availability. When pharmaceutical companies succeeded in formulating a coronavirus vaccine, concerns emerged among developing countries about the practice of developed countries stockpiling vaccines. Practices like this have given rise to strong reactions from developing countries, including Indonesia. In a recorded speech for the 2021 UN General Assembly session, President Jokowi slammed the politicization and discrimination toward vaccines and hoped for improvements to a fair global health system, especially for vaccine distribution (Cabinet Secretariat of the Republic of Indonesia 2021). To meet the need for vaccines, Indonesia reached out to many parties. For example, Indonesia is a key player in the GAVI COVAX Advance Market Commitment (AMC) collaboration initiative formed by the Coalition for Epidemic Preparedness Innovations (CEPI), WHO, and UNICEF. This initiative aims to distribute two billion vaccines to developing and least-developed countries. Indonesia proposed a vaccine request to the COVAX Facility in December 2020 and, through Minister of Foreign Affairs Marsudi, Indonesia was appointed as one of the initiative's co-chairs (Septiari 2021). This leadership role demonstrates the international community's trust in Indonesia's commitment to equitable vaccine distribution. Marsudi also voiced Indonesia's rejection of vaccine nationalism in other international forums. In a meeting organized by the World Economic Forum (WEF), for example, she said that equitable distribution of vaccines is a humanitarian issue, not a political issue, and therefore, vaccine nationalism must be stopped.

Indonesia aggressively carried out vaccine diplomacy with various parties through its foreign ministry to ensure vaccine supplies are met. Due to its persistence, Indonesia received assistance amounting to 2.5 million doses of the AstraZeneca vaccine from Australia, 4 million doses of the Moderna vaccine from the US through the COVAX Facility, and 1 million doses of the AstraZeneca vaccine from Japan (Editorial Board 2021). In providing vaccine assistance, China is the country that provides the most significant number of vaccines to Indonesia. The Sinovac vaccine from China first entered Indonesia with 3 million doses in December 2020. Since January 2022, approximately 200 million Sinovac vaccines have been distributed to Indonesia (Rakhmat and Purnama 2022). The distribution of the Sinovac vaccine plays a significant role in achieving herd immunity in Indonesia. The assistance that Indonesia receives from external parties is not just vaccines. Several Chinese investors provided assistance totaling 40 tons of medical equipment ranging from test kits, N95 masks, surgical masks, and Personal Protective Equipment (Afifa 2020).

With its status as a developing country with the largest economy, Indonesia also shows that it is more than passively receiving aid. Indonesia has also sent assistance to several countries to handle COVID-19. In May 2021, Indonesia sent humanitarian grant assistance through 200 units of oxygen concentrates for handling COVID-19 to India (BPMI Setpres 2021). Previously, the Indonesian government and industrial players had sent 1400 oxygen cylinders to Indonesia's partner countries. Meanwhile, in September 2021, Indonesia sent USD 200,000 in COVID-19 relief assistance through the ASEAN Coordinating Center for Humanitarian Assistance on Disaster Management (AHA Centre). The assistance provided includes medical equipment made in Indonesia, such as masks and gloves (Kementerian Luar Negeri 2023).

We can see that the Indonesian Foreign Affairs Ministry has played a very significant role in the recovery from COVID-19 in Indonesia. Without the hard work of Foreign Affairs Minister Marsudi and her team in negotiating with foreign countries and international institutions, Indonesia would not be able to meet the availability of vaccines and medical equipment. However, even though Marsudi and the ministry she leads play a very strategic role, she is separate from the special task force formed by the Indonesian government for COVID-19 recovery. As previously mentioned, the Jokowi government, which focuses on economic growth, formed a task force filled with most ministers in the financial sector. Even the head of the task force is led by the Coordinating Minister for Economic Affairs rather than the Minister of Health. This condition is very ironic considering that Indonesia encourages health integration into a country's foreign policy, and transnational health issues are one of the significant challenges in health diplomacy.

9.6 Conclusion

This article shows us how the Indonesian government worked during COVID-19. The discussion begins by explaining the development of health diplomacy, which increasingly gained ground after the Cold War ended. The discussion continued by reviewing the Indonesian health system, which changed when Indonesia experienced political changes in 1998. Even though Indonesia experienced economic improvement after the political transformation in 1998, the Indonesian health system was not strong enough to deal with health issues such as pandemics. The COVID-19 period in Indonesia was marked by the government's ineffectiveness in dealing with the coronavirus. Jokowi's government, which prioritizes the economy, has taken several contradictory steps to combat the spread of the coronavirus. The following discussion concerns the role of the Foreign Affairs Ministry as the government office that deals directly with the international community. Marsudi as the Minister of Foreign Affairs demonstrated her strategic role in ensuring that Indonesia receives assistance in coping with COVID-19, starting from vaccines and supplies of medical equipment. We conclude this article with some recommendations for policymakers. We strongly recommend that the government take health diplomacy seriously, especially since Indonesia was one of the countries that pushed the norm of integrating health into the scope of foreign policy. As one of the countries with the largest population in the world, the Indonesian government should take health issues seriously, particularly since Indonesia still needs a qualified health system. The Indonesian government should adopt a more unison pandemic preparedness policy to anticipate the next pandemic. The COVID-19 experience in Indonesia has shown the need for more preparedness and trust in the scientific community in recommending the pandemic policy.

References

Adams, V., T. E. Novotny, and H. Leslie. 2008. Global Health Diplomacy. *Medical Anthropology* 27 (4): 315–323. https://doi.org/10.1080/01459740802427067.

Afifa, L. 2020, March 27. *40 Tons of COVID-19 Medical Kits from China Arrive in Indonesia.* Tempo. https://en.tempo.co/read/1324664/40-tons-of-covid-19-medical-kits-from-china-arrive-in-indonesia.

Al Jazeera. 2021, September 5. *Rich Countries to have 1.2bn Surplus COVID Vaccine Doses.* https://www.aljazeera.com/news/2021/9/5/rich-countries-will-have-1-2bn-vaccine-dose-sur plus-data-firm.

Almeida, C. 2020. Global health diplomacy: A theoretical and analytical review. In *Oxford Research Encyclopedia of Global Public Health.* https://oxfordre.com/publichealth/display/ https://doi.org/10.1093/acrefore/9780190632366.001.0001/acrefore-9780190632366-e-25.

Ananta, A., D. R. W. W. Utami, and N. B. Handayani. 2016. Statistics on Ethnic Diversity in the Land of Papua, Indonesia. *Asia & the Pacific Policy Studies* 3 (3): 458–474. https://doi.org/10.1002/app5.143.

Asmara, C. G. 2020, March 31. *Maaf Pak Anies, Permintaan Lockdown DKI Ditolak Jokowi*. CNBC Indonesia. https://www.cnbcindonesia.com/news/20200331085509-4-148631/maaf-pak-anies-permintaan-lockdown-dki-ditolak-jokowi.

Barron, L. 2020, March 13. *Coronavirus Lessons From Singapore, Taiwan and Hong Kong*. https://time.com/5802293/coronavirus-covid19-singapore-hong-kong-taiwan/.

BBC. 2020, December 9. *Rich Countries Hoarding Covid Vaccines, says People's Vaccine Alliance*. https://www.bbc.com/news/health-55229894.

BBC. 2021, July 5. *Indonesia Faces Oxygen Crisis Amid Worsening Covid Surge*. https://www.bbc.com/news/world-asia-57717144.

Bieber, F. 2022. Global Nationalism in Times of the COVID-19 Pandemic. *Nationalities Papers* 50 (1): 13–25.

Bond, L., J. B. Carlin, L. Thomas, K. Rubin, and G. Patton. 2001. Does Bullying Cause Emotional Problems? A Prospective Study of Young Teenagers. *BMJ* 323 (7311): 480–484.

BPMI Setpres. 2021, May 12. Pemerintah RI Kirim Bantuan Hibah untuk Penanganan Covid-19 di India. Presiden RI. https://www.presidenri.go.id/siaran-pers/pemerintah-ri-kirim-bantuan-hibah-untuk-penanganan-covid-19-di-india/.

BPS. 2023. Profil Kemiskinan di Indonesia Maret 2023. https://www.bps.go.id/id/pressrelease/2023/07/17/2016/profil-kemiskinan-di-indonesia-maret-2023.html.

BPS. 2024, June 28. Laju Pertumbuhan Penduduk (Persen), 2021–2023. https://www.bps.go.id/id/statistics-table/2/MTk3NiMy/laju-pertumbuhan-penduduk.html

Cabinet Secretariat of the Republic of Indonesia. 2021, September 23. Address of the President of the Republic of Indonesia at the General Debate Session of the 76th General Assembly of the United Nations on 23 September 2021. Sekretariat Kabinet Republik Indonesia. https://setkab.go.id/en/address-of-president-of-the-republic-of-indonesia-at-the-general-debate-session-of-the-76th-general-assembly-of-the-united-nations-23-september-2021/.

Çetin, M. 2020. *The Rise of Populism and Nationalism and its Implications for COVID-19 Pandemic: A Social Constructivist Perspective* [Master's Thesis, Tekirdağ Namık Kemal Üniversitesi]. https://acikerisim.nku.edu.tr/xmlui/handle/20.500.11776/4366.

Dabla-Norris, E., A.-M. Gulde-Wolf, F. Painchaud, I. M. F. Asia, and P. Department. 2020, June 29. *Vietnam's Success in Containing COVID-19 Offers Roadmap for Other Developing Countries*. IMF. https://www.imf.org/en/News/Articles/2020/06/29/na062920-vietnams-success-in-containing-covid19-offers-roadmap-for-other-developing-countries.

Djalante, R., J. Lassa, D. Setiamarga, A. Sudjatma, M. Indrawan, B. Haryanto, C. Mahfud, M. S. Sinapoy, S. Djalante, and I. Rafliana. 2020. Review and Analysis of Current Responses to COVID-19 in Indonesia: Period of January to March 2020. *Progress in Disaster Science* 6: 100091.

Dyer, O. 2021. *Covid-19: Indonesia Becomes Asia's New Pandemic Epicentre as Delta Variant Spreads*. British Medical Journal Publishing Group. https://www.bmj.com/content/374/bmj.n1815.

Editorial Board. 2021, July 14. Retno's Vaccine Diplomacy—Editorial. *The Jakarta Post*. https://www.thejakartapost.com/academia/2021/07/13/retnos-vaccine-diplomacy.html.

Ekawati, L. L., A. Arif, I. Hidayana, A. Nurhasim, M. Z. Munziri, K. D. Lestari, A. Tan, F. Ferdiansyah, F. Nashiruddin, and Q. E. Adnani. 2022. Mortality Among Healthcare Workers in Indonesia During 18 Months of COVID-19. *PLOS Global Public Health* 2 (12): e0000893.

Fadli, and A. Arbi. 2020, February 27. Indonesia pressured to do more to detect coronavirus amid zero reported cases. *The Jakarta Post*. https://www.thejakartapost.com/news/2020/02/26/indonesia-pressured-to-do-more-to-detect-coronavirus-amid-zero-reported-cases.html.

Fidler, D. P. 2006. Health as Foreign Policy: Harnessing Globalization for Health. *Health Promotion International* 21 (Suppl_1): 51–58.

Fisher, D., Y. Y. Teo, and D. Nabarro. 2020. Assessing National Performance in Response to COVID-19. *The Lancet* 396 (10252): 653–655.

Foreign Policy and Global Health. 2020, May 18. *Foreign Policy and Global Health Joint Ministerial Statement COVID-19: Turning this Unprecedented Crisis into an Opportunity to Achieve Affordable Health Care for All.*

Gorbiano, M. I. 2020, March 2. BREAKING: Jokowi announces Indonesia's first two confirmed COVID-19 cases. *The Jakarta Post*. https://www.thejakartapost.com/news/2020/03/02/breaking-jokowi-announces-indonesias-first-two-confirmed-covid-19-cases.html.

Hameiri, S. 2014. Avian Influenza, 'Viral Sovereignty', and the Politics of Health Security in Indonesia. *The Pacific Review* 27 (3): 333–356.

Ihsanuddin, and K. Erdianto. 2020, April 1. Jokowi Akhirnya Blak-blakan soal Alasan Tak Mau Lockdown. *KOMPAS.com*. https://nasional.kompas.com/read/2020/04/02/05405561/jokowi-akhirnya-blak-blakan-soal-alasan-tak-mau-lockdown.

Jenne, E. K. 2022. Varieties of Nationalism in the Age of Covid-19. *Nationalities Papers* 50 (1): 26–44.

Kemenko Bidang Perekonomian. 2020, July 20. Pemerintah Bentuk Komite Penanganan Covid-19 dan Pemulihan Ekonomi Nasional—Kementerian Koordinator Bidang Perekonomian Republik Indonesia. https://ekon.go.id/publikasi/detail/389/pemerintah-bentuk-komite-penanganan-covid-19-dan-pemulihan-ekonomi-nasional.

Kemenko PMK. 2020, October 8. Pemerintah Dukung Pengembangan Obat Herbal Di Masa Pandemi Covid-19 | Kementerian Koordinator Bidang Pembangunan Manusia dan Kebudayaan. https://www.kemenkopmk.go.id/pemerintah-dukung-pengembangan-obat-herbal-di-masa-pandemi-covid-19.

Kementerian Luar Negeri. 2020, September 4. Pimpin Sidang Foreign Policy And Global Health Menlu Retno Kembali Dorong Akses Setara Dan Berkeadilan Terhadap Vaksin | Portal Kementerian Luar Negeri Republik Indonesia. https://kemlu.go.id/portal/id/read/1651/berita/pimpin-sidang-foreign-policy-and-global-health-menlu-retno-kembali-dorong-akses-setara-dan-berkeadilan-terhadap-vaksin.

Kementerian Luar Negeri. 2023, September 20. Kontribusi Indonesia Senilai Usd 200000 Untuk Bantuan Kemanusiaan Asean Kepada Myanmar Mulai Didistribusikan Akhir September 2021 | Portal Kementerian Luar Negeri Republik Indonesia. https://kemlu.go.id/portal/id/read/2943/berita/kontribusi-indonesia-senilai-usd-200000-untuk-bantuan-kemanusiaan-asean-kepada-myanmar-mulai-didistribusikan-akhir-september-2021.

Kickbusch, I., G. Silberschmidt, and P. Buss. 2007. Global Health Diplomacy: The Need for New Perspectives, Strategic Approaches and Skills in Global Health. *Bulletin of the World Health Organization* 85: 230–232.

Kissinger, H. 2014. Diplomacy. In *Geopolitics*, 114–115. Routledge. https://www.taylorfrancis.com/chapters/edit/ https://doi.org/10.4324/9780203092170-21/diplomacy-henry-kissinger.

Lowe, C. 2019. Viral sovereignty: Security and mistrust as measures of future health in the Indonesian H5N1 Influenza outbreak. *Medicine Anthropology Theory*, *6*(3). https://www.medanthrotheory.org/article/view/4962.

Mahendradhata, Y., L. Trisnantoro, S. Listyadewi, P. Soewondo, T. Marthias, P. Harimurti, and J. Prawira. 2017. *The Republic of Indonesia Health System Review*, ed. K. Hort and W. Patcharanarumol, vol. 7. World Health Organization.

Maliszewska, M., A. Mattoo, and D. Van Der Mensbrugghe. 2020. The potential impact of COVID-19 on GDP and trade: A preliminary assessment. *World Bank Policy Research Working Paper, 9211*. https://papers.ssrn.com/sol3/papers.cfm?abstract_id=3573211.

Morgenthau, H. J. 1993. *Politics Among Nations The Struggle for Power and Peace*. Mc Graw Hill.

Mukaromah, V. F., and J. Akbar. 2020, August 9. *5 Provinsi dengan Jumlah Kasus Covid-19 Tertinggi dan Terendah di Indonesia*. KOMPAS.com. https://www.kompas.com/tren/read/2020/08/09/204500565/5-provinsi-dengan-jumlah-kasus-covid-19-tertinggi-dan-terendah-di-indonesia.

Nabbs-Keller, G. 2013. Reforming Indonesia's Foreign Ministry: Ideas, Organization and Leadership. *Contemporary Southeast Asia: A Journal of International and Strategic Affairs* 35 (1): 56–82.

panrb. n.d. *5 Prioritas Kerja Presiden 2019–2024*. Kementerian Pendayagunaan Aparatur Negara dan Reformasi Birokrasi. Retrieved July 20, 2024, from https://www.menpan.go.id/site/tentang-kami/tentang-kami/5-prioritas-kerja-presiden-2019-2024.

Pibulsonggram, N., C. Amorim, P. Douste-Blazy, H. Wirayuda, J. G. Støre, and C. T. Gadio. 2007. Oslo Ministerial Declaration—Global Health: A Pressing Foreign Policy Issue of Our Time. *Lancet* 369 (9570): 1373–1378.

Rakhmat, M. Z., and Y. Purnama. 2022, November 16. China's Expanding Health Diplomacy in Indonesia. *The Diplomat*. https://thediplomat.com/2022/11/chinas-expanding-health-diplomacy-in-indonesia/.

Repucci, S., and A. Slipowitz. 2020. *Democracy under Lockdown*. Freedom House. https://freedomhouse.org/report/special-report/2020/democracy-under-lockdown.

Rina, R. 2020, February 11. *Terawan: Corona tak Masuk RI, Itu Berkat yang Maha Kuasa*. CNBC Indonesia. https://www.cnbcindonesia.com/news/20200211162800-4-137044/terawan-corona-tak-masuk-ri-itu-berkat-yang-maha-kuasa.

Ruckert, A., R. Labonté, R. Lencucha, V. Runnels, and M. Gagnon. 2016. Global Health Diplomacy: A Critical Review of the Literature. *Social Science & Medicine* 155: 61–72.

Rüland, J. 2017. Democratizing Foreign-Policy Making in Indonesia and the Democratization of ASEAN: A Role Theory Analysis. *TRaNS: Trans-Regional and-National Studies of Southeast Asia* 5 (1): 49–73.

Septiari, D. 2021, February 1. Indonesia against 'vaccine nationalism'. *The Jakarta Post*. https://www.thejakartapost.com/paper/2021/02/01/indonesia-against-vaccine-nationalism.html.

Smith, R., and R. Irwin. 2016. Measuring Success in Global Health Diplomacy: Lessons from Marketing Food to Children in India. *Globalization and Health* 12 (1): 28. https://doi.org/10.1186/s12992-016-0169-5.

Umar, A. R. M. 2023. The Rise of the Asian Middle Powers: Indonesia's Conceptions of International Order. *International Affairs* 99 (4): 1459–1476.

United Nations. n.d. *COVID-19 to slash global economic output by $8.5 trillion over next two years*. United Nations; United Nations. Retrieved June 25, 2024, from https://www.un.org/en/desa/covid-19-slash-global-economic-output-85-trillion-over-next-two-years.

United Nations. 2008, November 26. *Resolution adopted by the General Assembly on 26 November 2008 63/33 Global health and foreign policy*.

United Nations. 2021, September 23. *Decrying Covid-19 Vaccine Inequity, Speakers in General Assembly Call for Rich Nations to Share Surplus Doses, Patent Waivers Allowing Production in Low-Income Countries*. https://press.un.org/en/2021/ga12367.doc.htm.

Wang, D., and Z. Mao. 2021. A Comparative Study of Public Health and Social Measures of COVID-19 Advocated in Different Countries. *Health Policy* 125 (8): 957–971.

WHO. 2020a, January 2020. Statement on the second meeting of the International Health Regulations (2005) Emergency Committee regarding the outbreak of novel coronavirus (2019-nCov). https://www.who.int/news/item/30-01-2020-statement-on-the-second-meeting-of-the-international-health-regulations-(2005)-emergency-committee-regarding-the-outbreak-of-novel-coronavirus-(2019-ncov).

WHO. 2020b, June 28. *Coronavirus Disease (COVID-19) Situation Report—160*. https://www.who.int/docs/default-source/coronaviruse/situation-reports/20200628-covid-19-sitrep-160.pdf.

WHO. 2020c, June 29. *Listings of WHO's response to COVID-19*. https://www.who.int/news/item/29-06-2020-covidtimeline.

WHO. 2022, May 5. *14.9 million excess deaths associated with the COVID-19 pandemic in 2020 and 2021*. https://www.who.int/news/item/05-05-2022-14.9-million-excess-deaths-were-associated-with-the-covid-19-pandemic-in-2020-and-2021.

World Bank. n.d. *GDP growth (annual %)—Indonesia*. World Bank Open Data. Retrieved August 9, 2024, from https://data.worldbank.org.

Wu, J., W. Cai, D. Watkins, and J. Glanz. 2020, March 22. How the Virus Got Out. *The New York Times*. https://www.nytimes.com/interactive/2020/03/22/world/coronavirus-spread.html.

Xue, S., and K. He. 2022. Indonesia's Foreign Policy since the Reformasi. In *The Routledge Handbook of Diplomacy and Statecraft*, 221–232. Routledge. https://www.taylorfrancis.com/chapters/edit/ https://doi.org/10.4324/9781003016625-22/indonesia-foreign-policy-since-reformasi-song-xue-kai.

Yuniar, R. W. 2020, July 7. Covid-19: Kalung "antivirus" buatan Kementan kini disebut "kalung aromaterapi" dan "aksesoris kesehatan", pakar farmakologi sebut berpotensi "misleading." *BBC News Indonesia*. https://www.bbc.com/indonesia/indonesia-53311692.

Yessi Olivia works as a lecturer of international relations at the Department of International Relations, Universitas Riau. She received her doctoral degree from Flinders University in 2019 with a thesis focused on politics of human rights in Indonesia. Her research interests are politics of human rights, democracy in Southeast Asia, and international norms and institutions.

Yusnarida Eka Nizmi is a senior lecturer at the Department of International Relations, Universitas Riau. She received her doctoral degree from Universitas Padjadjaran, Indonesia, in 2018 with a dissertation focused on subregional economic cooperation. Her research interests are regionalism, globalization and transnationalism, and gender studies.

Umi Oktyari Retnaningsih is a senior lecturer at the Department of International Relations, Universitas Riau. She received her doctoral degree from Ohio State University, USA, in 2019 with a dissertation focused on global education. Her research interests are gender studies, post-colonial, and cultural studies.

Purwasandi is a junior lecturer at the Department of International Relations, Universitas Riau. He received his master's degree from Universitas Pertahanan, Indonesia, in 2016, with a thesis focused on national security. His research interests are security and defense policy, peace and conflict studies, and contemporary issues in international relations.

Rahmi Yulia is a junior lecturer at the Department of International Relations, Universitas Riau. She received her master's degree from Universitas Pertahanan, Indonesia, in 2019 with a thesis focused on defense management. Her research interests are security and defense policy, gender studies, and contemporary issues in international relations.

Chapter 10
Digital Diplomacy and the Protection of Indonesian Citizens

Albert Triwibowo

Abstract Citizen protection is one of the most palpable tasks of diplomacy for every Ministry of Foreign Affairs (MoFA), including for Indonesian MoFA. In the digitalization era, the use of digital medium in citizen protection has created the possibility to address consular issues through dialogical communication using digital tools. The practice of digital diplomacy on citizen protection also demanded more responsibility and active participation from national citizens. However, digital diplomacy is experienced and implemented differently by every country and MoFA. Every MoFA faced their own challenges and opportunities. This article examines the conduct of digital diplomacy by Indonesian MoFA and argues that Indonesian MoFA has tried to be creative to face challenges and opportunities on implementing digital diplomacy in regard to citizen protection. Indonesian MoFA practices digital diplomacy to disseminate information on the issue of citizen protection which is intended not only for national citizens who are living abroad or outside the country, but also to inform people at home. Besides that, to face challenges, Indonesian MoFA is also mixing social media and the work of traditional diplomacy through Indonesian diaspora in the protection of Indonesian citizens. This article uses qualitative method with the support of primary and secondary data.

Keywords Digital diplomacy · Indonesia · Citizen protection

10.1 Introduction

> I believe that if used effectively, internet could become as powerful tool in protecting our citizens and interests abroad. (Retno Marsudi, as cited by Rahayu 2019)

Citizen protection is one of the most palpable tasks of diplomacy for every Ministry of Foreign Affairs (MoFA). As the world become more borderless and numbers of travelers keep increasing, the need of consular diplomacy is on the rise.

A. Triwibowo (✉)
Department of International Relations, Parahyangan Catholic University, Bandung, Indonesia
e-mail: albert.triwibowo@unpar.ac.id

I. Kartini and Kanisa (eds.), *Joko Widodo's Foreign Policy and Diplomatic Legacy*,
https://doi.org/10.1007/978-981-95-2085-5_10

Besides that, there are also pressures and expectations from public, parliament, press, principles within the foreign ministries and politicians to put consular work as a core task of government. Nowadays, the advancement of technology has created more pressure toward the task of consular diplomacy, especially toward citizen protection, in which the consular works is demanded to be up to date, provide fast-response action, and still not lacking human approach (Hocking and Melissen 2015; Uvais 2010). Here, the conduct of consular diplomacy is practiced through digital medium, which is popularly known as digital diplomacy. In the digitalization era, the use of digital medium in citizen protection has created the possibility to address consular issues through dialogical communication using digital tools. The digital tools, such as social media, have created the possibility to be present without being physically present and conduct two-ways communication. On the other hand, the practice of digital diplomacy on citizen protection also demanded more responsibility and active participation from national citizens. Digital tools have empowered citizens to be more active in communicating and providing actual situations to the MoFA (Melissen and Caesar-Gordon 2016). However, digital diplomacy is experienced and implemented differently by every country and MoFA. Every MoFA faced their own challenges and opportunities (Manor 2019). Indonesian citizens at home are considered as vocal as overseas citizens and the digital literacy is still considered to be relatively low. Besides that, Indonesian citizens are also more familiar and used to monologic rather than dialogic communication (Dharossa and Rezasyah 2020; Triwibowo 2023a). Indonesian MoFA has also experienced other challenges in the digital diplomacy, such as budgetary, infrastructure, and human resources (personal communication with Indonesian MoFA representative, 2020; Madu 2018; Dwikardana et al. 2017).

The discussion of digital diplomacy is mainly focused on advanced economies. Besides that, the focus is usually related to the conduct of public diplomacy (Pamment 2016; Bjola and Jiang 2015; Manor 2019). There is only limited source of academic discussion on the practice of digital diplomacy outside public diplomacy of non-advanced economies, for example on Indonesia's digital diplomacy on consular diplomacy. However, there are some important articles that have tried to discuss the implementation of ICT on Indonesian citizen protection. Mahfudzi (2020) discusses the governance of Indonesian digital diplomacy on citizen protection, while Dharossa and Rezasyah (2020) analyze the implementation of three applications in the issue of citizen protection in Malaysia. Another important study about the implementation of Indonesia's digital diplomacy on citizen protection is from Radiananti (2021). She explores how the approach of using digital medium is developed in Indonesian diplomacy, and how it is implemented by the Indonesian MoFA. Moreover, other researches discuss the issue from the point of view of effectivity, such as Dwikardana et al. (2018). They discuss the effectivity of digital diplomacy for Indonesian migrant workers.

Meanwhile, Surwandono and Nugroho (2022) discuss the issue using a bigger lens of citizen protection policy, in which it includes the use of digital medium, through the paradigm of duty of care. Furthermore, Triwibowo discusses citizen protection issues in a specific time of the pandemic (2023a; 2023b; 2025). All those researches focus mainly on the challenges and implementation of Indonesian digital diplomacy in the issue of citizen protection. Only few, or maybe none, of them connect the dots between challenges and the implementation. Here, it is important for the discussion to be linked, especially to look how Indonesian MoFA and diplomats execute digital diplomacy, face the challenges on the ground, and respond to it. By focusing on how digital diplomacy is executed and responds to challenges in the context of Indonesian diplomacy experience, this article tries to see not only how digital diplomacy is harnessed by a specific country but also how MoFA and diplomats have tried to maximize and adjust digital diplomacy based on the nation's characteristics.

This article examines the conduct of digital diplomacy by Indonesian MoFA in the issue of citizen protection, especially during the administration of Joko Widodo (Jokowi). This article argues that Indonesian MoFA Indonesian MoFA has tried to be creative in order to face challenges and opportunities on implementing digital diplomacy in regard to citizen protection. The use of Information and Communication Technology (ICT) for citizen protection, especially social media, is intensified during Jokowi administration. Besides that, from 2014 to 2023, Indonesian MoFA was also responsible to provide consular services for more than 200.000 Indonesian citizens living abroad (Naibaho 2024). Indonesian MoFA practices digital diplomacy to disseminate information on the issue of citizen protection which is intended not only for national citizens who are living abroad or outside the country, but also to inform people at home. As stated above by Indonesian Foreign Minister Retno Marsudi, Indonesian MoFA and citizens could take benefit from the use of digital tools for citizen protection issue (Rahayu 2019). In doing consular diplomacy, Indonesian MoFA was also mixing social media and the work of traditional diplomacy through Indonesian diaspora in the protection of Indonesian citizens. This article uses qualitative method with the support of primary and secondary data. Primary data is obtained through interview with Indonesian citizens living abroad and Indonesian diplomats and officials who are familiar with the issue of citizen protection. Secondary data from journals and news will also be used to support the analysis.

To discuss the issue, the article will be divided into five main parts. After the introduction, the article will discuss the basic concepts of digital diplomacy briefly. The discussion will be followed with discussion over Indonesian digital diplomacy on the issue of citizen protection during Jokowi presidency and then a short discussion on how the digital diplomacy is being practiced. Thereafter, there will be a conclusion and also short projection on what can be achieved in the future.

10.2 Digital Diplomacy and Citizen Protection: Basic Concepts and the Debate

The early use of ICT in diplomatic practice has introduced the term ediplomacy, which is defined as "the use of the internet and new Information Communications Technologies to help carry out diplomatic objectives" (Hanson 2012, p. 2). Later, the phenomenon is popularly known as digital diplomacy and defined by Bjola as "the use of social media for diplomatic purposes, could change practices of how diplomats engage in information management, public diplomacy, strategy planning, international negotiations or even crisis management" (Bjola 2015, p. 4). In similar vein, Holmes defined digital diplomacy as "a strategy of managing change through digital tools and virtual collaboration" (Holmes 2015, p. 15). Here, some important aspects for digital diplomacy can be underlined. Those aspects are: information dissemination, digital medium, interaction, international event, and diplomatic objectives (Triwibowo 2023b, p. 8).

In consular diplomacy, digital tool is considered to be one of the medium that can support the implementation of government "duty of care," especially in time of crisis, like during natural disaster, conflict or in time of the pandemic situation. The duty of care is mainly related to the citizen's expectation about the state's duty to care for its citizens beyond state borders from both traditional and non-traditional security threats (Græger and Lindgren 2018). According to Leira and Græger (2020, p. 9), duty of care signifies "a contractual relationship between unequal partners, usually, but not exclusively, between state and citizens, rooted in a normative community." Duty of care can also involve people on the move like tourist, students, and diaspora communities. In times of a crisis, digital channels are being used to provide urgent and important information, to protect and assist citizens, such as by giving actual and accurate information, advice for return flights as well as repatriation process. Besides that, it also can be used to show solidarity and coordination for joint responses (Melissen and Caesar-Gordon 2016, p. 2; Bjola and Manor 2020).

Moreover, the use of ICT for consular diplomacy has created three important trends. First is related to citizens' expectation for quick solution and the use of technology in consular services. Second, there is an expectation that the citizens will also be more communicative and responsible in contacting government representatives. Lastly, there is a trend collaboration between public and private in creating consular innovation (Hocking and Melissen 2015). In practice, digital medium is used in consular diplomacy to provide up to date information and gather information from citizens. To grasp attention, hashtags are used and encouragement to forward the message is posted to expand the outreach (Manor 2019, p. 106). Digital medium, like social media, is considered helpful in helping the MoFAs allocate limited communication resources. Apart from providing quick response, consular diplomacy is also intended to project state's image in borderless world (Bjola and Manor 2020; Uvais 2010).

In academic discussion, the use of digital channels for consular issues has raised debate. The debate is about the impact of digitalization to citizen services in diplomatic work. From the point of view of the duty of care concept, citizen services are considered as part of domestic affairs. However, as it is stated by Melissen and Caesar-Gordon (2016, pp. 2–3), the consular services are increasingly intersect with foreign policy in recent development, especially with the globalization and the use of ICT. The conduct of consular diplomacy through digital medium then has created a more complex relations between domestic public and the government in term of public diplomacy (Melissen and Caesar-Gordon 2016, p. 9), as digital diplomacy is also understood as part of public diplomacy (Melissen 2013). Meanwhile, Manor stated that, digital diplomacy is related to domestic agenda in which MoFA and diplomats cannot ignore opinions from people at home (Manor 2019, p. 85).

Here, the issue of citizen protection always attracts attention and thus creating a big responsibility for MoFA and diplomats to deal with it (personal communication with Indonesian diplomat, 2021; Melissen and Caesar-Gordon 2016, p. 2). On the other hand, practitioners and diplomats have stated that the popularity of citizen protection in digital diplomacy has created more expectation in which it has slight connection with foreign policy in general. They argued that citizen protection issue is a domestic agenda and thus the issue has domesticated the MoFA itself. Besides that, the issue of citizen protection is usually highly political and the use of digital channels is only creating more responsibility for diplomats. Another fact is that the citizens usually not responsible enough to join digital initiatives which is proposed by diplomats and MoFA (personal communication with Indonesian diplomat 2024; personal communication with Indonesian diplomat, 2021; Melissen and Caesar-Gordon 2016, pp. 2–3).

As a result, consular diplomacy nowadays is using hybrid communications to tackle consular needs of national citizens. This approach is to accommodate people with traditional way of thinking and prepare a new way of communication in the future. As stated by Græger and Lindgren (2018, p. 191), the duty of care in the diplomatic work is oftentimes solved on the ground which requires more coordination between actors and digital communication. Hence, it is important to see how diplomats and MoFA face the challenges and find the hybrid form of communication to cater between analogue and digital citizens.

10.3 Indonesian Digital Diplomacy and Citizen Protection: Joko Widodo's Legacy

Citizen protection is considered as one of the most obvious diplomatic objectives for Indonesian MoFA from time to time (personal communication with a representative from Indonesian MoFA, 2020). The objective is always apparent in the Ministry's strategic plan, although it appeared in different name: the protection of Indonesian citizens (WNI/*Warga Negara Indonesia*), improvement of consular and protection for Indonesian citizens, protection of Indonesian migrant workers to protection diplomacy (*Diplomasi Perlindungan*). The theme of citizen protection is always central in the diplomatic objectives of Indonesian MoFA and always appeared in every administration. The focus on citizen protection is much related to the mandate stated in the Indonesian 1945 Constitution. In the preamble of the 1945 Constitution, it is clearly stated that one of the duties of the state is to protect the entire Indonesian nation and homeland (Dewan Perwakilan Rakyat Republik Indonesia, n.d.).

The practice continues in the era of Jokowi. In January 2024, Foreign Minister Retno Marsudi even stated that Indonesian MoFA always strengthen the system on citizen protection in the last nine years (Kementerian Luar Negeri Republik Indonesia

2024). In Jokowi's first administration (2014–2019), Indonesian foreign policy was guided with the vision of "down to earth diplomacy" or "diplomacy for the people" (*diplomasi untuk rakyat*). In the MoFA's 2014 Strategic plan, the issue of citizen protection was clearly stated in the strategic target of having an excellent service and protection for Indonesian citizens, legal entity, and diaspora (Kementerian Luar Negeri Republik Indonesia 2015, p. 26). Later in 2020, the issue of citizen protection is appeared in the name of protection of Indonesian citizens abroad and excellent public services (Kementerian Luar Negeri Republik Indonesia 2020, pp. 81–82). In the issue of citizen protection, Jokowi administration is responsible in developing the awareness and opening the conduct of digital diplomacy in consular diplomacy.

Although the use of ICT already started in the tenure of Foreign Minister Dr. Hasan Wirajuda, the effort to maximize digital tools to reach overseas Indonesian citizens was intensified during Jokowi and Foreign Minister Retno Marsudi. As stated by one representative from Indonesian MoFA in 2020 that "The attention to digital diplomacy happened in the last five to six years, especially during the Jokowi era" (personal communication with a representative from Indonesian MoFA, 2020; Triwibowo 2023b, p. 20). She added that the first initiative to incorporate digital tools for Indonesian diplomacy was started during the leadership of Foreign Minister Hasan Wirajuda but get intensified during the administration of Jokowi and Foreign Minister Retno Marsudi (personal communication with Indonesian MoFA representative, 2020: Triwibowo 2023b, 20). The initiative on using digital tools in consular diplomacy was at first mainly influenced by the need to communicate easily with Indonesian citizens, especially Indonesian migrant workers. The idea was then discussed during a briefing between the President, Foreign Minister, and the Head of Indonesian Missions. The decision later was to start using digital tools to support the citizens' protection matters. One of the initial actions is the use of social media, like Twitter and Facebook, to reach Indonesian migrant workers and citizens (personal communication with Indonesian MoFA representative, 2020; Triwibowo, 2025). Another impetus to use ICT in citizen protection was the need to response toward the situation of crisis or conflict. In time of crisis, cell phone and social media were considered as the fastest and easiest medium to share information (Dwikardana et al. 2017, p. 53; Meganingratna and Nur 2022, pp. 93–94).

In regard to citizen protection, Jokowi's initiative to embrace digital diplomacy in citizen protection was marked, at least, in two important contributions. First, Indonesian MoFA encourages and strengthens the use of digital diplomacy in consular services through new regulations. In the words of Bjola, this aspect is much related to policy innovation, especially on how digital aspect is translated in the policy level and regulations (Bjola 2015, p. 5). On general level, Indonesian MoFA started to address digital diplomacy in 2017 through Decree No. 84/2017. The words digital diplomacy explicitly stated as one of the objectives in the 2017 Strategic Plan. The objective was stated as "Digital Diplomacy and Systematic and Strong Management of Foreign Relations Information" (Dwikardana et al. 2017, p. 47). Thereafter, the Indonesian MoFA issued more specific regulation about digital diplomacy: The Regulation of the Minister of Foreign Affairs Number 10 Year 2018 about the management of digital media. These initiatives have underlined the Ministry's focus on digital diplomacy.

However, the utilization of digital channels in consular diplomacy was addressed and regulated further in different regulations and initiatives.

In 2018, the MoFA issued the Regulation of the Minister of Foreign Affairs which started to underline the utilization of digital tools in consular diplomacy. The new regulation specifically addressed the issue of citizen protection and is named differently compared the last regulation in 2004. Before, the regulation on citizen protection was titled "Services to Indonesian Citizens at Indonesian representatives Abroad." In 2018, the regulation was revised into the Regulation of the Minister of Foreign Affairs Number 5 Year 2018 regarding the protection of Indonesian citizens abroad. This means that the protection of Indonesian citizens becomes the focus in consular diplomacy during the administration of Jokowi and Foreign Minister Retno Marsudi. More importantly, the use of digital channels was addressed in the new regulation. The regulation in 2018 firmly stated that citizen protection services are managed through information system. Furthermore, the regulation stated clearly two important digital applications, namely: Portal Peduli WNI and Safe Travel (the Regulation of the Minister of Foreign Affairs Number 5 Year 2018).

The development of Safe Travel application and Portal Peduli WNI also marked another contribution of Jokowi on the issue of citizen protection. In this regard, Jokowi has attempted to adapt and develop appropriate infrastructure to support digital diplomacy on citizen protection. The initiative to use ICT for consular diplomacy has started even before the specific regulation was issued. In December 2014, the Indonesian MoFA has started to work with five national cellular providers to provide information through SMS Blast program. The information is about the nearest Indonesian representatives abroad when Indonesians first activate their cell phone overseas. This effort has intended to help overseas Indonesians (Dharossa and Rezasyah 2020, pp. 110–111; Detik Travel 2015). Before the SMS Blast program was initiated, citizen protection services were mainly based on a database, named *e-perlindungan*, developed by Indonesian overseas representatives.

The Indonesian MoFA started to utilize more ICT and introduced Safe Travel application in a beta version since 2017 (Madu 2018, p. 14). The application was intended to give a quick response toward the emergency needs of Indonesian citizens abroad. In Foreign Minister Press Statement in 2017, Retno Marsudi stated that "With this application, the Ministry of Foreign Affairs can find out the distribution, location and identity of Indonesian citizens abroad in real time and provide quick response in emergency situations" (Kementerian Luar Negeri Republik Indonesia 2017). The application is fully launched in 2018, and nowadays has been updated with more features. One of the most important updates was the panic button. This feature is mainly intended to give easy access for Indonesian citizens in times of emergency (Cinti 2019). Another new feature is Indonesia Nearby, which is to locate the presence of the nearest Indonesian people (Dwikardana et al. 2018, pp. 31–33). Meanwhile, Portal Peduli WNI is an online portal that serves as a single database, mainly for Indonesian citizens who are planning or staying abroad for a long time. There are three main features: self-report (lapor diri), services (*pelayanan*), and complaints (*pengaduan*). The portal not only serves as a database to protect citizens in times of emergency but also to provide consular assistances (Kementerian Luar Negeri Republik Indonesia, n.d.; Dharossa and Rezasyah 2020, pp. 112–113).

For digital diplomacy initiatives in general, Jokowi and Foreign Minister Retno Marsudi are also responsible for the establishment of Digital Command Center (DCC) in 2017. The DCC plays an important role to support Indonesian digital diplomacy and to watch as well as monitor public trends toward Indonesian foreign policy both domestically and internationally. More importantly, the DCC plays crucial role in analyzing and providing real time data in time of crisis or conflict (Kementerian Luar Negeri Indonesia 2017; Madu 2018, pp.13–14). Apart from the self-developed applications, Indonesian MoFA also used social media to easily engage and communicate with Indonesian citizens abroad, such as using X (previously twitter), Facebook, and Instagram. To attract attention, especially during crisis, hashtags are being used so that the information can be easily found. There are five main hashtags encouraged by the Indonesian MoFA, #IniDiplomasi as hashtag for general initiatives, #NegaraMelindungi for citizen protection, #IndonesiaUntuk Dunia as representation of Indonesia's initiative and participation on global issues, #RintisKemajuan for economic issues, and #DemiNKRI for sovereignty issues. Additionally, Indonesian representatives could also use #IndonesianWay to show Indonesia's diplomatic initiatives (Direktorat Jenderal Informasi dan Diplomasi Publik, n.d.; Kurniawati et al. 2020, p. 89; Madu 2018, p. 14; Triwibowo 2025, p. 51). In citizen protection issues, the hashtag #NegaraMelindungi was often used by the Indonesian MoFA, representatives, and diplomats on social media to show the effort from the Government and Ministry to assist and protect Indonesian citizens. The hashtag is also to show the presence of the state for the citizens.

10.4 Indonesian Digital Diplomacy and Citizen Protection: Practices and Reality

Indonesian MoFA initiatives on consular diplomacy, particularly in incorporating digital channels, cannot be underestimated. New regulations and proper infrastructures signed the ability of the Ministry to adapt and take benefit on digital diplomacy. The question then is, how does the implementation of digital diplomacy on citizen protection? And, what can be learnt from Indonesian experiences?

The main task of consular diplomacy using digital diplomacy is to provide assistance and protection for citizens through digital channels, for everyday life and particularly during crisis time. The phenomenon of digital diplomacy in consular diplomacy has then coined three important trends: quick solutions, more responsible citizens, and collaboration between public and private in consular innovation. However, the task and those trends in utilizing digital channels for citizen protection issues are experienced differently based on country's own characteristics. Here, Indonesia characteristics on digital diplomacy is related to the, still, limited understanding of digital diplomacy, sporadic and ad hoc basis of digital diplomacy, and influenced by domestic interests (Triwibowo 2023a; 2025). These characteristics play role in the use of digital channels on citizen protection, especially since the practice of digital diplomacy on citizen protection is mainly influenced by the responsibility of the citizen to take part in the Government "duty of care" (Melissen and Caesar-Gordon 2016). In this case, all effort in consular diplomacy using digital channels will mean nothing without the active participation from citizens' side.

In doing digital diplomacy on citizen protection, Indonesian MoFA and diplomats must face some challenges, both from the internal MoFA and also from Indonesian citizens. Internally, Indonesian digital diplomacy in general has to face the lack of resources, for example human and financial resources. In the beginning of digital diplomacy practice, there is no dedicated team which was dealing with the social media and other digital channels (personal communication with Indonesian diplomat, 2020). Although Indonesian MoFA has already built DCC to conduct a proper digital diplomacy, it is designed to cover many areas in diplomacy. As a result, there is still no specific designated digital team who focuses on citizen services. In regard to citizen protection, it also needs a proper infrastructure, for example servers and data center. As explained by Radiananti (2021, pp. 56–57), the main challenge in the development of infrastructure is due to the budget. Nowadays, Indonesia still doesn't have any proper server and data center to support citizen services as the budget is not yet fixed and waiting for approval.

Nowadays, conducting digital diplomacy often require focus and time consuming while the representatives only have a limited number of people. Thus, the treatment of citizen protection usually requires more time (Dwikardana et al. 2017; personal communication with Indonesian citizen in Germany, 2024; personal communication with Indonesian citizen who lived in the UK, 2024). Another internal hurdle is the interoperability of the application which is developed by the Directorate of Protection of WNI. So far, the Safe Travel application is developed separately for its Android and iOS version. Besides that, the application is mainly worked using Google Maps to support the positioning features, in which it doesn't work in some countries. In terms of organization, Indonesian MoFA still has to face human resources issue.

Challenges come not only internally from the Ministry itself, but also from the citizens' side. One of the attributes of Indonesian citizens living abroad is their passive attitude. Here, Indonesian citizens living abroad is considered to have low participation rate when dealing with digital application (Dharossa and Rezasyah 2020, pp. 113–116; personal communication with Indonesian citizen in Germany, 2024). The application of Portal Peduli WNI, for instance, can work properly with the support from Indonesian citizens itself. In this regard, the success of Portal Peduli WNI relies on the responsibility of Indonesian citizens to fill the online form when they first move for a long time abroad. This portal is used by Indonesian representatives to locate and monitor Indonesian citizens, especially in times of important events or crisis situation, for example during the national elections, natural disasters, war, or pandemic. In 2022, the estimation number of Indonesian citizens living abroad is around 9 million people. However, only one third of those people are recorded in the database of Indonesian MoFA (Yahya and Meiliana 2022). In 2023, there is only around 2.7 million people of Indonesian citizens living abroad who is officially registered (Direktorat Perlindungan WNI 2024). The situation confirmed the argument by Surwandono and Nugroho (2022, p. 272), in which they underline the low participation rate of Indonesian citizens using Indonesian MoFA digital application. They further stated that for the Safe Travel application, there are only 50.000 people who downloaded the application. This fact once again show little attention and participation of Indonesian citizens in using digital diplomacy for citizen protection issue.

The limited understanding and responsibility of Indonesian citizens to be active in the matters of citizen protection can also be seen from the familiarity of using digital communication. In terms of digital communication, Indonesian citizens are more familiar and used to one-way communication, in which it doesn't need to give any reaction unless it is needed. Besides that, Indonesian people are used to a more personal and direct communication, such as via WhatsApp, direct phone call, or message (Dharossa and Rezasyah 2020, pp. 113–114; Triwibowo 2025, pp. 142–143, personal communication with Indonesian citizen in Germany, 2024). This attribute has also contributed to the implementation of digital diplomacy on citizen protection, especially regarding the sporadic and ad hoc initiatives. As consequence, SMS Blast is considered as the most successful digital initiative on citizen protection, because the initiative doesn't need two-way communication or reaction from the citizen side (Dharosa and Rezasyah 2020, pp. 113–116). From the side of Indonesian representatives and diplomats, the absence of dialogue on online platform will decrease the possibility for unnecessary debate. When further communication needed, then personal and direct communication will be conducted (Triwibowo 2023a;2023b [forthcoming]). This phenomenon at the end also leads to a sporadic and ad hoc initiatives, in which diplomats' initiative will play a key role when further consular services are needed for Indonesians citizens. One of the examples of the passive attribute of Indonesian citizens can be seen from the participation on the Portal Peduli WNI.

Furthermore, Indonesian MoFA also faces other challenges when it comes to the issue of citizen protection, such as the attention from domestic people at home. The issue of violence against Indonesian migrant workers, Indonesian citizens getting stranded abroad, having problems, or waiting for repatriation usually receive massive attention at home. The demand for Indonesian MoFA and representatives as well as diplomats to solve the problems is always raising (Hardono 2005, p. 730). The Covid-19 pandemic is the latest example. The repatriation process becoming national headline and receive more attention form the MoFA when there are support and push for a quick repatriation process. The push and attention are coming not only from citizens at home, but also from politicians, member of House of Representatives, political parties, and religious groups. (Triwibowo 2023a; Metro TV News 2020a, 2020b; Junaedi 2020; Kumparan News 2020a, 2020b; MPR 2020). Therefore, information on citizen protection through digital channels is also intended to inform people back at home. In this regard, effort to provide information for people back at home is also considered as an attempt of accountability. With digital diplomacy, people in Indonesia will know that Indonesian representatives and diplomats are working to provide consular services that protect Indonesian people abroad (personal communication with Indonesian diplomat, 2021).

Because of these challenges, the use of digital channel in citizen protection is limited. To respond toward the situation, Indonesian digital diplomacy on citizen protection has tried to be creative by combining digital channels and the work of traditional diplomacy through Indonesian diaspora. Here, Indonesian representatives and diplomats use a more personal communication, such as WhatsApp, to contact and communicate with Indonesian diaspora. At the same time, digital applications such as Portal Peduli WNI and Safe Travel as well as social media are also used to maximize the information dissemination. One important aspect that needs to be underlined in the citizen protection is the role of Indonesian diaspora. Here, Indonesian diaspora communities include Overseas Indonesian Students Association (*Persatuan Pelajar Indonesia*/PPI) and other specific communities who have interest on Indonesia.

One of the examples that represent all the above explanation is the work of PPI in Rostock, Germany, during the crisis time of Covid-19 pandemic in 2020. The group is active as a diaspora community of Indonesian students and migrants that shares information about Indonesia in Rostock, Indonesia in Germany and news back at home. In times of crisis, the group actively shares new information and assistance from the Indonesian representatives. When the outbreak became worse in 2020, Indonesian diplomats who are based in Berlin (the location of Indonesian Embassy) contacted the representative of PPI Rostock, asking about Indonesian community in the location, such as the total number, logistic and health situation, as well as sharing emergency contact. When any group member needs consular support, the representative from PPI Rostock can also contact the emergency number. Moreover, when the representatives of Indonesian Embassy are unreachable, other members can also contribute to solve the problem while waiting for the support from the Indonesian representatives. Besides having regular meeting, the group also uses a WhatsApp application and Instagram to share information. In time of the pandemic, the representatives from Indonesian Embassy in Berlin send logistical supported couple of times and also visited Rostock to conduct consular services. Here, apart from Portal Peduli WNI and Safe Travel, WhatsApp is becoming central in the consular diplomacy for Indonesian living abroad, as it can bridge the needs and characteristics of Indonesian citizens living abroad and Indonesian representatives. In this regard, Indonesian citizens are more used to a personal communication and interaction as it can be, in many cases, more responsive rather than rely on digital application (personal communication with Indonesian citizen in Germany, 2024).

10.5 Digital Diplomacy and Indonesia's Citizen Protection: What's Next

Under the presidency of Joko Widodo and Foreign Minister Retno Marsudi in 2014–2023, Indonesian digital diplomacy has gone through remarkable development. Although Indonesian diplomacy will still develop in the future, the current development cannot be underestimated. Indonesian diplomacy has adapted with current trends and has tried to grasp opportunities to fulfill the national interest through updating the regulations to infrastructure to support the practice of digital diplomacy. In terms of consular diplomacy, citizen protection is becoming the focus during Jokowi administration, and the use of digital media is already incorporated in the Regulation of the Minister of Foreign Affairs Number 5 Year 2018. The application of Portal Peduli WNI and Safe travel complement the use of social media to strengthen the protection of Indonesian citizens.

However, there are some challenges in the practice of Indonesian digital diplomacy in citizen protection, for example the passive attitude of Indonesian citizens and the problem of resources. Therefore, Indonesian MoFA, representatives, and diplomats are working with Indonesian diaspora to ensure the works of citizen protection. Nevertheless, Indonesia's digital diplomacy on citizen protection has potential to develop in the future, especially when it can be more open minded and maximize the potential of Indonesian diaspora amidst the lack of resources that needs to be faced.

Thus, there are some important things that can be done to improve the conduct of Indonesian digital diplomacy, particularly on citizen protection issues. First, there is a need to develop a specific digital team for citizen protection or consular services in order to maximize all the digital ideas. Second, to overcome the problem of human and financial resources, Indonesian MoFA can work more closely with Indonesian diaspora communities, especially in the issue of consular diplomacy. Here, Indonesian diaspora has a potential to contribute in the issue of consular diplomacy, especially when Indonesian representatives abroad still experience challenges in terms of human and financial resources. The presence and support from Indonesian diaspora can reduce the burden of Indonesian representatives and also strengthen the network of citizen protection mechanism. Here, Indonesian diaspora can help in providing and updating data in the field (personal communication with Indonesian citizen in Germany, 2024). When Indonesian MoFA has more budget and resources, Indonesian MoFA can add more features for Indonesian diaspora in the current available digital channels, such as in Portal Peduli WNI and Safe Travel application. Another possible initiative is to develop a designated application for Indonesian diaspora. In time of crisis, this application could contribute in the effort to protect Indonesian citizens by providing more accurate data and communicating with Indonesians living abroad.

Third, Indonesian MoFA needs to be more creative and open minded toward the use of new social media. One of the challenges in digital diplomacy is the high cost in developing a new and specific application. Thus, one of the strategies is to be creative in maximizing what is available and popular in the market. Besides that, Indonesian

MoFA, representatives, and its diplomats need to be more communicative and take benefit from the possibility of more dialogical communication in digital channels, such as in social media or even in its own application. A dialogical communication will help the ministry to solve and address the issue of citizen protection faster. Lastly, Indonesian MoFA needs to update its strategy and also regulations in order to be more accommodative toward new trends. So far, Indonesian MoFA is acting more as a follower instead of a leader. Here, in terms of regulation, Indonesian MoFA needs to formulate grand strategy of Indonesian digital diplomacy, in which it needs to include the aspect of consular diplomacy. Nowadays, Indonesian digital diplomacy focuses more on the conduct of digital diplomacy for public diplomacy. Therefore, the current regulation in digital media is not addressing specifically about citizen protection. Although the regulation of Indonesian citizens did address the use of digital channels, the separation between the two means that the main intention is different and conducted differently. In fact, collaboration between the two could maximize the practice of Indonesian diplomacy. All in all, what has been achieved in the administration of Joko Widodo can be an important milestone to develop a more digital citizen protection in the future even though there are still many things that need to be done.

References

Bjola, C., and L. Jiang. 2015. Social Media and Public Diplomacy: A Comparative Analysis of the Digital Diplomatic Strategies of the EU, US and Japan in China. In *Digital Diplomacy. Theory and Practice*, ed. C. Bjola and M. Holmes, 71–88. Routledge.

Bjola, C., and I. Manor. 2020. Digital Diplomacy in the Time of the Coronavirus Pandemic. *Center on Public Diplomacy*, March 31. Retrieved August 20, 2024, from https://www.uscpublicdiplomacy.org/blog/digital-diplomacy-time-coronavirus-pandemic.

Bjola, C. 2015. Introduction: Making Sense of Digital Diplomacy. In *Digital Diplomacy. Theory and Practice*, ed. C. Bjola and M. Holmes, 1–9. Routledge.

Cinti, P.I.R. 2019. Aplikasi Ini Bikin Kamu Lebih Aman Saat Traveling di Luar Negeri, November 29. Retrieved August 18, 2024, from https://travel.detik.com/travel-news/d-4803407/aplikasi-ini-bikin-kamu-lebih-aman-saat-traveling-di-luar-negeri.

Detik Travel. 2015. SMS Blast, Cara Melindungi Traveler Indonesia di Luar Negeri. Retrieved August 18, 2024, from https://travel.detik.com/travel-news/d-2835510/sms-blast-cara-melindungi-traveler-indonesia-di-luar-negeri.

Dewan Perwakilan Rakyat Republik Indonesia. n.d. Undang-Undang Dasar Negara Republik Indonesia Tahun 1945 [The 1945 Constitution of the Republic of Indonesia]. Retrieved August 20, 2024, from https://www.dpr.go.id/jdih/uu1945.

Dharossa, T., and T. Rezasyah. 2020. Upaya Perlindungan WNI oleh Pemerintah Indonesia melalui Pendekatan Diplomasi Digital (2014–2019) [Efforts to Protect Indonesian Citizens by the Government of Indonesia through a Digital Diplomacy Approach (2014–2019)]. *Padjadjaran Journal of International Relations* 2 (1): 105–118. https://doi.org/10.24198/padjir.v2i1.26055.

Direktorat Jenderal Informasi dan Diplomasi Publik. n.d. Diplomasi Digital dan Penggunaan Media Sosial di Kementerian Luar Negeri [Digital Diplomacy and The Use of Social Media in the Ministry of Foreign Affairs]. Unpublished presentation of Directorate General of Information and Public Diplomacy, Ministry of Foreign Affairs of the Republic of Indonesia.

Direktorat Perlindungan WNI. 2024. Data Persebaran WNI di Luar Negeri Tahun 2023. Retrieved August 17, 2024, from https://sdi.kemlu.go.id/dataset/pwni-persebaran-2023.

Dwikardana, S., S. Djelantik, A. Triwibowo, A. Valerisha, and J. Martha. 2017. *Transformasi Strategi Diplomasi di Era Digital: Identifikasi Postur Diplomasi Digital di Indonesia* [Diplomatic Strategy Transformation in the Digital Era: Identification of Digital Diplomacy Posture in Indonesia] (Research Report). Lembaga Penelitian dan Pengabdian Masyarakat Universitas Katolik Parahyangan. https://repository.unpar.ac.id/handle/123456789/4719.

Dwikardana, S., A. Valerisha, and S. Yazid. 2018. *Implementasi Diplomasi Digital Indonesia: Studi Kasus Pekerja Migran Indonesia di Luar Negeri* [Implementation of Indonesia's Digital Diplomacy: A Case Study of Indonesian Migrant Workers] (Research Report). Lembaga Penelitian dan Pengabdian Masyarakat Universitas Katolik Parahyangan. https://repository.unpar.ac.id/bitstream/handle/123456789/7727/lpdsc243_Sapta_Implementasi%20Diplomasi%20Digital-p.pdf.

Græger, N., and W. Y. Lindgren. 2018. The Duty of Care for Citizens Abroad: Security and Responsibility in the in Amenas and Fukushima Crises. *The Hague Journal of Diplomacy* 13 (2): 188–210. https://doi.org/10.1163/1871191X-11302009.

Hanson, F. 2012. Baked in and Wired: eDiplomacy@State. *Brookings Institute Foreign Policy Paper Series* (30). https://www.brookings.edu/wp-content/uploads/2016/06/baked-in-hansonf-5.pdf.

Hardono, D. 2005. Perlindungan Warga Negara Dalam Perbandingan [Citizen Protection in Comparison]. *Jurnal Hukum Internasional* 2 (4): 727–746. https://download.garuda.kemdikbud.go.id/article.php?article=2347470&val=22631&title=Perlindungan%20Warga%20Negara%20dalam%20Perbandingan.

Hocking, B., and J. Melissen. 2015. *Diplomacy in the Digital Age* (Clingenadael Report). Clingenadael. https://www.clingendael.org/sites/default/files/pdfs/Digital_Diplomacy_in_the_Digital%20Age_Clingendael_July2015.pdf.

Holmes, M. 2015. Digital Diplomacy and International Change Management. In *Digital Diplomacy. Theory and Practice*, ed. C. Bjola and M. Holmes, 13–32. Routledge.

Junaedi, E. 2020. Muhammad Iqbal: Pemerintah Harus Segera Evakuasi WNI di China, Jangan Diam! January 28. Retrieved August 17, 2024, from https://timesindonesia.co.id/peristiwa-daerah/248527/muhammad-iqbal-pemerintah-harus-segera-evakuasi-wni-di-china-jangan-diam.

Kementerian Luar Negeri Republik Indonesia. n.d. Pelayanan dan pelindungan WNI di Luar negeri [Services and Protection of Indonesian Citizens Abroad]. Retrieved August 17, 2024, from https://peduliwni.kemlu.go.id/beranda.html.

Kementerian Luar Negeri Republik Indonesia. 2015. Rencana Strategis 2015–2019 [Strategic Plan 2015–2019]. Kementerian Luar Negeri Republik Indonesia. Retrieved August 16, 2024, from https://kemlu.go.id/portal/id/read/2033/akip/rencana-strategis.

Kementerian Luar Negeri Republik Indonesia. 2017. Pernyataan Pers Tahunan Menteri Luar Negeri RI Tahun 2017 [2017 Annual Press Statement of the Minister of Foreign Affairs of the Republic of Indonesia]. Retrieved August 17, 2024, from https://kemlu.go.id/portal/id/read/757/pidato/pernyataan-pers-tahunan-menteri-luar-negeri-ri-tahun-2017.

Kementerian Luar Negeri Republik Indonesia. 2018a. *Peraturan Menteri Luar Negeri Republik Indonesia Nomor 5 Tahun 2018 Tentang Pelindungan Warga Negara Indonesia Di Luar Negeri* [Regulation of the Minister of Foreign Affairs of the Republic of Indonesia Number 5 of 2018 Concerning Protection of Indonesian Citizens Abroad]. Retrieved August 17, 2024, from https://peraturan.bpk.go.id/Home/Details/139502/permenlu-no-5-tahun-2018.

Kementerian Luar Negeri Republik Indonesia. 2018b. *Peraturan Menteri Luar Negeri Republik Indonesia Nomor 10 Tahun 2018 Tentang Pengelolaan Media Digital Kementerian Luar Negeri Dan Perwakilan Republik Indonesia* [Regulation of the Minister of Foreign Affairs of the Republic of Indonesia Number 10 of 2018 Concerning Digital Media Management of the Ministry of Foreign Affairs and Representatives of the Republic of Indonesia]. Retrieved August 17, 2024, from https://peraturan.bpk.go.id/Home/Details/139509/permenlu-no-10-tahun-2018.

Kementerian Luar Negeri Republik Indonesia. 2020. Rencana Strategis 2020–2024 [Strategic Plan 2020–2024]. Kementerian Luar Negeri Republik Indonesia. Retrieved August 16, 2024, from https://kemlu.go.id/portal/id/read/2033/akip/rencana-strategis.

Kementerian Luar Negeri Republik Indonesia. 2024. Selama Sembilan Tahun Terakhir Kemlu Terus Perkuat Sistem Pelindungan WNI [Over the Past Nine Years, the Ministry of Foreign Affairs Has Continued to Strengthen the Indonesian Citizen Protection System]. Retrieved August 18, 2024, from https://kemlu.go.id/portal/idAkun/read/5669/berita/selama-sembilan-tahun-terakhir-kemlu-terus-perkuat-sistem-pelindungan-wni!

Kumparan News. 2020. Muhammadiyah Minta Pemerintah Evakuasi Mahasiswa di Wuhan. Retrieved August 17, 2024, from https://kumparan.com/kumparannews/muhammadiyah-minta-pemerintah-evakuasi-mahasiswa-di-wuhan-1sirUnpxndU/full.

Kurniawati, E., I. Rachmawati, and M. A. Dewi. 2020. @KemluRI: Diplomasi Publik Digital? [@KemluRI: Digital Public Diplomacy?]. *Andalas Journal of International Studies* 9 (1): 83–99. https://doi.org/10.25077/ajis.9.1.83-99.2020.

Leira, H., and N. Græger. 2020. Introduction: The Duty of Care in International Relations. In *The Duty of Care in International Relations. Protecting Citizens Beyond the Border*, ed. N. Græger and H. Leira, 1–17. Routledge.

Madu, L. 2018. Indonesia's Digital Diplomacy: Problems and Challenges. *Jurnal Hubungan Internasional* 7 (1): 11–8. https://doi.org/10.18196/hi.71121.

Mahfudzi, N. e. 2020. Tata Kelola Perlindungan Warga Negara Indonesia dalam Melakukan Peran Diplomasi Digital [Governance of Protection of Indonesian Citizens in Carrying Out the Role of Digital Diplomacy]. Jurnal Hubungan Internasional XIII (1): 19–38.

Manor, I. 2019. *The Digitalization of Public Diplomacy*. Palgrave Macmillan.

Meganingratna, A., and A. Nur. 2022. Implementasi Aplikasi Safe Travel Terhadap Perlindungan Warga Negara Indonesia (WNI) di Luar Negeri [Implementation of Safe Travel Application for the Protection of Indonesian Citizens (WNI)]. *NeoRespublica: Jurnal Ilmu Pemerintahan* 3 (2): 87–106. http://ojs.uho.ac.id/index.php/NeoRespublica.

Melissen, J., and M. Caesar-Gordon. 2016. "Digital Diplomacy" and the Securing of Nationals in a Citizen-Centric World. *Global Affairs*: 1–10.https://doi.org/10.1080/23340460.2016.1239381.

Melissen, Jan. 2013. Public Diplomacy. In *The Oxford Handbook of Modern Diplomacy*, ed. A. Cooper, J. Heine, and R. Thakur, 436–452. Oxford: Oxford University Press.

Metro TV News. 2020. Keluarga Minta Pemerintah Segera Evakuasi WNI di Wuhan, January 29. Retrieved August 17, 2024, from https://www.metrotvnews.com/play/bw6C57Lq-keluarga-minta-pemerintah-segera-evakuasi-wni-di-wuhan.

MPR. 2020. Hidayat Nur Wahid: "Segera Evakuasi WNI di China, Selamatkan Mereka Dari Ancaman Virus Corona", January 27. Retrieved August 17, 2024, from https://www.mpr.go.id/berita/hidayat-nur-wahid-segera-evakuasi-wni-di-china-selamatkan-mereka-dari-ancaman-virus-corona.

Naibaho, R. 2024. Selama 9 Tahun, Kemlu Selesaikan 218.313 Kasus WNI di Luar Negeri [Over 9 Years, the Ministry of Foreign Affairs Resolves 218,313 Cases of Indonesian Citizens Abroad], January 8. Retrieved August 19, 2024, from https://news.detik.com/berita/d-7130815/selama-9-tahun-kemlu-selesaikan-218-313-kasus-wni-di-luar-negeri.

Pamment, J. 2016. *British Public Diplomacy & Soft Power. Diplomatic Influence & Digital Disruption*. Palgrave Macmillan.

Radiananti, B.D. 2021. Diplomasi Digital dan Implementasi Aplikasi Safe Travel di Kementerian Luar Negeri Republik Indonesia. *PESIRAH: Jurnal Administrasi Publik*, 2 (1), 50–60. https://doi.org/10.47753/pjap.v2i1.1.

Rahayu, L.S. 2019. Menlu Bicara Pentingnya Diplomasi Digital untuk Kerja Sama Ekonomi [Foreign Minister Talks about the Importance of Digital Diplomacy for Economic Cooperation]. *Detik News*, September 10. Retrieved August, 20, 2024, from https://news.detik.com/berita/d-4699957/menlu-bicara-pentingnya-diplomasi-digital-untuk-kerja-sama-ekonomi.

Surwandono and A. Nugroho. 2022. Mengevaluasi Kebijakan Diplomasi Perlindungan WNI melalui Paradigma "Duty of Care". *Politica* 13 (2). https://doi.org/10.22212/jp.v13i2.3499.

Triwibowo, A. 2023a. The Characteristics of Indonesian Digital Diplomacy. *Journal of ASEAN Studies* 11 (1): 167–196. https://doi.org/10.21512/jas.v11i1.8525.

Triwibowo, A. 2023b. The Prospect and Limitations of Digital Diplomacy: The Case of Indonesia. *The Hague Journal of Diplomacy* 19 (1): 1–46. https://doi.org/10.1163/1871191x-bja10176.

Triwibowo, A. 2025. *Digital Diplomacy in Indonesia's Foreign Policy*. Routledge.

Uvais, I. 2010. *Consular Diplomacy: A Conceptual Framework for Analysis and Decision-making* [Master Dissertation: University of Malta]. Library of University of Malta. https://www.um.edu.mt/library/oar/handle/123456789/76984.

Yahya, A.N., and D. Meiliana. 2022. Kemenlu Perkirakan Ada 9 Juta WNI di Luar Negeri, Hanya 3 Juta yang Tercatat [Ministry of Foreign Affairs Estimates There Are 9 Million Indonesian Citizens Abroad, Only 3 Million Are Recorded],, July 13. Retrieved August 18, 2024, from https://nasional.kompas.com/read/2022/07/13/11200371/kemenlu-perkirakan-ada-9-juta-wni-di-luar-negeri-hanya-3-juta-yang-tercatat.

Albert Triwibowo is an Assistant Professor at Parahyangan Catholic University, Bandung, Indonesia. He is also a researcher at Parahyangan Center for International Studies (PACIS). He received his doctoral degree form Universität Rostock and was supported by DAAD. His research interests are diplomacy, digital diplomacy, foreign policy, Indonesia, East and Southeast Asia regionalism.

Chapter 11
Indonesian Border Diplomacy Under Joko Widodo's Presidency: Land Boundary Negotiations Between Indonesia–Timor-Leste and Indonesia–Malaysia

Indriana Kartini, Awani Irewati, and Agus R. Rahman

Abstract Protecting the integrity of the Unitary State of the Republic of Indonesia received special attention during Jokowi's Presidency. During the first and the second periods of Jokowi's administration, sovereignty diplomacy was prioritized with its key aspects being state border resolution through negotiations. Given that several segments of the land boundary remain unresolved, this chapter examines Indonesian border diplomacy under Jokowi's administration by specifically looking at the land boundary negotiations between Indonesia andTimor-Leste, particularly in the Noel Besi–Citrana segment, and the land boundary negotiations between Indonesia and Malaysia, especially in the Sebatik Island segment. This chapter finds that although Jokowi's administration had a strong commitment to resolving several segments of the land boundary between Indonesia–Timor-Leste and Indonesia–Malaysia, up until his last period, Jokowi's administration has not finalize the land boundary negotiations in the conflictual segment between Indonesia and Timor-Leste. Only in the conflictual segment between Indonesia and Malaysia, in Sebatik (West pillar to East pillar) it has been finalized and under MoU preparation. This became the challenge for the next administration under Prabowo Subianto.

Keywords Border diplomacy · Jokowi's administration · Indonesia · Timor-Leste · Malaysia · Naktuka · Sebatik Island

---

I. Kartini (✉) · A. Irewati · A. R. Rahman
National Research and Innovation Agency (BRIN), Research Center for Politics, Jakarta, Indonesia

I. Kartini and Kanisa (eds.), *Joko Widodo's Foreign Policy and Diplomatic Legacy*,
https://doi.org/10.1007/978-981-95-2085-5_11

11.1 Introduction

During President Joko Widodo's (Jokowi) administration, protecting the integrity of the Unitary State of the Republic of Indonesia received special attention. It was prioritized as the top focus in sovereignty diplomacy during the first period of Jokowi's administration and as the third priority in the second period. A key aspect of this sovereignty diplomacy is the resolution of state borders, both land and sea, through negotiations. The 2024 Annual Press Statement of the Minister of Foreign Affairs of the Republic of Indonesia outlines the state boundary agreements, highlighting both those that have been completed and those still/ remain pending.

The sea boundary negotiations include: (1) The EEZ boundary delimitation agreement with Vietnam, which, after 12 years of negotiations, became the first dual-line agreement among Southeast Asian countries; (2) The Indonesia–Malaysia agreement on two segments of the maritime boundary in the Sulawesi Sea and the Southern Malacca Strait, which was completed in June 2023 after 18 years of negotiations; (3) The agreement between Indonesia and the Philippines on the Principles and Guidelines for the Boundary of the Continental Shelf in October 2022, which will serve as an important reference in the ongoing negotiations to complement the 2014 EEZ Boundary agreement in the Sulawesi Sea.

Meanwhile, the land boundary negotiations include: (1) The agreement on three land boundary segments with Malaysia in Kalimantan-Sabah, reached during the 2017–2019 period. Additionally, three other segments—**the Sebatik Island segment**, **the Sinapad-Sesai segment**, and **West Pillar-AA 2** in Kalimantan-Sabah—are targeted for completion after 24 years of negotiations; (2) The land boundary agreement between Indonesia and Timor-Leste in **the Subina–Oben** and **Noel Besi–Citrana segments**, which was negotiated for 19 years. Given that several segments of the land boundary remain unresolved, this study focuses on border diplomacy at the Indonesia–Timor-Leste border, particularly in the Noel Besi–Citrana segment, and the Indonesia–Malaysia border, especially in the Sebatik Island segment. The framework for analysis emphasizes that a nation's territorial sovereignty is a key priority in the pursuit of national interests, essential for maintaining territorial integrity, which underpins sovereignty. The realist approach is primarily employed in addressing land border issues through diplomatic efforts, focusing on the power, security, and strategic positioning of the country.

When it comes to addressing land border disputes, the country consistently engages in vigorous diplomatic negotiations. However, if these negotiations become stalled, the government may resort to coercive actions, including military deployment, to maintain control over contested border regions. For instance, the border conflicts between China and India, as well as between Thailand and Cambodia, illustrate the use of military force alongside negotiation efforts to resolve issues.

11.2 Literature Review

Numerous studies have focused on the borders between countries, particularly those between Indonesia and Malaysia, as well as between Indonesia and Timor-Leste. However, research that highlights the significance of Ordinary Boundary Problems in these two areas during the period leading up to the end of President Joko Widodo's term (October 20, 2024) has emerged as a fresh addition to earlier studies.

Research on border diplomacy in Nunukan was discussed by Anandra and Kusumawardhana in an article published in The Journalish: Social and Government (Vol. 4, No. 4, 2023). The study focuses on border issues in North Kalimantan (specifically in Nunukan) from political, security, and socio-economic viewpoints. Border diplomacy is employed as a primary strategy to address these border challenges. This institutional approach is used to handle border issues between North Kalimantan and Sabah.

Regarding the study on the settlement of land boundaries between Indonesia and Timor-Leste, Nomleni (2019) analyzes the resolution of border conflicts using a cultural approach, focusing on the settlement of boundaries in the Naktuka neutral zone. By employing the customary meeting communication model "Nekaf Mese Ansaof Mese Atoin Pah Timor" (one heart, one soul as Dawan people), the study highlights the reasons for utilizing customary meetings, the models used, and the impressions of these meetings between the East Amfoang community (Indonesia) and the indigenous community of Timor-Leste. The findings indicate that the communication processes within the customary rituals of the Timorese people, combined with conciliation and deliberation approaches, resulted in a customary agreement accompanied by a customary oath. This agreement serves as one of the options for resolving the Naktuka territorial boundary dispute between Indonesia and Timor-Leste (Nomleni 2019, p. 117).

Another study related to boundary disputes is Mangku's (2017) analysis of the resolution of the boundary dispute between Indonesia and Timor-Leste from an international law perspective. Mangku highlights the 2005 Provisional Agreement, specifically Article 6, point (b), which emphasizes the involvement of local communities in resolving border disputes peacefully, as stipulated in Article 8 of the same agreement. According to Mangku (2017, p. 63), the people living in West Timor (Indonesia) share the same socio-cultural background as those in East Timor (Timor-Leste), which ensures that the customary legal system is similar for both groups. This customary legal system substantively regulates land issues and customary boundaries. Mangku's study (2017, p. 63) underscores the significant role of customary leaders in negotiations to resolve boundary disputes. Both Nomleni's and Mangku's studies emphasize the potential role of traditional leaders in offering solutions to territorial boundary disputes in the Naktuka neutral zone through customary and legal communication approaches. However, neither study examines the process of land boundary negotiations between Indonesia and Timor-Leste, which this study aims to address.

Nurdin et al. (2023) examined the issue of the Outstanding Boundary Problems (OBP) on Sebatik Island in their work titled "The Unsettled Agreement of the 2019 Re-demarcation of the Indonesia-Malaysia Boundary Line's Impact." This study investigates the consequences of the 2019 boundary re-measurement on the inhabitants and the nation of Sebatik, particularly in relation to the transfer of land that was originally Malaysian territory to Indonesia, as well as the transfer of Indonesian land to Malaysia. While this research shares thematic similarities with our article, it diverges in focus; our article does not concentrate on the repercussions of re-demarcation but rather examines the diplomatic negotiation processes undertaken during President Joko Widodo's administration, aimed at resolving all OBPs in the Eastern region, including Sebatik.

11.3 Theoretical Framework

Ullah and Kumpoh (2018, p.3) argue that borders have become a contested subject in the international arena, drawing renewed attention in recent years. They note that traditional diplomacy and foreign policy focused on issues of war and peace, but the scope of border-related agendas has expanded significantly over the past few decades, with notable changes in border controls and security perceptions. Additionally, Nordquist (2010) asserts that boundary demarcation is a complex process involving both domestic and international legal frameworks. While boundary agreements between states are considered international legal documents, their creation and ratification are inherently domestic political processes. Similarly, the delimitation phase involves negotiations that precede the signing of agreements, which can be intricate and contentious, potentially leading to serious disagreements among the involved parties (Nordquist 2010).

According to Shrestha (2021), borders define a state's territorial boundaries and uphold its sovereignty, yet their relational nature can simultaneously challenge that sovereignty. As such, borders can reflect internal disputes and serve as a potential source of interstate conflicts. Given their complex and conflict-prone nature, preventive diplomatic measures through negotiation are crucial to de-escalating boundary disputes globally. These diplomatic efforts operate within the framework of international law, which prioritizes peaceful conflict resolution. In the modern era, the use of force in disputes is widely regarded as inappropriate, making peaceful settlement a foundational principle for fostering confidence in international law's dedication to global peace. Chetail (2003) notes that International law plays a key role in resolving disputes by employing arbitration or tribunals, which depoliticize conflicts by entrusting technical experts. In this context, the International Court of Justice (ICJ) has proven to be an effective and beneficial arbitration mechanism for addressing disputes in contemporary times.

Border diplomacy entails employing transborder collaboration, discussions, alliances, policies, or occasionally other coercive measures to resolve issues impacting state borders in relation to the external environment. It amicably resolves border-related disputes through demarcation, regulation, or alternative methods (Shrestha 2021). Ogunnubi and Awosusi (2021) define border diplomacy as the diplomatic strategy of utilizing a state's geostrategic border advantages (land, airspace, and seaport) to promote transborder cooperation or address territorial disputes with neighboring states or other relevant cross-border entities.

Another aspect highlighted by the Constructivism approach is the significance of national identity and socio-cultural connections. It is important to note that border regions are deeply intertwined with the relationships among ethnic groups, which are closely linked to the interactions between border communities. The same essence of the approach taken by some scholars believes that the constructivism-based approach emphasizes four dimensions that exclusively support flexible and mutually acceptable migration standards: (i) collaborative transborder security and policing, (ii) joint rehabilitation of deteriorated and permeable transboundary regions, (iii) bilateral alignment of immigration laws and policies, and (iv) safeguarding immigrants' social rights (Iwuoha and Mbaegbu 2021). Another scholar emphasizes the approach of using the concepts of identity, norms, and ideas to explore development cooperation among states in South Asia (Malhotra 2022).

The border disputes involving Malaysia–Indonesia and Indonesia–Timor-Leste are inherently connected to these relationships. Efforts to address border issues through diplomacy become more adaptable, which can lead to a better alignment of social and cultural dynamics within the communities. However, this flexibility may also risk escalating tensions. Thus, it is crucial to preserve the local community ties between the two nations as part of their diplomatic efforts. Once the border issues are resolved, future steps will depend on the relationships among local communities in both countries.

11.4 The Process of Land Boundary Negotiations Between Indonesia and Timor-Leste

Since the Permanent Court of Arbitration (PCA) decision on June 25, 1914, which resolved the territorial dispute over the border enclave of Oecusse in favor of the Netherlands, the potential for border conflict between the Netherlands and Portugal was reduced. The two countries then shifted their focus to a more pressing issue—the looming threat of World War I. During World War II, both nations witnessed the colonization of Timor by Japan.

Indonesia's independence in 1945 meant that, as the successor state to the Netherlands' colonial territory, Indonesia inherited the international border with Portuguese Timor on the island of Timor. While there was a strong desire to integrate Portuguese Timor into the Republic of Indonesia, the Indonesian government initially focused on the reintegration of the western part of Papua Island. This area, defended by the Netherlands in 1949, only became part of the Republic of Indonesia in 1963 under the name West Irian, later renamed Irian Jaya, and finally Papua. The 1974 revolution in Portugal and the civil war in Portuguese Timor provided an opportunity for Indonesia to integrate East Timor into the Republic of Indonesia in 1975.

Although the international community never formally recognized the integration of East Timor into the Republic of Indonesia, de facto incorporation did occur. The former international border between Indonesia and Portugal was reclassified as an administrative division in 1975, separating East Timor Province from East Nusa Tenggara Province (West Timor). The unified border, which had previously separated the two regions, remained unchanged from 1975 to 1999, and East Timor continued to exist as a distinct territorial entity.

The UN referendum held in 1999 resulted in a majority decision by the people of East Timor to choose independence and separate from the Republic of Indonesia. This referendum changed the political history of both Indonesia and East Timor. The 1999 referendum showed that nearly 80% of East Timorese voted for independence. Interestingly, however, according to Deeley (2001, 12), the results in the East Timor enclave of Oecusse showed a preference for integration with Indonesia. Despite this, the overall result determined the victory of the independence option, leading to East Timor's independence from Indonesia.

The relevant border delimitation for the Timor-Leste–Indonesia border dates back to 1904 for most of the boundary, and to 1914, when the PCA settled territorial disputes related to the Oecusse enclave. A significant portion of the boundary follows rivers and streams, with the deepest point of the river (thalweg) used as the definitive boundary marker. However, the maps produced during the 1904 and 1914 delimitations were limited in accuracy and precision. Despite this, these maps are considered official documents and hold significant value in legal proceedings. In addition to the maps created by the Netherlands and Portugal, there are also maps produced by Indonesia during East Timor's integration. Although these maps reflect a high level of state agreement on territorial boundaries, differences in scales, projections, and geodetic references can lead to discrepancies in significant territorial treaties. Therefore, these maps should be used with caution, as the use of contradictory maps by opposing parties could be a potential source of disputes and conflicts (Deeley 2001, p. 14).

To establish and assert the boundaries, the governments of the two countries re-examined the historical boundaries that once existed between West Timor and East Timor. Since Portugal declared East Timor as its colony in 1701, various agreements and activities related to the delimitation of land boundaries between Indonesia and East Timor have taken place, including the following:

a. *The Treaty of Demarcation and Exchange of Some Portuguese and Dutch Dominions on the Solor and Timor Archipelago* between Portugal and the Netherlands was signed on April 20, 1859, and ratified on August 13, 1860, in Lisbon, Portugal.
b. A *joint boundary commission surveys the Island of Timor's boundaries, and agreement is reached on most of its length.* A complete boundary survey of the

Ocusse enclave is prevented by the actions of hostile tribesmen, in the period between 1898 and 1899.

c. *Convention for the Demarcation of Portuguese and Dutch Dominions on the Island of Timor* was signed on October 1, 1904, in The Hague and ratified on October 29, 1908.
d. *1914 Arbitral Award Rendered in Execution of Compromise* which was signed in The Hague on April 3, 1914.
e. The Oil Poli Document signed on February 9, 1915, on the partial description of the Oecusse boundary as surveyed after the Arbitration decision, including a list of boundary pillars.
f. The Mota Talas document signed on April 22, 1915, describes the pillars installed on the border in the eastern sector (main border) following the survey activities in Article 10 of the 1904 Treaty.

Since the transition period in 1999, before Timor-Leste formally became independent, Indonesia worked together with UNTAET to discuss the issue of land boundaries on the island of Timor. In 2001, the two parties formed the Technical Subcommittee on Border, Demarcation, and Regulation (TSC-BDR), and in a 2002 meeting, they agreed to carry out a joint demarcation survey program. When Timor-Leste officially became an independent country on May 20, 2002, TSC-BDR RI-RDTL began operating in cooperation to resolve land boundary issues. In the same year (2002), unresolved segments were identified during the survey following the second TSC-BDR meeting.

In addition to the TSC-BDR, the RI-UNTAET Joint Border Committee (JBC) Forum was also formed, which agreed to use the 1904 Treaty as one of the determinants for the land boundary between RI and RDTL. The process of negotiating the land boundary between Indonesia and Timor-Leste is conducted at three levels: the Joint Ministerial Commission (JMC) level, the Joint Border Committee (JBC) level, and the Technical Subcommittee on Border, Demarcation, and Regulation (TSC-BDR) level (see Fig. 11.1).

At the JMC level, the negotiation process aims to resolve issues that arose after the 1999 polls, such as those concerning Indonesian government assets, Indonesian citizens left behind in Timor-Leste, the settlement of civil servants' rights, the rights of former East Timorese TNI members and police officers, the problem of former East Timorese refugee children, border-related issues (e.g., illegal border crossings, smuggling of merchandise, and other illegal economic activities), and the determination of land and sea boundaries. At the JBC level, the negotiation process focuses on resolving border problems such as cross-border issues involving people and goods, security concerns, order, river management at the border, and the delimitation and demarcation of land boundary lines. Meanwhile, at the TSC-BDR level, the focus is on the technical aspects of determining land boundaries, including discussions and reviews of the 1904 Treaty document, the 1914 PCA document, maps, and other related documents. Additionally, joint work planning and scheduling, joint reconciliation surveys, joint delineation surveys, mapping implementation, and the documentation of related data and information are conducted (Anggitta 2014, p. 36).

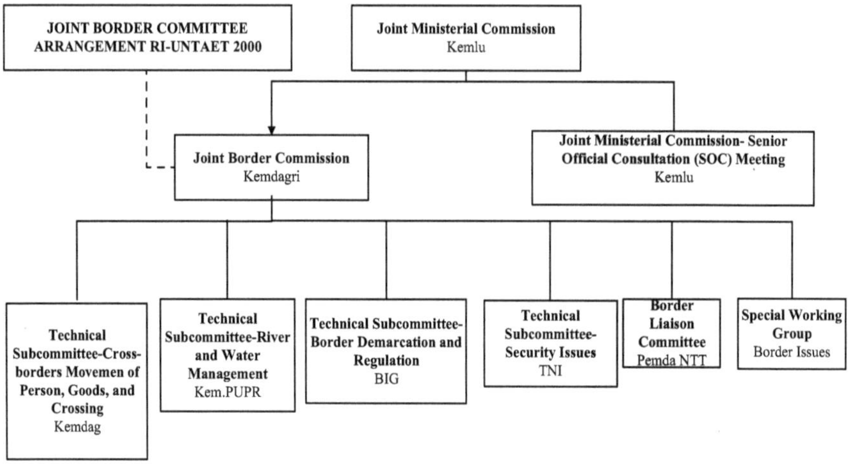

Fig. 11.1 Organizational structure for Indonesia–Timor-Leste Land Boundary Management. *Source* Presentation Materials of the Ministry of Foreign Affairs on the Development of Cooperation in Handling RI-RDTL Land Borders at the Coordination Meeting for Handling Land Boundary Areas, August 2022

Meanwhile, the Special Working Group (SWG) was formed in 2002 to handle unresolved segments. The Senior Officials' Consultation (SOC) was established later, in 2017, and has held five meetings up until 2019. The five SOC meetings include: (1) The 1st SOC, held in Bali in March 2017; (2) The 2nd SOC, held in Dili in April 2017; (3) The 3rd SOC, held in Bandung in October 2017; (4) The 4th SOC, held in Dili in December 2018; and (5) The 5th SOC, held in Bali in July 2019. The SOC was established as a forum for resolving unresolved segments through political means, as settlement at the technical level through discussions in the TSC-BDR and SWG forums had reached an impasse, requiring additional efforts.

Before the 5th SOC, a Joint Field Visit (JFV) was conducted by the TSC-BDR Sub-Committee from April 29 to May 15, 2019. Based on information from BNPP, at the time of the 5th SOC, the Chairman of the TSC-BDR Sub-Committee reported the purpose of the Joint Field Visit, which was:

1. For the Noel Besi–Citrana Segment, the purpose of the Joint Field Visit is: (a) to identify the elements specified in Article III.1. of Treaty 1904, namely the Noel Besi thalweg, the mouth of the Noel Besi River and the azimuth point to Fatu Sinai/Batek Island, including Nono Tuinan (the old Noel Besi river); (b) to collect relevant social information to support the creation of Special Arrangements.
2. For the Bijael Sunan–Oben Segment, the objectives of the Joint Field Visit are: (a). To follow up on the delineation process based on the results of the 2015 UAV Indonesia survey according to a common understanding of the general description of what constitutes the boundary line; (b) to collect relevant social information to support the creation of Special Arrangements.

Based on information from the Regional Director of the Directorate General of Strahan, Ministry of Defense of the Republic of Indonesia in 2022, the following is the chronology of the formation of the SOC RI–RDTL:

- On November 3, 2016, in Jakarta, a meeting took place between Xanana Gusmão, the Coordinating Minister for Politics, Law, and Security, and the Minister of Foreign Affairs of the Republic of Indonesia. During the meeting, both parties agreed to resolve the unresolved segments.
- On November 11, 2016, in Jakarta, a meeting was held between the Coordinating Minister for Politics, Law, and Security and the Minister of Foreign Affairs of the Republic of Indonesia. The meeting discussed options for resolving Unresolved Segments to be presented at the meeting between Xanana Gusmao and the Coordinating Minister for Politics, Law, and Security scheduled for November 15, 2016.
- On November 25, 2016, the Coordinating Minister for Politics, Law, and Security of the Republic of Indonesia sent a letter to Xanana Gusmao. The content of the letter states that the Indonesia side offers the following options for resolving unresolved segments:

    (1) For the Noel Besi–Citrana segment, a new line needs to be drawn between the Overlapping Claim Noel Besi.
    (2) For the Bijael Sunan–Oben segment, it is necessary to make a Special Arrangement (SA) to provide access for indigenous peoples.

- On December 6, 2016, the Ministry of Home Affairs of the Republic of Indonesia c.q. The Directorate General of BAK held a follow-up meeting on the Letter of the Coordinating Minister for Politics, Law, and Security to Xanana Gusmao. The meeting forum underlines several things:

    (1) Technical aspects of negotiations, agreements, and legal instruments of the Special Arrangement (SA), with the Ministry of Foreign Affairs as the focal point;
    (2) Economic, social, and cultural aspects that will be followed up in the Special Arrangement, with the Ministry of Home Affairs/BNPP as the focal point. The meeting forum then concluded with several important notes, including:
        (1) The basis of the withdrawal of the Noel Besi–Citrana line is an equal portion-based median line, based on the precedent of the completion of Dilumil–Memo in 2013;
        (2) The settlement of Bijael Sunan/Manusasi–Oben is a line based on the Boundary Convention of the 1904 Tract, with a Special Arrangement respecting the customary rights of the local community.

- On January 12, 2017, a meeting was held to discuss the completion of the Unresolved Segments RI–RDTL at the Ministry of Foreign Affairs. The meeting forum concluded with several important notes, including:

    (1) The establishment of a SOC so that it can be realized immediately for the two countries to meet shortly;

(2) The Ministry of Foreign Affairs coordinated the formation of a technical team divided into two groups (Technical Delineation and Special Arrangement) and tentatively prepared the SOC agenda, including the time and venue for negotiations. On January 16, 2017, the Ministry of Foreign Affairs issued letter Number 01026/HI/01/2017/53, regarding the report on the meeting to discuss the settlement of unresolved segments. The results of the meeting decided that:
    (1) the Indonesian side will form two teams to complete the unresolved segments of Noel Besi–Citrana and Bidjael Sunan/Manusasi–Oben;
    (2) a delegation team will be formed, consisting of the Chairman, Head of Regional Boundary Mapping from BIG, and representatives from the Directorate of Topography of the Indonesian Army;
    (3) A Special Arrangement team will be formed, consisting of the Chairman, Director of Legal and Regional Agreements from the Directorate General of HPI, Ministry of Foreign Affairs, and representatives from the Directorate General of Waskoban and the Directorate General of BAK from the Ministry of Home Affairs.

- On December 13–14, 2018, a decree was issued by the Director-General of Strahan of the Ministry of Defense, Number: Sprin 1466/VI/2019, regarding an official trip to Dili, Timor-Leste, as a member of the Indonesian delegation for the 4th meeting of SOC on the unresolved segments of the RI-RDTL land boundary. The meeting results were an agreement between the two parties to conduct a Joint Field Visit (JFV) in both unresolved segments following the schedule and Standard Operating Procedures.
- On July 1–3, 2019, a decree was issued by the Director-General of Strahan of the Ministry of Defense, Number: Sprin 1466/VI/2019, regarding an official trip to Dili, Timor-Leste, as a member of the Indonesian delegation for the 5th meeting of SOC on the unresolved segments of the RI-RDTL land boundary. During the meeting, the settlement of two segments of the land boundary between RI and RDTL—Noel Besi (RI)–Citrana (RDTL) and Bijael Sunan/Manusasi (RI)–Oben (RDTL)—was discussed. The meeting's results focused on efforts to reach an agreement by gathering materials for higher leadership's consideration in negotiations related to unresolved segments.
- On March 19, 2021, a meeting was held at the Mason Pine Hotel in Padalarang, West Java, which resulted in several outcomes, including:

(1) For the Noel Besi (RI)–Citrana (RDTL) segment, the use of the Simplified Median Line was confirmed as the final decision, and the Indonesian side no longer needs to discuss technical matters internally;
(2) For the Bijael Sunan/Manusasi (RI)–Oben (RDTL) segment, this segment was surveyed bilaterally regarding the 1904 Treaty between the Netherlands and Portugal, confirming that it belongs to Timor-Leste (Portugal).

However, during the transition period before the Provisional Agreement in 2005, there was an agreement regarding the TNI patrol line that crossed the area. Indonesian citizens assumed the line at the border to be within Indonesian territory. This territory was divided proportionally in half, leading to dissatisfaction among Indonesian citizens at the border, who rejected the survey results and caused the area to be classified as an unresolved segment. Consequently, the Indonesian side requested a Special Arrangement to allow Indonesian citizens to jointly manage the segment area.

11.4.1 Negotiation Process for the Determination of the Land Boundary Between Indonesia and Timor-Leste in 2019

Based on information from the Geospatial Information Agency, the TSC-BDR Technical Team from Indonesia conducted an exercise from April 29 to May 15, 2019, in preparation for the 5th Senior Officials' Consultation (SOC) meeting. The Technical Team proposed several options for resolving the Noel Besi–Citrana segment, including:

1. Completion options using the Median Line

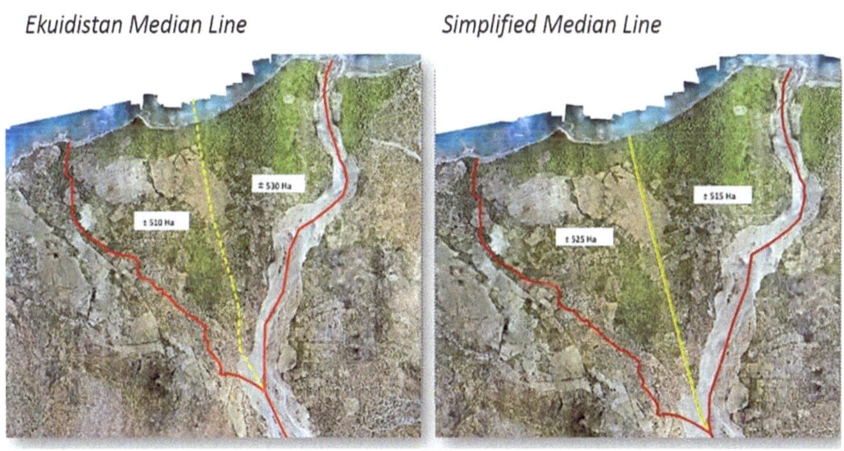

Source: BIG. (2021). Development of the Unresolved Status of the Noel Besi–Citrana Segment. Bogor: Center for Boundary Mapping of the Geospatial Information Agency (BIG).

2. Completion options using nature features

   The identified natural features are a combination of small rivers, rice paddies, and land cover boundaries, as shown in the aerial photos and DEM from the 2015 UAV acquisition.

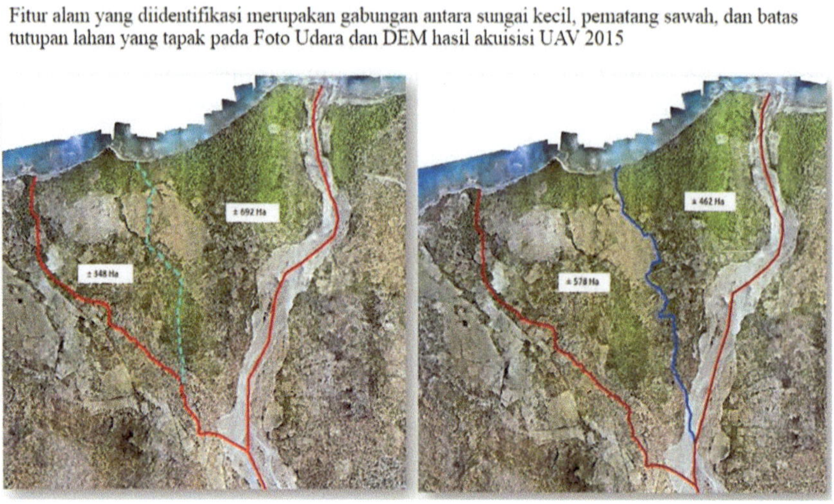

Source BIG. (2021). Development of the Unresolved Status of the Noel Besi–Citrana Segment. Bogor: Center for Boundary Mapping of the Geospatial Information Agency (BIG)

Based on the results of the 5th SOC, under agenda item Number 5, it was stated that both countries are committed to resolving two unresolved segments—the Noel Besi–Citrana segment and the Bijael Sunan–Oben segment—by September 2019. As a follow-up, a meeting was held on July 22, 2019, between the Chairman of SOC, the Coordinating Minister for Politics, Law and Security Wiranto, and Xanana Gusmão at the office of the Coordinating Ministry for Politics, Law and Security of the Republic of Indonesia. This meeting resulted in the Agreed Principles on the Final Settlement of the Land Boundary, outlining how the teams from Indonesia and Timor-Leste will proceed in addressing the unresolved segments (BNPP 2021). The agreement is to be finalized by the SOC and incorporated into Addendum No. 2 of the 2005 Provisional Agreement between the Government of the Republic of Indonesia and the Government of the Democratic Republic of Timor-Leste on the Land Boundary. It will later be included in a comprehensive agreement between Indonesia and Timor-Leste. Both sides also agreed that, as a priority, the continuation of this process requires intensive cooperation through the Joint Border Committee (JBC) between the two countries (https://polkam.go.id/ri-timor-leste-sepakat-selesaikan-batas-darat/, 2019, accessed on May 6, 2020).

The Agreed Principles on The Final Settlement of Land Boundary between Indonesia and Timor-Leste in 2019 states that "the two parties have agreed, based on the reference to the '1904 Agreement between the Netherlands and Portugal', with a mutual understanding for the final settlement of several unresolved territories as follows":

a. **Noel Besi–Citrana Segment**

This segment will be completed by applying a simplified median line drawing approach, drawn from the north starting at the Bokos point with coordinate points to be determined later.

b. **Bijael Sunan–Oben Segment**

This segment will be completed following the 1904 Treaty between the Netherlands and Portugal.

c. **Haumeniana–Passabe Segment**

This segment will be determined by adjusting the line along the road in order of both sides whose coordinate points will be determined later.

d. **Mota'ain–Batu Gade Area**

The SOC will define the coordinate point and the endpoint (*Terminus Point*).

e. **Special Arrangements**

Taking into account the cultural bond between the communities living on the border of the two countries, both sides agreed to make special arrangements that fully respect the traditional, cultural, and religious activities of the people living in the Noel Besi–Citrana and Bijael Sunan–Oben segments, including access to Batek/Fatu Sinai Island for Timor-Leste citizens residing in the Noel Besi–Citrana segment.

Based on the Principles of the Agreement on the Final Settlement of the Land Boundary between Indonesia and Timor-Leste, both parties agreed that, by referring to the 1904 Treaty between the Netherlands and Portugal, the Noel Besi–Citrana segment would be resolved by applying the approach of drawing a simplified median line, starting from the north at the Bokos point, with coordinate points to be determined later. The establishment of the regional boundary based on the common midpoint is a political settlement, whereby the regional boundary in the Noel Besi–Citrana segment, which has been divided into two regions, is a political line that will be logically converted into a technical line by drawing a straight line down from the common midpoint. This approach is similar to the method used by the Indonesian government and the Papua New Guinea government when determining the territorial boundary between Indonesia and Papua New Guinea by drawing a straight line following the flow of the Fly River in Papua New Guinea (see Fig. 11.2) (https://www.youtube.com/watch?v=uB2weUW34R0&t=2502s, 2020, accessed on July 1, 2021).

After the implementation of the 5th SOC, a Desktop Exercise Meeting was held in Bali on August 14–15, 2019. During this meeting, the midpoint and endpoint coordinates for the Noel Besi–Citrana segment were agreed upon. On this occasion, the Timor-Leste side submitted their version of the claim. Then, in December 2019, the Desktop Exercise Meeting was held again; however, the Timor-Leste side did not attend.

Based on this explanation, it can be noted that after a long period following the signing of the Provisional Agreement in 2005, the Agreed Principles regarding the determination of the territorial boundary between the governments of Indonesia and Timor-Leste were finally signed in July 2019. The agreement stipulated the use of a simplified median line approach based on equal portions as an option for

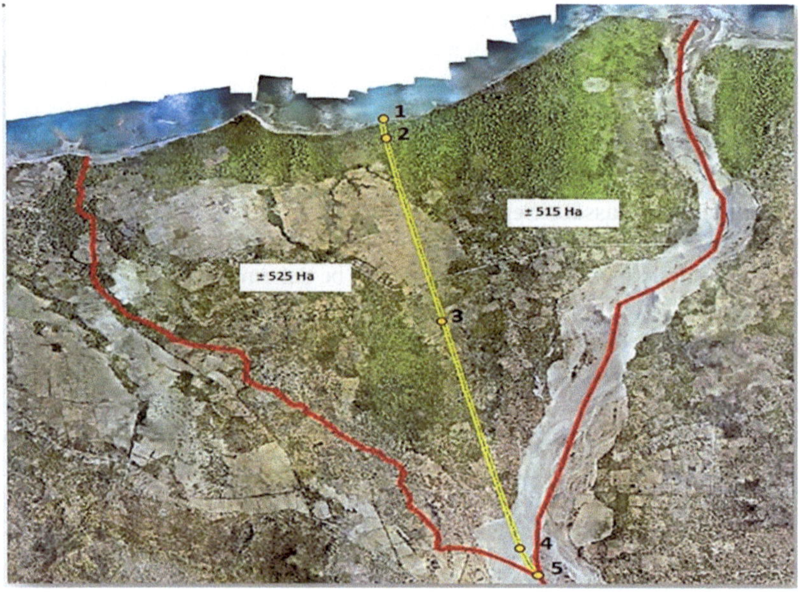

Fig. 11.2 Regional distribution map in the Noel Besi–Citrana segment with a simplified median line approach. *Source* BIG. (2021). Development of the Unresolved Status of the Noel Besi–Citrana Segment. Bogor: Center for Boundary Mapping of the Geospatial Information Agency (BIG)

settling the land boundary in the Naktuka (Noel Besi–Citrana) segment. Shortly thereafter, the Indonesian government conducted an exercise on mapping the territorial boundary based on the common midpoint following the Agreed Principles. This exercise included the possibility of mapping the boundary according to the Timor-Leste government's version, which follows the features of roads and rice paddies while considering the settlements in Timor-Leste. This exercise was undertaken by the Indonesian government as a precautionary step during renegotiations with the Timor-Leste government.

11.4.2 Negotiation Process After the Determination of the Land Boundary of Indonesia and Timor-Leste in 2019

After the signing of the Agreed Principles on the Final Settlement of the Land Boundary on July 22, 2019, a series of developments occurred in the efforts to settle the unresolved segments in Noel Besi–Citrana by the two countries, as follows:

- On August 7, 2019, Xanana Gusmão sent a letter to the Coordinating Minister for Politics, Law, and Security of the Republic of Indonesia, conveying a proposal for the Noel Besi–Citrana settlement line. Xanana proposed a settlement line option that allocates a larger portion of territory to Timor-Leste, with approximately 74% for Timor-Leste and 26% for Indonesia."

- On December 13, 2019, the Coordinating Minister for Politics, Law, and Security of the Republic of Indonesia sent a reply letter in response to a letter from Xanana Gusmão.
- On February 4, 2020, Xanana Gusmão visited Indonesia and met with the Coordinating Minister for Politics, Law, and Security of the Republic of Indonesia.
- On February 5, 2020, the Chairman of SOC RI sent a reply letter in response to Xanana Gusmão's letter.
- In March 2020, Timor-Leste postponed the VI SOC due to the COVID-19 pandemic.
- In December 2020, Xanana Gusmão sent a letter to the Coordinating Minister for Politics, Law, and Security of the Republic of Indonesia, outlining plans to resume negotiations in the first quarter of 2021.
- Throughout 2021, the Indonesian Ministry of Foreign Affairs continued to encourage the resumption of negotiations with Timor-Leste.
- In February 2021, the Indonesian Minister of Foreign Affairs sent a letter to Xanana Gusmão proposing to resume negotiations early in the year.
- In March 2021, the SOC Chairman held an informal virtual meeting, resulting in an agreement to hold the 6th Desktop and SOC in April 2021.
- In May 2021, the Chairman of SOC RI sent a letter to the Chairman of SOC RDTL, requesting the implementation of a technical team (Desktop) in August 2021. That same month, Xanana sent a letter to the Coordinating Minister for Politics, Law and Security Affairs of the Republic of Indonesia, proposing to begin exploratory negotiations on the RI-RDTL sea boundary in October 2021.
- In August 2021, the Indonesian side proposed that the Desktop and SOC be held in September 2021.
- In September 2021, RDTL again postponed the 6th Desktop and SOC due to the pandemic and upcoming elections, proposing to continue in 2022.
- On April 4, 2022, an informal one-on-one meeting was held between the Chairman of SOC RI and RDTL.
- On July 18–20, 2022, RDTL President Ramos Horta paid a state visit to Indonesia.

During an informal one-on-one meeting between the chairmen of SOC RI-RDTL in April 2022, the Chairman of SOC Timor-Leste conveyed the following points:

1. Timor-Leste is willing to hold a Technical Team meeting only if the Republic of Indonesia accepts the proposal outlined in Xanana Gusmão's letter dated August 7, 2019.
2. The only acceptable final solution is the one proposed in Xanana Gusmão's letter.
3. If the Government of Indonesia is unwilling to consider the proposal, Timor-Leste will not participate in further Technical Team meetings.
4. Timor-Leste will revert to the position outlined in the 1904 Treaty map agreed upon by the Netherlands and Portugal, which follows the old Noel Besi river on the west side.

In response, the Chairman of SOC Indonesia stated Indonesia's position as follows:

1. Indonesia cannot accept Timor-Leste's proposal.
2. The settlement of the Noel Besi–Citrana segment must align with the Agreed Principles on The Final Settlement of Land Boundary agreement of July 22, 2019, which stipulates that the final boundary is a simplified median line, reflecting an equitable solution for both parties.

Indonesia's position was conveyed in a letter from the Coordinating Minister for Politics, Law, and Security of the Republic of Indonesia to Xanana Gusmão on December 13, 2019, and reiterated in a letter from the Chairman of SOC RI on February 5, 2020. However, the Chairman of Timor-Leste's SOC continues to reject Indonesia's position for the following reasons:

1. The map in the 1904 Treaty depicts the boundary line following the old Noel Besi River (Nono Tuinan) on the west side.
2. The Timorese side faces difficulties in explaining the situation to their parliament and constituents.
3. Timor-Leste believes that the Indonesian government should accept the compromise line proposed by Timor-Leste, which they consider a win–win solution for both parties.

In response, the Indonesian government maintains that Xanana Gusmão's proposed line disregards the proportional division of territory negotiated by the Technical Team. Both Indonesia and Timor-Leste remain steadfast in their respective views on the settlement of unresolved segments. Indonesia has detected indications that Timor-Leste might seek to bring the land boundary issue in the Noel Besi–Citrana segment to a third party, potentially shifting the negotiation strategy toward a political determination of the boundary without Technical Team consideration.

To anticipate these developments, Indonesia, following the direction of the Coordinating Minister for Politics, Law and Security Affairs on August 12, 2022, has adopted the following stance:

1. Indonesia will defend its last position and request Timor-Leste to consider the proposed line by the Chairman of SOC RI.
2. Indonesia will be flexible if Timor-Leste is also willing to show flexibility.
3. It is necessary to hold the 6th SOC Meeting with the Technical Team present.

Meanwhile, if Timor-Leste withdraws the proposed Xanana line based on the August 2019 Xanana Letter and reverts to its original position following the 1904 Treaty, both countries would retract their commitments and undo the progress made at the technical level.

Based on the explanation above, it is clear that the process of negotiating the land boundary between Indonesia and Timor-Leste has been ongoing since 2001, continuing up to the time this research was conducted in 2022. Negotiations preceding the determination of the 2019 land boundary began in 2001 between the Indonesian government and UNTAET, leading to the formation of the Technical Subcommittee on Border, Demarcation, and Regulation (TSC-BDR). When Timor-Leste officially became a sovereign state in 2002, unresolved segments were identified during surveys following the 2nd TSC-BDR meeting.

Since then, territorial boundary disputes have emerged in several segments (unsurveyed and unresolved), namely in the Subina–Oben segment, the Bijael Sunan–Oben (Manusasi) segment, the Dilumil–Memo segment, and the Noel Besi–Citrana segment. The Dilumil–Memo segment was resolved in the determination of land boundaries in 2013, with a decision to use the principle of the median line based on equal portions. The other three segments—Subina–Oben, Bijael Sunan–Oben (Manusasi), and Noel Besi–Citrana—were addressed in the determination of the land boundary in 2019 as a result of the meeting between the Chairmen of SOC Indonesia and Timor-Leste on July 22, 2019, which was outlined in the Agreed Principles on The Final Settlement of Land Boundary.

Specifically for the Noel Besi–Citrana segment, the Agreed Principles decided to use the principle of a simplified median line based on equal portions, similar to the settlement of the Dilumil–Memo segment in 2013. However, shortly after the 2019 land boundary determination, political dynamics in Timor-Leste led to a demand for a larger portion of the territory (approximately 74% for Timor-Leste and approximately 26% for Indonesia), rather than the equal portion agreed upon in the Agreed Principles. These demands from the Timor-Leste government resulted in a deadlock in the determination of land boundaries in the Noel Besi–Citrana segment. The Indonesian government rejected these demands and called for both parties to adhere to the 2019 Agreed Principles.

11.5 The Process of Land Boundary Negotiations Between Indonesia and Malaysia

Indonesian territory is the continuation of the colonial space of the Netherlands in Southeast Asia named the Dutch Indies. Its border with Malaysia covers a border region consisting of a land border in Kalimantan and Sebatik and a maritime border in the Malacca Straits and the Sulawesi Sea. Both parties accepted the result of the British and the Netherlands agreement on those border areas in 1824. They finally signed a bilateral border agreement in 1969. This treaty intends to resolve their dispute over the British occupation of Dutch colonial possessions in Southeast Asia (Wright 1950; Webster 1998).

The colonial legacy sprinkled border issues after both countries had taken independence from their colonial power. They involved border territorial disputes both in the land border area in Kalimantan and Sebatik islands and the sea border area in the Malacca Straits, the South China Sea, and Sulu-Sulawesi Seas (Ruhana and Karim 2024, pp. 1–32). Both parties had participated in a long negotiation series to demarcate their land border in disputed areas. They successfully put basic agreement as the strategic step of further negotiations when they finally signed a bilateral border agreement in 1969 (Djalal 2012, pp. 15–27).

The land border between Indonesia and Malaysia in Kalimantan is being re-measured by both nations after an 18-year pause. The boundary, originally set during the colonial period by the Dutch for Indonesia and the British for Malaysia, was initially satisfactory. However, both countries now interpret these historical boundaries differently. Specifically, the Kalimantan border (both West and North) stretches 2,062 km, marked by 20,329 boundary markers and monitored by 113 security posts (*Kompas*, 22 December 2022). An intriguing and challenging aspect is the interdependence between communities on both sides of the border, a relationship that has existed since the colonial era and persists today due to social, cultural, and economic ties. However, this situation also allows for illicit activities, which must be addressed under existing laws.

Nonetheless, border issues would be greatly reduced if the boundary agreement between Indonesia and Malaysia were finalized with clarity and certainty. This is technically challenging, as both sides conduct their own measurements, which must be aligned through joint efforts in the field. This process requires repeated collaboration until a mutual agreement is reached. The measurement process is also complicated by the landscape and topography of the border area, which includes rugged terrain and natural obstacles. Under President Jokowi's leadership, there has been a notable dedication to addressing border disputes that have lingered unresolved for nearly 18 years. However, before examining the particular areas that still need resolution, it's helpful to first grasp the overall profile of the Indonesia–Malaysia land border.

11.5.1 Profile of the Land Border Area

The Indonesia–Malaysia border context includes two main types: the terrestrial/land border (between West and East Kalimantan and Sarawak and Sabah) and the maritime boundaries (between Sumatra and Peninsular Malaysia and between North-East Kalimantan and Sabah). Both borders frequently experience fluctuating issues. These border problems often parallel the relationship between the two countries, resulting in dynamic border interactions.

This analysis focuses on the land border issues, which, when examined in detail, reveal strong social, economic, and cultural dynamics. The border areas also have landscapes and topographies that, empirically, are more complex than one might imagine.

Before exploring this land border in greater detail, here is a comprehensive overview of the land border in Kalimantan, including the areas it traverses. This information provides a foundational understanding or serves as a refresher for those familiar with these border issues.

The Kalimantan border stretches for 1,773.6 km, with 723.4 km in West Kalimantan, 223.6 km in East Kalimantan, and 826.7 km in North Kalimantan. This extensive route spans from the westernmost point at Temajuk to the easternmost point at Malinau. A parallel border road has been constructed adjacent to the national boundary, along with access roads leading to Border Crossing Posts (PLBN). These access roads are connected to the parallel border road, facilitating easier mobility for the local communities.

Table 11.1 presents various types of Border Crossing Posts in Kalimantan, classified into three categories: A, B, and C. These categories are determined based on six indicators: (1) **the monthly** number of people crossing, (2) **the daily** number of vehicles crossing, (3) the weight of each vehicle that crosses, (4) the size of the area, (5) the size of the core building zone, and (6) the size of the supporting zone. This article primarily focuses on land borders, in accordance with its main theme.

The Map 1 (**the left side**) shows that the yellow line running from the left end (Aruk) to the right end (Nanga Era) indicates the parallel border road. The blue line, on the other hand, marks the border access road that connects this parallel road to five Border Crossing Posts (PLBNs). As noted earlier, West Kalimantan hosts five PLBNs: Aruk, Jagoi Babang, Entikong, Jasa Sungai Kalik, and Nanga Badau. With these developments, road infrastructure significantly enhances the movement of people, goods, and services, and boosts trade between the two nations. Although there are no PLBNs in East Kalimantan (the right side), a national road has been

Table 11.1 Types and profiles of cross border post in Kalimantan

Types of CBP	Number of people passing by	Number of vehicles passing	Load per vehicle	Land area	Core zone building area	Area of supporting zona building
Land Border Type A	>7.500	<100	< = 40 Ton	10 Ha	> = 9.000 m^2	> = 5.000 m^2
Sea Border Type A	>7.500	–	–	–	As per the Port Master Plan	–
Land Border Type B	>1.000–<7.500	>30–<100	<=20 Ton	10 Ha	>=9.000 m^2	>=9.000 m^2
Sea Border Type B	>1.000–<7.500	–	–	–	As per the Port Master Plan	–
Land Border Type C	<1.000	<30	< = 20 Ton	10 Ha	>=9.000 m^2	> = 9.000 m^2
Sea Border Type C	<1.000	–	–	–	As per the Port Master Plan	–

Source https://peraturan.bpk.go.id/Details/264853/peraturan-bnpp-no-1-tahun-2021
Kepala Badan Nasional Pengelola Perbatasan Republik Indonesia, Peraturan BNPP No. 1 Tahun 2021 Tentang Tipologi Pos Lintas Batas Negara

Fig. 11.3 Map 1 of Border Routes in West Kalimantan (the left side) and East Kalimantan (the right side). *Source* Direktorat Jenderal Binamarga, Kementerian Pekerjaan Umum dan Perumahan Rakyat, 19 August 2021 retrieved from https://x.com/jalur5_/status/1428256840052207618/photo/3

constructed to connect with the parallel border road from **Tering** to **Long Pahangai** (see Fig. 11.3 indicated by the blue lines as National Road to border).

In North Kalimantan, there are four PLBNs: Long Nawang, Long Midang, Labang, and Seipancang. In total, there are nine PLBNs across Kalimantan. On the North Kalimantan side, the sea boundary between Indonesia and Malaysia is marked by several small islands near North and East Kalimantan. These islands include Sebatik, Nunukan, and Tarakan. Sebatik Island is the only one where a clear demarcation line separates the territories, with the northern portion belonging to Malaysia and the southern portion to Indonesia.

In the map of North Kalimantan above (see Fig. 11.4), the parallel border road from Long Boh to Malinau has yet to reach the northernmost point of North Kalimantan. Ideally, this road should extend from Malinau to the Labang Border Crossing Post (PLBN) and the Sei Pancang PLBN via border access routes. However, the road leading to the Long Midang PLBN is already accessible through border roads (refer to Map 3, marked in blue line). Several YouTube videos indicate that the road conditions between Malinau and Krayan are still incomplete. Notably, the Labang PLBN in North Kalimantan can be accessed by the river to reach Sabah (as shown in one of the Malinau Channel videos). The Integrated Labang PLBN is situated in a forested area on a hill, with an attractive design. During a recent visit, BNPP (BNPP channel, 1 month ago) mentioned that the Labang PLBN is planned to be developed as a key center for border area development, similar to other PLBNs.

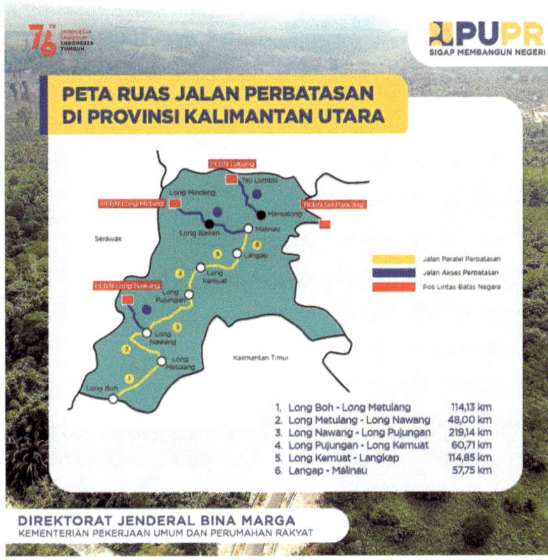

Fig. 11.4 Map 2 of border routes in North Kalimantan. *Source* Direktorat Jenderal Binamarga, Kementerian Pekerjaan Umum dan Perumahan Rakyat, Retrieved August 19, 2021 from https://x.com/jalur5_/status/1428256840052207618/photo/3

11.5.2 Outstanding Border Segments to Address

The most recent and difficult issue at the border is the possible shifting of border markers on the Kalimantan side. If this occurs, Indonesia would undoubtedly be disadvantaged. To prevent this, border officers will inspect the markers in the field and verify that they align with the designated coordinate points.

The biggest question on our minds of course concerns where the resolution process for the Outstanding Border Problem (OBP) between Indonesia and Malaysia. The goal is to create clear legal lines on the border and in theory reduce illegal crossings. We will get you up to date on what occurred and when it happened in the following timeline:

1. Seven unresolved border sections in 2020. In 2020, the Minister of Home Affairs who simultaneously shot BNPP Tito Karnivan announced that there are seven segments left deep enough. The latter became 3 matters to be settled in 2020 and the last four during 2021 and 2024. Additionally, two segments had already been agreed upon in November 2019 (Republika.news, March 11, 2020).
2. Following 2020, Harmen Batubara, noted that Malaysia had recognized the Simantipal segment and Segment C 500–C 600 in the Eastern Sector OBP. As a result, the issues in these segments have been resolved.
3. At the time, three OBP segments remained unresolved on the eastern side: 1. Sebatik Island, Sinapad-Sesai River, 2. Point B 2700, and 3. Point B 3100 near Taulumbis village in Lumbis subdistrict, Nunukan regency (see the explanation of these OBP in the next subsection). Concerning Sebatik Island, Foreign Minister Retno Marsudi stated that the land boundary, particularly from the West

Pillar to AA-2 (on Sebatik Island), was nearing completion in terms of technical measurements (Antara, July 3, 2024). The most recent update of the issue (based on an interview with BIG on September 25, 2024) indicates that this segment has been finalized.

According to BIG, resolving these segments involve a lengthy process of negotiations and agreements between the parties. To achieve a consensus, technical collaborations, such as surveys and measurements, are conducted by the respective institutions. For instance, BIG is tasked with mapping boundaries, including national, provincial, district/city, and village borders (BIG, https://www.big.go.id, July 4, 2024).

11.5.3 Snapshot of the Land Boundary Problem

The ongoing border disputes between these two nations are closely related to the legal agreements established by their historical colonial rulers, specifically the Netherlands and England. As explained by the Directorate of Regional Boundary Mapping and Toponymy in an interview on September 25, 2024, there are three legal documents created by these colonial powers that pertain to the land border between Indonesia and East Malaysia, including: (1) The Boundary Convention, which was signed in London on June 20, 1891; (2) The Boundary Agreement between the Netherlands and England signed in London on September 28, 1915; (3) The Boundary Convention between the Netherlands and England signed in The Hague on March 26, 1928.

Each of the legal bases mentioned above was established based on the efforts of re-translation (from 1891 to 1915 to 1928) by each party at that time, in relation to the possibility of boundary changes due to natural condition changes. The differences in interpretation of the legal provisions have led to issues concerning the 9 segments of the OBP. The resolution must be based on an agreement between the two countries, both politically and technically. To conduct the re-measurement, both countries utilized and applied a measurement system that is significantly more accurate than the colonial measurements (based on Astronomy). However, both the Dutch and British colonial powers, at that time, did not establish coordinate points for the division of border areas, yet both had already considered the possibility of determining those points for the future.

There are 9 OBP segments faced by both countries, and these 9 segments are divided into two, namely the State of Sarawak with West Kalimantan, and the State of Sabah with a small part of East Kalimantan and North Kalimantan (see the map below). Four segments are located in West Kalimantan, namely OBP D 400, OBP Gunung Raya, OBP Gunung Jagoi, and OBP Batu Aum (indicated by the yellow column). Meanwhile, five segments are located in North Kalimantan, namely OBP B2700-B3100, OBP Sinapad and Sesai, OBP Simantipal, OBP C500-C600, and OBP Sebatik (indicated by the blue column). During Joko Widodo's administration, the priority for the completion of these OBP segments located in the Eastern Sector has been established. Technically, the Eastern segments face a higher level of technical difficulty compared to the Western segments, which still have sparse populations and dense forests (as stated by BIG).

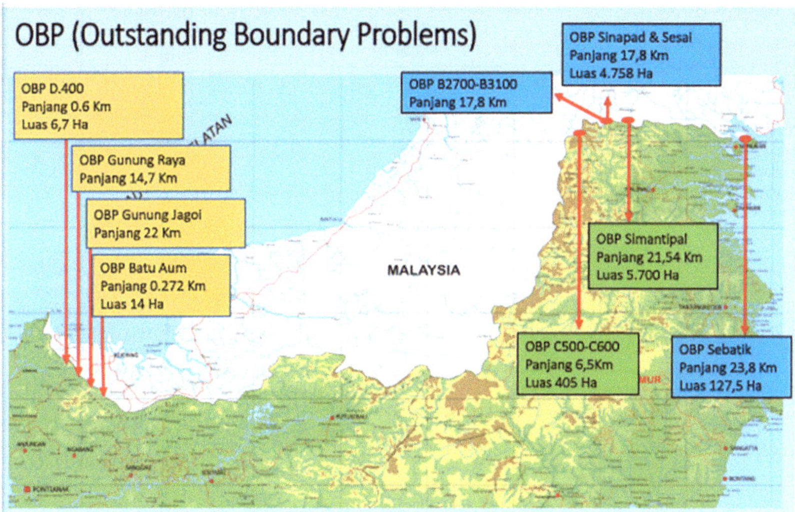

Source The Geospatial Information Agency (BIG), 2024. This map was directly acquired during a comprehensive interview with the Head of BIG in Cibinong on September 25, 2024

Out of the five segments, two (Simantipal and C500-C600) were agreed upon and completed in 2018. A technical agreement for the Sebatik area was established in 2019, while the other two segments (OBP B2700-B3100 and OBP Sinapad & Sesai) were technically finalized in 2023 and are now pending formal signing, expected to take place around November this year. According to BIG, the technical discussions with Malaysia have been more intensive than those with other countries (BIG Cibinong, September 25, 2024). The OBP segments in Sebatik extend from West to East, with Sinapad located at the easternmost point. Consequently, all five OBP segments on the Eastern side have been completely finished (OBP C500-C600 and OBP Simantipal in 2023), while the three segments OBP B2700-B3100, OBP Sinapad & Sesai, and OBP Sebatik have been technically completed and are now just waiting for the MoU to be signed by both countries, which will be done by the Foreign Ministers and witnessed by the Heads of State.

Addressing the Land Boundary Issues between Indonesia and Malaysia, number of related institutions of Indonesia have been involved and handled, both politically and technically actions. Figure 11.5 illustrates Indonesia's Organizational Framework for resolving issues through coordinated actions at the top, middle, and first levels. The Joint Border Committee (JBC), led by the Secretary General of the Ministry of Home Affairs, includes members such as BIG. The Joint Technical Committee (JTC), reporting to the JBC, is chaired by the Directorate of Defense Territorial Affairs from the Ministry of Defense and also includes BIG among others. The JTC collaborates

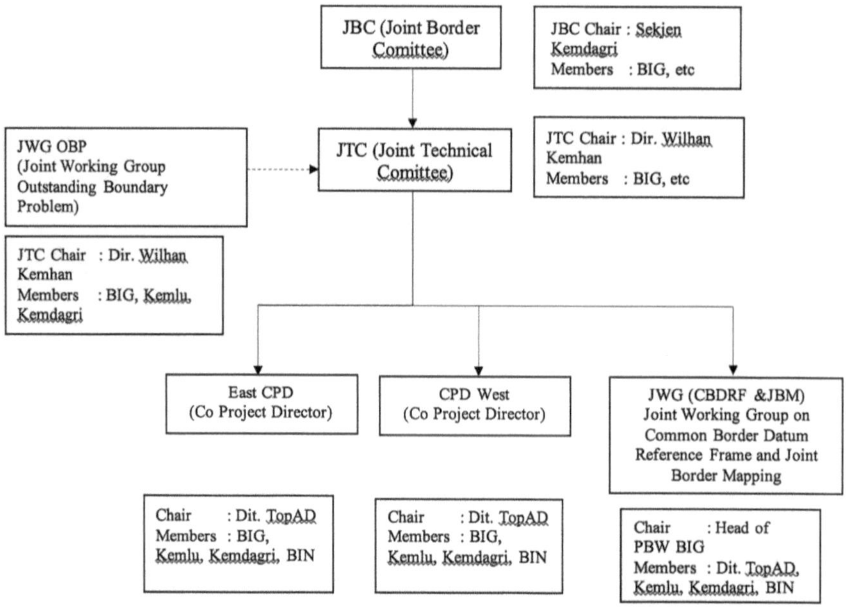

Fig. 11.5 Organizational framework for addressing the Indonesia–Malaysia Land border issues. *Source* The Geospatial Information Agency (BIG), 2024. This map was directly acquired during a comprehensive interview with the Head of BIG in Cibinong on September 25, 2024

with the Joint Working Group on OBP matters. Below the JTC, Co-Project Directors for the eastern and western segments, along with their teams, handle technical issues on an ongoing basis.

Despite the completion of the technical aspects and the ongoing preparations for the signing of the Memorandum of Understanding (MoU), it is imperative to address the socio-economic challenges present in the designated area, particularly in Sebatik. For example, there exists a residence where the backyard, which contains the kitchen, is situated within Malaysian territory, while the front yard lies in Indonesia. The boundary in Sebatik is delineated by a straight line that follows the 4°10' North Latitude, with the eastern side being densely populated. This boundary was established based on astronomical calculations conducted in 1928, and it is fortunate that the existing boundary markers are still well maintained. Both nations have reached an agreement to adhere to the 4°10' line as indicated by these markers. According to BIG, this North Latitude line is not actually straight but curves when re-measured using GPS due to differing coordinate systems. However, the reference points are based on specific pillars, known as the west pillar and the east pillar (see the map on the lower left). Between these two type A pillars, which are about 22 km apart, there are smaller type B and C pillars (Fig. 11.6).

The Map 4 on the right side illustrates the outcomes of the joint measurements conducted by Malaysia and Indonesia on Sebatik Island, indicating changes in land areas on both sides, with some regions experiencing increases and others decreases.

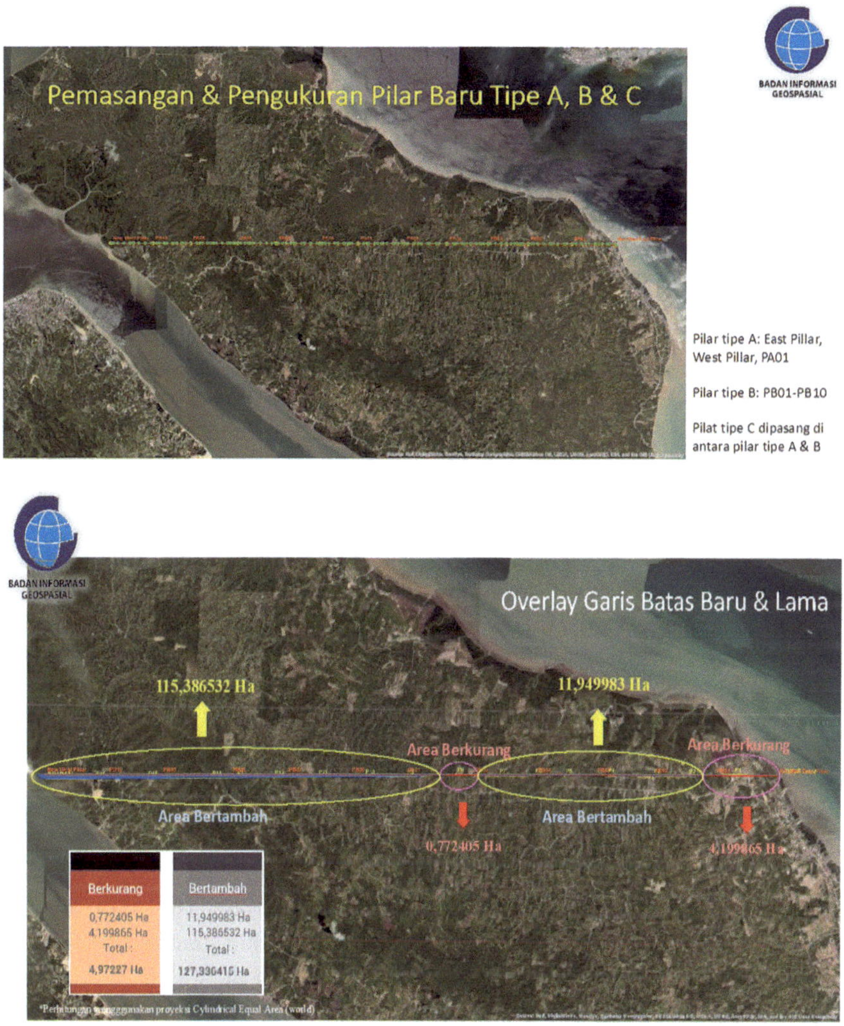

Fig. 11.6 Map 4 Land Boundary Line on Sebatik Island between Indonesia and Malaysia and the Agreement on Defining Land Boundaries. *Source* The Geospatial Information Agency (BIG), 2024. This map was directly acquired during a comprehensive interview with the Head of BIG in Cibinong on September 25, 2024

From Indonesia's viewpoint, there is a loss of land amounting to 4.97227 hectares in the central and right sections of Sebatik (marked by the red circle). In contrast, Indonesia has gained an additional 127.336415 hectares on the left and right sides of Sebatik (marked by the yellow circle). The implications of these joint measurement results extend beyond this, as the government, particularly the Ministry of Home Affairs, needs to address compensation for the land loss, which requires the establishment of legal regulations. The government is currently working on the presidential decree for this purpose. It is anticipated that this decree regarding legal regulations will be finalized before the end of President Joko Widodo's term. In Sebatik, the area with a high population density is situated on the eastern coast, while the western side is predominantly agricultural land. While the increase in land for Indonesia is not problematic, the reduction in land could potentially create socio-economic challenges for the residents of Sebatik.

11.6 Conclusion

Settling land boundaries between Indonesia–Timor-Leste, and Indonesia–Malaysia, is one of the goals of border diplomacy pursued by the Joko Widodo administration. The Indonesian government has made various diplomatic efforts to finalize several problematic segments of land boundaries, such as the Naktuka segment on the Indonesia–Timor-Leste border and the Sebatik Island segment on the Indonesia–Malaysia border.

For the Naktuka segment, a political dynamic emerged after the determination of the land boundary in 2019, influenced by Timor-Leste's domestic politics. The Timor-Leste government revealed that its parliament objected to the 2019 land boundary agreement, which used a simplified median line approach. They proposed instead that a larger portion of the territory—approximately 74%—be allocated to Timor-Leste, leaving around 26% for Indonesia. However, the Indonesian government opposed this proposal and remained committed to the simplified median line approach. As a result, a deadlock occurred, preventing the boundary determination in the Naktuka segment from advancing to the demarcation stage, in the form of installing boundary markers and preparing special arrangements.

Similarly, the Sebatik Island segment, one of the Outstanding Border Problems (OBP) on the Indonesia–Malaysia border, has been finalized. Therefore, it can be said that the Indonesian government during the Jokowi administration has completed technically the five OBP in the Eastern segment. One of these OBP, only Sebatik (West pillar to East pillar) is under MoU preparation. This could be signed in the end of 2024 under the Administration of the New Presiden Republic of Indonesia, Prabowo Subianto.

Tackling these issues is difficult, especially due to the technical challenges present on the ground, which are closely linked to possible future social impacts. Muh Aris Marfai, the Head of BIG, points out that one significant concern is the social consequences that may emerge following the establishment of the border demarcation between Indonesia and Malaysia. For example, the matter of several hectares of Indonesian land that has been incorporated into Malaysia needs to be addressed. It is crucial to have a legal framework in place to manage the compensation process for these situations. Caution is necessary in this approach, as the intricate nature of overlapping land ownership between the two countries could result in conflicts if not backed by suitable legal provisions. Therefore, it can be said that resolving problematic land boundary segments remains a task and challenge that should be addressed by the new government under the leadership of Prabowo Subianto.

References

Anggitta, Mutti. 2014. Kesepakatan Batas Darat RI-Timor Leste: Sebuah Kajian Diplomasi Perbatasan RI. *Jurnal Penelitian Politik* 1(1).

Anandra, Sahid Fadhil dan Kusumawardhana, Indra. 2023. Tapal Batas SangGaruda: Pendekatan Indonesia Dalam Diplomasi Dan Konflik Perbatasan Dengan Malaysia. *The Journalish: Social and Government* 4 (4). https://doi.org/10.55314/tsg.v4i4.686.

Chetail, Vincent. 2003. The contribution of the International Court of Justice to International Humanitarian Law. *International Review of the Red Cross (IRRC), 58*(850).

Deeley, Neil. 2001. The International Boundaries of East Timor. *Boundary and Territory Briefing* 3 (5).

Djalal, Hasyim. 2012. Border Diplomacy. In *Maritime Border Diplomacy*, ed. Myron H. Nordquist and John Norton Moore. Martinus Nijhoff.

Iwuoha, Victor Chidubem, and Casmir Chucwuka Mbaegbu. 2021. Border Governance and Its Complications in West Africa: What Can Be Learned from Constructivism? *Society*, vol. 58, September 3. Springer.

Malhotra, Aditi. 2022. Constructivist Approaches to the International Relations of South Asia. In *Routledge Handbook of the International Relations of South Asia*, 1st ed.

Mangku, Dewa Gede Sudika. 2017. Implementasi *Joint Border Committee* (JBC) untuk Penyelesaian Sengketa Perbatasan Darat antara Indonesia dan Timor Leste. *Jurnal Yuridis* 5 (1).

Nomleni, Kristin E.J. 2019. Dinamika Resolusi Konflik Wilayah Perbatasan Berbasis Budaya (Studi Kasus pada Aktivitas Komunikasi Pertemuan Adat "Nekaf Mese Ansaof Mese Atoin Pah Timor" Masyarakat Amfoang Timur Kabupaten Kupang NTT)". Thesis. Padjadjaran University.

Nurdin, Amin, Sudjito, Arsana, I Made Andi. 2023. The Unsettled Agreement of the 2019 Re-demarcation of the Indonesia-Malaysia Boundary Line's Impact. *Indonesian Journal of Multidisciplinary Science*. E-ISSN: 2808-6724. ijoms.internationaljournallabs.com.

Nordquist, K.-A. 2010. *Boundary Conflicts and Preventive Diplomacy*. Sweden.

Ogunnubi, O., and E.O. Awosusi. 2021. Nigeria's Border Diplomacy: Rhetoric or Substance for Regional Hegemonic Leadership? *Global Society* 36 (4).

Ruhana, Siti, and Tun Abdul Karim. 2024. Indonesia vs Malaysia: The Battle for Border Territory Resolved. *International Law Discourse in Southeast Asia* 3 (1).

Shrestha, B.N. 2021. Diplomacy in the Perspective of Boundaries. *Journal of Foreign Affairs* 1.

Ullah, A.K., and A.A.-Z. Kumpoh. 2018. Are Borders the Reflection of International Relations? Southeast Asian Borders in Perspective. *Journal of Asian Security and International Affairs* 3 (3).

Webinar Border History: Melacak Batas Indonesia–Timor Leste Hingga ke Portugal. 2020. https://www.youtube.com/watch?v=uB2weUW34R0&t=2502s.

Webster, Anthony. 1998. *Gentlement Capitalists: British Imperialism in Southeast Asia*. Tauris Academic Studies.

Wright, H.R.C. 1950. The Anglo-Dutch Dispute in the East, 1814–1824. *Economic Historical Review* 3 (2).

Indriana Kartini is a Senior Researcher at the Research Centre for Politics, the National Research and Innovation Agency (BRIN), Indonesia. She was the Cluster Coordinator for Foreign Policy and International Issues. She obtained her Master of International Politics from the Faculty of Arts, The University of Melbourne, Australia in 2008. She received her Ph.D from the Political Science Study Program, at the University of Indonesia in 2023. Her research interests are international politics, international border, sub-regionalism and geopolitics. In 2017, she was awarded a Research Grant from the Sumitomo Foundation for a project on Japan's public diplomacy in Indonesia. She is the editor and contributor of academic books, including Potret Politik Luar Negeri Indonesia di Era Reformasi (Indonesian Foreign Policy's Portrait in the Reform Era) (Yayasan Pustaka Obor, 2019); Penguatan Konektivitas Lintas Batas dalam Kerja sama Ekonomi Subregional (Strengthening Cross Border Connectivity in Subregional Economic Cooperation (Yayasan Pustaka Obor, 2021)

Awani Irewati is a Senior Researcher at the Research Center for Politics, the National Research and Innovation Agency (BRIN), Indonesia. She earned her Bachelor's degree in International Relations from the Faculty of Social and Political Sciences at Airlangga University in 1987, where she wrote a scientific paper on Japan's Military Defense for her final project. She completed her Master's degree in Asia and International Studies at Griffith University, Queensland, Australia in 1994, with a thesis on Japan's role in APEC. Her research has primarily focused on inter-country borders, especially within ASEAN and Indonesia. Awani has coordinated and edited several research books, including Border Dispute between Thailand and Cambodia (ANDI Publisher, 2015), Dynamics of Subregional Cooperation in Southeast Asia: Greater Mekong Subregion (LIPI Press, 2018), and Mechanisms for Resolving Border Disputes in Southeast Asia (Mahara Publishing, 2017). She has authored numerous scientific papers for journals and international seminars and is a frequent speaker on border issues. In 2021-2022, she was awarded the Sumitomo Foundation program, during which she led a project on "the development of an Integrated Marine and Fisheries Center to support the resilience and sovereignty of the Natuna Islands", focusing on Japan's assistance in building the SKPT in Selat Lampa, Natuna. She authored a book chapter titled "Resilience and Preparedness Among Coastal Communities on the Natuna Islands", with Sjarief Widjaja and Imam Syafi'i as co-authors. This chapter was included in the book Climate Crisis, Social Responses, and Sustainability, published by Springer in 2024 and edited by Uttam Mukhopadhyay, Subhasis Bhattacharya, et al.

Agus R. Rahman worked as a prospective researcher in 1988 when the organization was still called the Indonesian Institute of Sciences (LIPI) and has continued his role until now in 2025 as a Principal Expert Researcher at the Research Center for Politics, the National Research and Innovation Agency (BRIN), Indonesia. In 2019, he also served as the Coordinator of the International Research Group. Between 2000 and 2019, he was a non-permanent lecturer in the International Relations Study Program at the Faculty of Social and Political Sciences, Prof. Dr. Moestopo University (Beragama). His research over the last five years has focused on border connectivity in subregional cooperation within the Southeast Asian region.

Chapter 12
Containing the Internationalization of Papua and Protecting Indonesian National Integrity

Johni R. V. Korwa, Nicholas D. Metherall, Petrus K. Farneubun, and Made Selly Dwi Suryanti

Abstract This chapter examines how Papua's internationalization was managed throughout the ten years of President Jokowi's administration. Within this chapter, the concept of internationalization is defined as a process in which awareness of issues within Papua is expanded beyond a domestic scale to an international scale. By expanding awareness overseas, the internationalization of issues in Papua has served as a springboard for advocates seeking to call upon international support for Papua's quest for self-determination, which we conceptualize as "traditional internationalization," as well as progress toward the wider quality of life objectives for the Indigenous people of Papua, which we call "contemporary internationalization." The conceptual frameworks of internationalization can be understood as an active or politically motivated phenomena and thus distinct from wider theoretical frameworks under globalization. Under the Jokowi administration, diplomacy has sought to contain internationalization (both traditional and contemporary) through constant advocacy in multilateral fora—such as the United Nations General Assembly—by calling upon other countries to adhere to the norms of state sovereignty and non-interference in Indonesia. At the regional level, since joining the Melanesian Spearhead Group as an associate member in 2015, Jokowi's administration has focused on strengthening foreign relations and obtaining support for Indonesia's national integrity over Papua. Jokowi's government has also enhanced its bilateral ties with Papua New Guinea (PNG) to gain support for its efforts to protect the territorial integrity of the Indonesia–PNG border. At the domestic level, Jokowi's regime has undertaken several reforms frequently highlighted in contemporary internationalization, such as infrastructure, social and economic development, the partitioning of Papua (*pemekaran*), and the amendment of the Special Autonomy Law. Despite these reforms, his administration has faced criticism for its inability to address human rights

J. R. V. Korwa (✉) · P. K. Farneubun · M. S. D. Suryanti
Cenderawasih University, Papua, Indonesia

N. D. Metherall
The Australian National University, Canberra, Australia

The University of the South Pacific, Suva, Fiji

I. Kartini and Kanisa (eds.), *Joko Widodo's Foreign Policy and Diplomatic Legacy*,
https://doi.org/10.1007/978-981-95-2085-5_12

abuses, entrenched inequality, and the protracted conflict. Further critics have claimed that physical infrastructure and policies like *pemekaran* and special autonomy do not truly represent the will of Indigenous Papuans. These examples are seen as lingering and unanswered issues to be left for the next administration.

Keywords Internationalization of Papua · Indonesia · Jokowi · Territorial Integrity

12.1 Introduction

The process of internationalization of Papua has spanned from the earliest days of integration into the Dutch East Indies to the present day. Amidst the external pressures of internationalization, this chapter examines President Joko (Jokowi) Widodo's efforts to uphold Indonesia's national integrity over Papua.[1] The February 2024 presidential election signaled the closing of Jokowi's term as head of state. However, there are several lingering issues surrounding Papua that have not yet been fully addressed by the Jokowi Administration and thus remain a concern for the incoming Indonesian government; in particular, the phenomena of internationalization. Within this chapter, internationalization is defined as a process in which awareness of issues within Papua is expanded beyond a domestic scale to an international scale. By expanding awareness overseas, the internationalization of issues in Papua may serve as a springboard for advocates seeking to call upon international support for Papua's quest for self-determination. Thus, internationalization implies an active process whereby international awareness also increases international political pressure on Indonesia.

The political impetus driven by the increasing pressure of internationalization distinguishes it from the wider pervasive processes of globalization. We contend that the original goal of this internationalization of Papua was to gain international support for Papuan independence or *Merdeka.* Yet since *Reformasi* (Indonesia's reformation era), the process of internationalization has gradually broadened to encapsulate wider objectives. For example, international awareness of issues in Papua may also serve as a frame to encourage other countries and organizations to exert pressure on the Indonesian government, to make more incremental changes improving upon its track record of poor treatment of Indigenous Papuans. This could involve addressing several human rights violations, finding appropriate conflict resolution measures, de-escalation of conflict between the Indonesian military and the separatist movement, and fostering prosperity in the region. Thus, the chapter also conceptualizes the distinction between "traditional internationalization" for *Merdeka* or Independence from "contemporary internationalization" which focuses on addressing wider quality of life issues and specific problems faced by Indigenous Papuans. This shift in the

---

[1] In this chapter, the term "Papua" (also often referred to as "West Papua") refers to Indonesia's easternmost region, which has been administratively divided into six provinces: Papua, West Papua, Highland Papua, Central Papua, South Papua, and Southwest Papua.

political direction of internationalization has occurred over time as part of a contemporary re-orientation of both local and international advocates toward campaigning for more pragmatic and politically palatable piecemeal changes rather than a broader overarching goal of *Merdeka*.

On this matter, a series of questions emerge, "How did the Jokowi administration contain the internationalization of Papua both regionally and internationally?" and "What did Jokowi's government do during his tenure to maintain Indonesia's national integrity over Papua?" We take questions such as these as our starting point and discuss them under the conceptual framework of traditional and contemporary internationalization. It is our contention that despite the issues surrounding Papua gaining international attention, the Jokowi administration has been relatively effective in containing these issues on a domestic, regional, and global level over the ten years of his leadership and relative to preceding governments in Indonesia. This relative effectiveness can be judged from both the policies and programs implemented with implications for Papua as well as the relatively muted levels of international pressure on Indonesia in relation to issues in Papua when compared with previous more militant administrations. Jokowi's administration has indeed placed special emphasis on regional development in Papua. It has actively participated in multilateral and regional institutions to combat the internationalization of Papua and strengthened bilateral ties with Papua's only bordering neighboring country, Papua New Guinea (PNG), for the protection of its integrity. Yet, some critical issues, such as human rights abuse, poverty, inequality, and conflict resolution based on peaceful dialogue. These issues have not been adequately addressed throughout Jokowi's presidency. The chapter evaluates the administration of Jokowi in relation to the issues surrounding Papua during his tenure.

This chapter is outlined as follows: first it provides background on the internationalization of the issues in Papua since the New York Agreement. This is followed by an examination of the diplomatic measures taken by the Jokowi government to minimize international attention of the issues in Papua abroad. The chapter then examines how Jokowi's administration has also sought to protect Indonesia's sovereignty over Papua through strengthening relationships with neighboring PNG and discusses Jokowi's legacy in Papua within the broader context of Indonesia's national integrity.

12.2 Traditional and Contemporary Internationalization of Papua

The awakening of an international consciousness of issues in Papua could be attributed to the events precipitated from the New York Agreement in 1962. The Agreement set out the terms leading to the transfer of administration of Papua from the Netherlands to Indonesia. At the time, this transfer was designed to be part of a temporary stage of transition toward self-determination for the Papuan people. On 1 October 1962, the United Nations Temporary Administrative Authority (UNTEA) took over the interim government administration of Papua from the Netherlands. The UNTEA subsequently handed over its administration to Indonesia on 1 May 1963, after which the Indonesian government was responsible for preparing the Papuans to exercise self-determination by 1969—an event well known as the "Act of Free Choice," or *Pepera* (*Penentuan Pendapat Rakyat*) in Indonesia (Drooglever 2009).

Stipulated within the New York Agreement, *Pepera* would function as a form of referendum, overseen by the United Nations, to determine the final status of governance for Papua. Yet many critics argued that the UN oversight was not sufficient to ensure the fairness and democratic integrity of the Act of Free Choice (see Saltford 2003; Viartasiwi 2018; Webster 2013). While there have also been opposing views claiming that *Pepera* was sufficiently representative to provide a legitimate decision that was further solidified through audience and resolution with the UN General Assembly, most published works within the wider discourse, commonly expressed perceptions of an "Act of No Choice' with shortcomings including insufficient international oversight, coercion, compromised local representation and wider intransparency within *Pepera* (see Drooglever 2009; Saltford 2003; Webster 2013).

In this regard, the New York Agreement was perceived as a starting point for traditional internationalization of Papua. This form of traditional internationalization was often driven by anti-integration groups who sought to raise international awareness and gather international support for independence, as they vehemently opposed the agreement that transferred control of Papua from the Netherlands to Indonesia (Elisabeth 2006). Most importantly, these anti-integration groups initiated the internationalization of the Papua issue to challenge the results of the *Pepera*, which was considered unfair due to participants being pressured to vote unanimously in favor of Indonesian control (Chauvel 2004). It is noteworthy that one of the recommendations from the Second Papuan People's Congress in 2000 in Jayapura was to internationalize the Papua *Merdeka* issue by forming a team to lobby the international community, including requesting the UN Security Council to maintain peace in Papua until a permanent government was established (Elisabeth 2006). These events indicate how traditional internationalization of the Papua issue has been an integral part of the Papuans' long-standing struggle for independence from the beginning.

Over the past decades, since the New York Agreement and *Reformasi*, there has been an expansion in the scope of attention from the international community toward Papua. This has led to a shift beyond traditional internationalization for *Merdeka* toward contemporary internationalization pragmatically targeting more specific issues. In particular since the paradigm shifts in governance associated with *Reformasi,* there have been a wide range of examples of international criticisms of the management of Papua under Indonesian governance (Tebay 2005; Taliawo 2022). This mismanagement of Papua has even been characterized as "reactionary hypernationalism on the one hand, and a rhetorical anti-colonial internationalism on the other" (Wangge and Lawson 2023). Wider examples of international criticisms have included the human rights violations, inciting conflicts, heavy handed practices from the Indonesian military and Indonesia's Special Forces (Kopassus), environmental exploitation, and wider racism and inequality. Within the frame of contemporary internationalization, these criticisms functioned to exert pressure on the Indonesian government to pay greater consideration to addressing these complaints and in doing so, increasing the quality of life for Papuan peoples. Hence, the chapter distinguishes traditional internationalization from contemporary internationalization. While traditional internationalization has emerged since the New York Agreement, contemporary internationalization has expanded since the period of *Reformasi* and a clear response to this contemporary internationalization has been demonstrated under the leadership of the Jokowi Administration.

The criticisms and drivers of contemporary internationalization have originated from three major groups of international actors: (1) civil society, (2) nation-states, and (3) intergovernmental organizations. An alphabetically ordered list of some of the examples of international actors who have made at least one statement expressing concern over the issues of human rights and civil liberties in Papua is included in the tables below. While noting that this synthesis is not exhaustive, it already lists 29 civil society actors, 8 intergovernmental organizations, and at least 20 nation-states.

For instance, the calls for elected parliamentarians who have signed the Westminster Declaration relates to self-determination and the illegitimacy of *Pepera* (traditional internationalization) as well as human rights violation (contemporary internationalization). Building upon this, the International Parliamentarians for West Papua (IPWP) who have been gathering international support have called upon the Office of the United Nations High Commissioner for Human Rights (OHCHR) to visit Papua within the Brussels Declaration (IPWP 2024). These examples clearly demonstrate how internationalization has the capability to exert pressure on the Indonesian state. It is also worth noting the considerable role of the councils of churches, listed under civil society actors, with ties at the community level both in Papua and abroad. On 21 September 2002 on the UN International Day of Peace, faith leaders from all the main religions of Indonesia gathered in Jayapura to pray for peace and subsequently identified 5 February as an annual Papuan Day of Peace signaling how international networks may influence local activism (contemporary internationalization) (Tebay 2005).

The list of actors that have made statements advocating for *Merdeka* or Independence is far more limited indicating that international actors are more likely to advocate for smaller pragmatic issues (contemporary internationalization) rather than risk diplomatic ties through traditional internationalization calling for Papua's independence (traditional internationalization) (Tables 12.1, 12.2, and 12.3).

In the present day, civil resistance movements and advocacy for self-determination in Papua continue to demonstrate grassroots support. Many of these calls and demonstrations have garnered international attention and support from human rights organizations and solidarity groups. These movements, mainly driven by Papuan pro-independence activists, have also used social media such as Facebook, Twitter [X], YouTube, Instagram, and other online platforms to internationalize the Papua issue and gather international sympathy (Lantang and Tambunan 2020). Furthermore, there have been reports concerning the human rights abuses in Papua, such as those from Human Rights Watch (2023), Amnesty International (2018), and the U.S. Embassy and Consulates in Indonesia (2023). Some of these reports have gained international attention through referring to historic human rights violations. Yet these have

Table 12.1 Examples of Civil Society Organizations (CSOs) who have issued statements of concern about Papua

Civil Society Organizations (CSOs)
Amnesty International
Asia Pacific Centre for the Responsibility to Protect (University of Queensland)
Asian Human Rights Commission (AHRC)
Australia West Papua Association (AWPA)
Forest Peoples Programme (FPP)
Fransiscans International
Free West Papua Campaign
Human Rights Watch
International Academics for West Papua (IAWP)
International Crisis Group (ICG)
International Federation for East Timor (IFET)
International Lawyers for West Papua (ILWP)
International Parliamentarians for West Papua (IPWP)
International Service for Human Rights (ISHR)
International Work Group for Indigenous Affairs (IWGIA)
Pacific Islands Association of NGOs (PIANGO)
Papuan Student Alliance (AMP)
Papuan Voices
Peace Brigades International (PBI)
Rainforest Foundation (UK)
Survival International
TAPOL: UK based NGO
The Centre for Peace and Conflict Studies (Sydney University)
The Indonesian Human Rights Monitor (IMPARSIAL)
The International Coalition for Papua (ICP)
West Papua Action Network (WPAN)
West Papua Network (WPN)
West Papua Project (University of Woolongong)
World Council of Churches

rarely made headlines often remaining as 'media blind spots' as seen with cases of democide: Ballier (1977), Biak (1998), Theys Eluay (2001), communities in Wasior (2001), Wamena (2003), Paniai (2014) and others (Brundige et al. 2004; Widjojo 2006; McGibbon 2006; Leadbeater 2008; Purnomo 2022). In contrast, international attention to these issues has often peaked with greater headline-level attention when foreign nationals have been drawn into the conflict. One of the most recent stories to reach mainstream media was about a New Zealand pilot Phillip Mehrtens being

Table 12.2 Examples of Intergovernmental Organizations (IGOs) that have issued statements of concern about Papua

Intergovernmental Organizations
African, Caribbean and Pacific (ACP) Group of States
ASEAN Parliamentarians for Human Rights
European Parliament
Melanesian Spearhead Group (MSG)
Non-Aligned Movement (NAM)
Pacific Island Countries and Territories (PICTs)
Pacific Islands Forum (PIF)
United Nations (The Office of the High Commissioner for Human Rights)

Table 12.3 Examples of individual nation-states that have issued statements of concern about Papua

Individual nation-states and multilateral groupings of nation-states
Australia
Canada
New Zealand
United Kingdom
United States
15/22 individual nation statements from PICT members. The exceptions include the French Territories (New Caledonia, Wallis and Futuna, Tahiti, French Polynesia) and the US Territories (American Samoa, Guam, and the Northern Mariana Islands)

Notes The gray literature sources for Tables 12.1, 12.2, and 12.3 were derived from statements shared by each of the organizations on their websites or in reports

taken hostage by the National Liberation Army in Papua to be used as a bargaining chip to push for independence from Indonesia (Beazley 2024). Having spent over one and a half years in captivity in Indonesia's Papua region, and following critical diplomatic efforts by Wellington and Jakarta, Mr. Mehrtens was eventually released on 21 September 2024 (Fatima 2024). These examples signal how both perceived human rights abuses as well as widening conflicts and the capture of an NZ citizen have drawn significant international attention and criticism of Indonesia in relation to Papua. All of these concerns, such as human rights abuses and widening conflicts, have not only contributed to the internationalization of the Papua issue since the New York Agreement up to the present day, but have also drawn the attention of Indonesian presidents over time, including Jokowi. The next section highlights the efforts of the Jokowi administration to contain the internationalization of the Papua issue.

12.3 Jokowi's Diplomatic Efforts in Containing the Internationalization of Papua

Throughout the ten years of Jokowi's presidency, from 2014 to 2024, relative to previous administrations, Indonesia's foreign and domestic policy placed a greater emphasis on three areas: (1) development in Papua, (2) containing internationalization of Papua, and (3) suppressing the political independence movement. The nature of this policy agenda sought to appease local dissenters in Papua, contain internationalization of Papua and to ultimately defend territorial integrity and sovereignty against ongoing calls for self-determination. This marked a significant shift from Indonesia's previous policy, which neglected development in Papua.

Political dynamics in Papua compelled Indonesia to actively engage with foreign actors. Indonesia used a wide range of diplomatic strategies and participated in multilateral and regional institutions to reaffirm its sovereignty over Papua. The United Nations General Assembly (UNGA) served as a prominent platform for Indonesia to influence perceptions of its political stance toward Papua and refute any negative perceptions about the country. Citing the international principle of non-interference, state sovereignty, and territorial integrity, Indonesian leaders, particularly the President, Vice President, and Minister of Foreign Affairs, attempted to thwart the *Merdeka* internationalization of Papua. Article 2 (7) of the UN Charter specifies that states may not "intervene in matters which are essentially within the domestic jurisdiction of any state." Indonesia regularly invoked this clause to assert its sovereignty over Papua. For instance, although Jokowi did not address separatism and Papua during his speech at the UNGA, he did voice his profound concern about the frequent undermining of the sovereignty and territorial integrity principles of the UN Charter and international law (United Nations 2020a, 2021).

Jokowi was not a foreign policy president, and less active in global diplomacy, unlike his predecessor, President Yudhoyono, who frequently made appearances at international fora. Jokowi was frequently absent in giving speeches in such international fora and notably the UNGA. According to his diplomatic record, he only made two virtual appearances at the UNGA—during the 75th session in 2020 and the 76th session in 2021—during his presidential terms (both coinciding with COVID-19). Due to the epidemic, Jokowi was never physically present. This was also closely linked to his lack of training and lack of interest in the field. Jokowi, according to Connelly (2014, 2015), had little to no prior diplomatic experience. He was more preoccupied with domestic policies than with foreign policy.

Jokowi mostly delegated the responsibility for addressing the UNGA to First Vice president Jusuf Kalla, and his Minister of Foreign Affairs, Retno Marsudi. The Foreign Affairs Minister participated extensively in the Security Council and Human Rights Council sessions, in addition to UNGA debates. In their speeches at the UNGA, both representatives emphasized the need for mutual respect for states' sovereignty and territorial integrity. They did not specifically address the Papua *Merdeka* issue either (see United Nations 2018, 2019, 2022, 2023). For example, in one of his speeches, Kalla strongly emphasized his point that, "respect for sovereignty [of other countries] is the most fundamental commitment and the principle that underpins the relations among states. The commitment should be respected" (United Nations 2019), while along a similar vein Marsudi asserted "the fundamental principles of sovereignty and territorial integrity are non-negotiable" (United Nations 2022).

Despite not addressing the controversy around *Merdeka* in Papua explicitly, the reference to these norms implied the governments agenda to contain any forms of traditional internationalization emerging at these high-level talks. Indonesia has maintained that Papua has been an integral and indivisible part of Indonesia since 1969, and any necessary measures by foreign countries to support political independence of Papua has always been viewed as an infringement on Indonesia's domestic affairs and a violation of that state's sovereignty, as specified in the UN Charter. The Indonesian government's international diplomatic efforts toward Papua were evidently focused on safeguarding the country's territorial integrity and impeding its internationalization by referencing principles of state sovereignty and non-interference. If the country's President and Vice President did not mention the issue of Papua in their speeches, junior Indonesian diplomats brought up the issues in their remarks at the UNGA, maintaining Indonesia's sovereignty over Papua. To counter narratives that attempted to damage the government, Jokowi depended on these young diplomats. In fact, these young diplomats spoke for Indonesia in the majority of UNGA debates against foreign leaders.

One of the most frequently debated issues raised by other countries during UNGA sessions was human rights abuses in Papua. These countries were mostly Pacific nations, including Vanuatu, the Solomon Islands, the Republic of the Marshall Islands, Nauru, and Tuvalu. Indonesia vehemently disagreed with the Pacific countries' remarks about human rights violations, "We categorically reject the continuing insinuations in their statements, which clearly reflect an unfortunate lack of understanding of the history, current situation, and progressive developments in Indonesia."—a diplomat declared at the 71st UNGA session in 2016 (United Nations 2016). Another instance occurred on September 26, 2020, when an Indonesian diplomat exercised its right to a first reply at the 75th session of the UNGA in response to Vanuatu's statement on human rights violations in Papua. The diplomat denounced Vanuatu for what they described as political meddling in Indonesian domestic affairs. According to the diplomat, Vanuatu has violated the principle of non-interference as contained in the UN Charter and acted "under the guise of artificial human rights concerns" (United Nations 2020b).

Aside from the UNGA, another platform that Indonesia has actively used to affirm its sovereignty over Papua and contain Papua's internationalization is the Melanesian Spearhead Group (MSG). It has long been known that Papua belongs to Melanesia. However, as part of Indonesia's attempt to seek membership in the MSG in 2015, the Jokowi administration put forth a new claim that it possessed five Melanesian provinces: Papua, West Papua, Maluku, North Maluku, and East Nusa Tenggara (Lawson 2016). This assertion has undeniably altered the geopolitical landscape of Melanesian nations in the Pacific. This new claim also led to a historic decision, with Indonesia being granted Associate Member status at the 20th MSG Leaders' Summit in Honiara in 2015—just one year after Jokowi came into power. At the same time, the MSG leaders granted observer status to the West Papuan umbrella organization, the United Liberation Movement for West Papua (ULMWP) as representative of Papuans outside the country (Lawson 2016). As of 2024, the Jokowi government has established three new provinces in Papua, thereby bringing the total number of Indonesian Melanesian provinces to nine: Papua, West Papua, Highland Papua, Central Papua, South Papua, Southwest Papua, Maluku, North Maluku, and East Nusa Tenggara (Fig. 12.1).

As attention turned to the regional stage, Indonesia has engaged in economic diplomacy with practically every nation in the Pacific. Indonesia has developed strategic bilateral cooperation in a variety of critical areas, including trade, investment, and border management with the aim to garner support for its sovereignty over Papua. To increase engagement and show that it was dedicated to improving the standard of living for the communities in the region as a member of the Pacific families, Indonesia unveiled the Pacific Elevation vision in 2019 (Ministry of Foreign Affairs of the Republic of Indonesia 2022). According to Retno Marsudi, "Indonesia's diplomacy in the Pacific is entering a new era with the vision of Pacific Elevation" (Ministry of Foreign Affairs of the Republic of Indonesia 2024).

Furthermore, Indonesia adopted cultural diplomacy to maintain its presence and sustain its influence in the region by actively participating in the regional forums

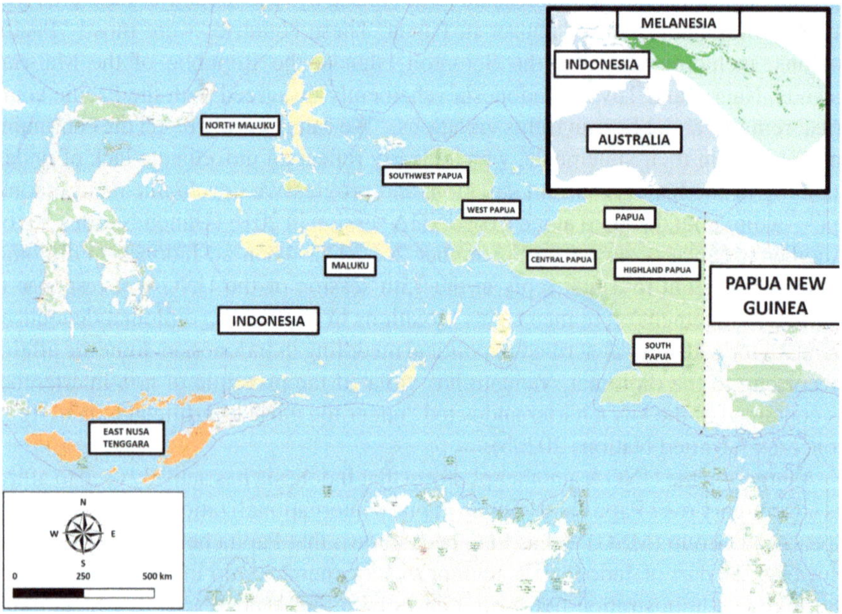

Fig. 12.1 Map of Melanesia with the addition of the nine Indonesian provinces considered for inclusion in the Melanesian Spearhead Group within the proposal from the Jokowi Administration. *Source* Digitized by authors using Quantum Geographic Information Systems (QGIS) Software

such as the Melanesian Spearhead Group (MSG), the Pacific Islands Forum (PIF), and Organisation of African Caribbean and Pacific States (OACPS). Furthermore Indonesia has sponsored Melanesian festivals. Indonesia is a dialogue partner in the PIF. Indonesia's roles in these fora have served to contain the influence Papuan separatism (traditional internationalization). Given Indonesia's claim over its numerous Melanesian provinces (Fig. 12.1), its active involvement in these fora was closely tied to this claimed geo-cultural identity. This cultural affiliation has an advantage for Indonesia as it may enhance Indonesian political bargaining power in the Pacific. Under the Jokowi administration, Indonesia often used this cultural factor to seek collective support for its territorial integrity and limit the ability of the ULMWP to voice its political goal. Indonesia has benefited from MSG participation by preventing ULMWP from becoming a full member.

Another strategy to highlight in the Jokowi administration's efforts to contain the internationalization of Papua is humanitarian diplomacy. It was recently introduced by Indonesia to its repertoire of strategies to win over the nations of the Pacific. Although Vanuatu, the Solomon Islands, Tuvalu, and Nauru continued to support Papua's right to self-determination and criticized Indonesia for not doing enough to improve the region's human rights condition, Indonesia nonetheless gave financial help to these countries (Wangge and Lawson 2023). As part of its humanitarian diplomacy, Indonesia also dispatched aid to Vanuatu in 2023 to help with relief after the damage of tropical cyclones: Kevin and Judy (Ministry of Foreign Affairs of The Republic of Indonesia 2023a).

Indonesia's heightened engagement was prompted by the profound shift in the Pacific region's diplomatic approach, as countries gave precedence to decolonization diplomacy (Fry and Tarte 2015; Melanesian Spearhead Group 2009), as well as the inclusion of Papua on the agenda of regional fora. Thus, the two approaches (cultural and humanitarian diplomacy) are also the main goal of forging strong ties with the Pacific nations and participating in the regional fora has been to ensure support for Indonesia's territorial integrity and deter support for the Papuan separatist movement. The next section highlights Indonesia's efforts to maintain sovereignty over and contain the internationalization of Papua, by strengthening its relationship with the only Pacific nation that borders Indonesia.

12.4 Strengthening Indonesia's Sovereignty over Papua Through Its Neighbor: PNG

Papua remains the only region in Indonesia where security forces are still confronting an active separatist movement. Indonesia's treatment of Papua is also being closely monitored by the international community due to the rapid spread of media in today's globalized world, as well as reports from Papuan local churches, NGOs, and activists. This activism is also mirrored at the grassroots level in neighboring PNG, which has historically showed some selective support for Papua's struggle for independence

(Kabuni 2021). PNG is one of Indonesia's three neighboring countries, along with Malaysia and East Timor, that share land borders. PNG shares Melanesian culture and Christianity with Papua as part of the cultural and historical context of New Guinea Island. Indonesia's relationship with PNG has been attributed great importance by Jokowi's administration for three reasons.

Firstly, PNG has been one a key strategic destination for Papuan freedom fighters, crossing the porous border and often settling there. It is also considered a safe haven for Papuan civilians fleeing the conflict between Indonesian armed forces and the *Organisasi Papua Merdeka* (OPM—Free Papua Movement) (Chauvel 2007). Even though some Papuan civilians have returned to Papua over the past decade, either voluntarily or through repatriation, the border remains a sensitive region difficult to control as part of a wider military operation to suppress insurgency against Indonesian rule. Reinforcing recognition of the Papua–PNG border also serves as a geopolitical mechanism to support Indonesian territorial integrity. Therefore, fortifying the border with PNG has become a critical aspect of Jokowi's foreign policy. His decentralization policy of "*Nawa Cita*" envisions development of Indonesia from the periphery. This has extended his domestic policy to frame his foreign policy in the context of border relations. The construction of the *Pos Lintas Batas Negara* (PLBN—Transnational border posts) exemplifies his efforts to secure the border area, transform it into a center of economic growth, and foster connectivity between Indonesia and PNG (see Metherall et al. 2022; Mandibondibo et al. 2022). Currently, three PLBNs have been inaugurated in Papua: PLBN Skouw in Jayapura, PLBN Sota in Merauke, and PLBN Yetetkun in Boven Digoel with plans for additional PLBNs in the remaining two Papua regencies bordering PNG, including Keerom and Pegunungan Bintang.

Securing the border area has also been a focus of a recent New Defence Cooperation Agreement. The agreement, signed by both countries in 2010 in Port Moresby, was ratified by Indonesia in 2012 and by PNG only in 2024. This agreement may also be seen as an example of containment of traditional internationalization uses the language of mutual respect for each other's independence, sovereignty, territorial integrity, and non-interference in internal affairs (PNG National Information Centre 2024). A border patrol is also included in the agreement to counter insurgency by Papuan separatists against Indonesian rule for decades, and to combat transnational crimes along the border (Joku 2024).

By securing their shared border through joint patrols, Indonesia has expected its neighbor PNG to fully respect Indonesian sovereignty over Papua and ensure that PNG does not support any Papuan independence movement, both at home and abroad. The reason is that the grassroots local communities in PNG have historically supported the struggle for self-determination in Papua as part of the Melanesian brotherhood. Therefore, a state-level confirmation from PNG is needed to ensure peace in the region and Indonesia's sovereignty over Papua. The Indonesian political motive can be seen from the beginning of Jokowi's leadership when he visited Port Moresby in 2015 to strengthen economic ties. In response, the Prime Minister of PNG, Peter O'Neill has also expressed the desire to work closely with Indonesia to create peace and stability in Papua (Parlina 2015).

At the end of his presidency, Jokowi and his foreign policy representatives reaffirmed his commitment to the Indonesia–PNG relationship during his meeting with the Prime Minister James Marape, the successor of O'Neill, on the sidelines of the 43rd ASEAN Summit in Jakarta in 2023. Acknowledging PNG's support for Indonesia, "I greatly appreciate Papua New Guinea's support for Indonesia's sovereignty and integrity…" (Ministry of Foreign Affairs of The Republic of Indonesia 2023b). The meeting between the two leaders undeniably aims to strengthen their sovereignty, particularly Indonesia's over the Papua region. The aim to secure the shared border region could be seen as one way to achieve the goal of solidifying Indonesian territory over Papua. In short, the collaboration with PNG not only affirms Indonesia's sovereignty and its territorial integrity over Papua but also counteracts any Papua independence movements using the border to advance their interests in the region.

The second reason PNG is considered vital to the strengthening of Indonesia's sovereignty over Papua is that it has played a major role as a bilateral partner for Indonesia. Consequently, numerous meetings between PNG and Indonesian officials from different levels of government have taken place every year to discuss any issues and to strengthen bilateral relations, including the Border Liaison Meeting (BLM), Joint Border Committee, and Joint Ministerial Committee (JMC). On 8 May 2024, for instance, the Fourth Joint Ministerial Commission between Indonesia and PNG, hosted by PNG, was held in Jayapura (Ministry of Foreign Affairs Republic of Indonesia 2024). It was a historic day as for the first time both the PNG and Indonesian foreign ministers met in Jayapura, the capital of the Indonesian province of Papua.

As for Indonesia, it has been in the interest of Jokowi's administration to convey to other countries that Papua is a peaceful, developed, and integral part of Indonesia. More importantly, it demonstrates Indonesia's commitment to developing Papua and fostering relations with neighboring PNG. In addition, the Indonesian government has long been supportive of other programs aimed at enhancing mutual friendship and strengthening relations between the governments and societies of Papua and PNG. Examples include the sister city cooperation between Jayapura (Indonesia) and Vanimo (PNG), as well as between Jayapura (Indonesia) and Wewak (PNG), which has been ongoing since 2016. These partnerships focus on areas such as trade and tourism, education, sports and health, and agriculture and fisheries. PNG indeed plays a crucial role in its bilateral relationship with Indonesia, as its support for border controls signals recognition of Indonesia's territorial integrity reinforcing support for the sovereignty of Indonesia from Sabang to Merauke.

Finally, PNG has greatly facilitated Indonesia's connection with the Melanesian nations. Indeed, PNG—along with Fiji—have become proponents for Indonesia's inclusion within the MSG due to their vested interests in increasing investment, military cooperation, and trade with Indonesia. This stands in contrast with two other MSG member states, Vanuatu and the Solomon Islands, which have long supported Papua's quest for independence. Accordingly, PNG and Fiji actively restrict any Papuan movements such as ULMWP within the MSG that are not in Indonesia's interest while seeking to accommodate Indonesian interests in the forum. Since becoming an associate member of the MSG, the Jokowi administration has been highly active in engaging with Pacific communities and PNG's role has always been acknowledged. For example, during the Press Statement at the 4th Meeting of Indonesia-PNG Joint Ministerial Commission in Jayapura in 2024, Indonesian Foreign Minister Retno Marsudi stated that "as our closest neighbour in the Pacific, PNG is a critical partner for Indonesia in its engagement with its brothers and sisters in the Pacific" (Ministry of Foreign Affairs of the Republic of Indonesia 2024). In other words, without PNG's assistance, Indonesia may not have been able to engage with the Pacific as it does today. As such, PNG will not only serve as a bridge for smooth Indonesian influence in the Pacific—a region where Papua is receiving more attention than elsewhere—but will also play a key role in securing Indonesia's sovereignty over the Papuan border with PNG.

The efforts of Jokowi's government to contain the *Merdeka* internationalization of Papua in multilateral or regional forums, as well as to strengthen bilateral relations with its neighboring country, have been clearly demonstrated through these examples. Nevertheless, the issue of internationalization is not solely about achieving independence; contemporary internationalization also has the potential to influence Indonesia to improve its treatment of Papua. The next section will, therefore, highlight Jokowi's legacy within Papua during his presidency.

12.5 Jokowi's Legacy in Papua and Indonesia's National Integrity

Jokowi gained relative popularity in Papua, as indicated by the large majority of Indigenous and non-Indigenous Papuans, voting for him in both 2014 and 2019 general elections (Hakim and Meiliana 2019). This may also be due to the relative distrust of his opposing candidate Prabowo with a military track record of human rights violations. Jokowi has also been recorded as the president who visited Papua most frequently during his presidency, with 19 visits up to July 2024, more than any other president (Silas 2024).

Under the decade of his leadership, Jokowi's administration employed a "down-to-earth diplomacy," with the aim to offer tangible benefits to people. This diplomacy adopted a slightly inward-looking approach because it involved negotiation, mediation, and conflict management within a domestic context (Djumala 2019). As noted by the Senior Expert Staff of the Presidential Staff Office (KSP) Theofransus Litaay, "The President has implemented various breakthroughs, not only by decentralising development planning policies from a Java-centric to an Indonesia-centric approach, but also by implementing programs to become a Papua-centric" (Tangkelangan 2024). The application of Jokowi's down-to-earth diplomacy style in Papua could also be one of the strategies to address the root causes of several issues in the region and accelerate its development. There have been three noticeable changes in Papua during Jokowi's leadership, which may be conceptualized as part of a response to contain the internationalization of Papua. In particular, Jokowi pursued relatively progressive policies seeking to improve Indonesia's treatment of Papuan Indigenous people. Shifting from coercions to incentives could also be seen as part of a wider strategy to maintain Indonesia's national integrity over Papua.

First and foremost, Papua has experienced more infrastructure development under Jokowi than under any prior presidents. Jokowi's policies enhanced government and private sector investment in the infrastructure sector with the hope and aim of accelerating the Papuan economy and improving the quality of life of the Papuan people. It could also be seen to demonstrate to the world that Indonesia is committed to Papuan development, weakening international and local perceptions of inequality and grounds for Papua *Merdeka* internationalization. At the Inauguration of the Papua Youth Creative activities in 2023, for example, the President conveyed that "Indonesia's development is now not Java-centric, but Indonesia-centric and the land of Papua is prioritised for… development" (Cabinet Secretariat of Indonesia 2023).

Among the many infrastructural advances in Papua under the Jokowi government, are the construction of 3,462 kms of the Trans-Papua Road and 1,098 kms of border roads, as well as the 1.3-km Youtefa Bridge in Jayapura and the Domine Eduard Osok Airport in Sorong (Cabinet Secretariat of Indonesia 2023; Sudarto 2018). Several infrastructure assets have also been repaired, such as terminal revitalization work and the building of an airport in Wamena, Jayawijaya. Jokowi's effort is aligned with implementing the central, provincial, and regional government budgets, reaching 1,036 trillion Rupiah (Cabinet Secretariat of Indonesia 2023).

Another milestone during Jokowi's leadership is the revisions to Papua's Special Autonomy Law signed by the President in 2021. It was essentially a renewal of the original special autonomy back in 2001 with some critical revisions including the issue of special autonomy funds. Papua' special autonomy is granted to recognize and respect special regional government units, to appreciate the equality and diversity of the socio-cultural life of the Papuan people, and to ensure that natural resource management and government administration can achieve community welfare, thereby narrowing the gaps between Papua and other provinces (Wulandari 2024).

Under Jokowi's administration, the partitioning of Papua (*pemekaran*) has also become more rapid than ever before. Indeed, four new Papuan provinces (Highland Papua, Central Papua, South Papua, and Southwest Papua) were established in 2022, bringing the total number of provinces in Papua to six, including the former provinces of Papua and West Papua. Jokowi affirmed that *pemekaran* is an effort to promote equitable development in Papua (BPMI of Secretariat Presidential 2022). Meanwhile, the Indonesian Minister of Home Affairs Tito Karnavian asserted that *pemekaran* is intended not only to create equitable development in the region, but also to improve public service delivery, increase community welfare, and elevate the dignity of Indigenous Papuans (Dirgantara and Prabowo 2022).

However, the achievements under the Jokowi government may also be viewed from a more critical perspective. Adriana Elisabeth, a researcher from the Indonesian Institute of Sciences (LIPI), observed that most of Jokowi's infrastructural assets in Papua were largely enjoyed by non-Indigenous immigrants, rather than native Papuans who are mainly inhabitants of mountainous areas (BBC Indonesia

2019). The phenomenon of "elite capture" has been widely documented on the basis of both class and ethnic divides. This phenomenon is further intensified through historic transmigration policies and gentrification of urban centers by elite and non-Indigenous communities (see Korwa et al. 2021). It is undeniable that many infrastructure projects in Papua are concentrated in cities or urban areas inhabited by diverse ethnic groups, making it challenging to reach native Papuans. For instance, the number of native Papuans is half of the total population in Papua, allowing non-Papuans to greatly benefit from infrastructure development. As mentioned by Ribka Haluk, the Head of the Papua Province Social, Population, Women's Empowerment and Child Protection Service, that the population of Papua Province was 4,349,344 people as of 31 December 2019, while data for native Papuans based on clan was 2,386,048 (54.86%) people (Weya 2020).

Although there has been an increase in GDP of 2.2% since Jokowi's administration in 2014, the economic growth and associated benefits have not been equally distributed across Indonesia. In particular, Papua has not benefited to the level of Java. Papua Province's per capita income is mostly yielded from extractive industries including the gas sector and wider mining activities. This is proven by data on gross domestic income through mining such as PT. Freeport Indonesia which manages the Grasberg mine in Mimika Regency: a colossal scar in the earth widely described as a symbol of the extraction from Papua through partnerships between post-colonial governance and corporate imperial structures (Ahmed 2015; Anderson 2015). Further economic value yielded from Papua has been seen from the tourism sector through destinations such as Raja Ampat. Yet, despite the large income generated through these sectors, the impact of "trickle-down" benefits have had little to no impact on the lives of many of the people of Papua as evidenced in Papua's per capita expenditure (representing relative purchasing power) which has always remained below the national average (Ghifari 2023). Papua is even included in the category of poor province. Trends in annual per capita expenditure shows the relative inequality in expenditure between Indigenous Papuans and Indonesians despite the relatively high living costs in Papua compared to the average of wider Indonesia (Fig. 12.2). Such economic trends highlight the disjoint between Papua's rich natural resources and significant contributions to Indonesia's GDP against a backdrop of limited trickle-down benefits for local Indigenous Papuans and growing income inequalities. These figures may serve to worsen perceptions of a disproportionately extractive role of Indonesia in Papua:

In addressing these perceptions and the associated risks of contemporary internationalization, the Jokowi Administration has had to consider various avenues. Regarding the revisions to Papua's Special Autonomy Law, there is a competing perception between Jakarta and Papua (Farneubun and Korwa 2020). Jakarta believes that special autonomy remains the best solution to address Papua's economic inequality on the one hand, while some Papuans, on the other hand, argue that it has failed to improve their living conditions. This has also been the case for *pemekaran*, which is seen as the will of Jakarta rather than the will of the Papuans. Even though *pemekaran* was introduced with the mission statement of encouraging equitable development in the region, looking back at its evolution in the context of Papua

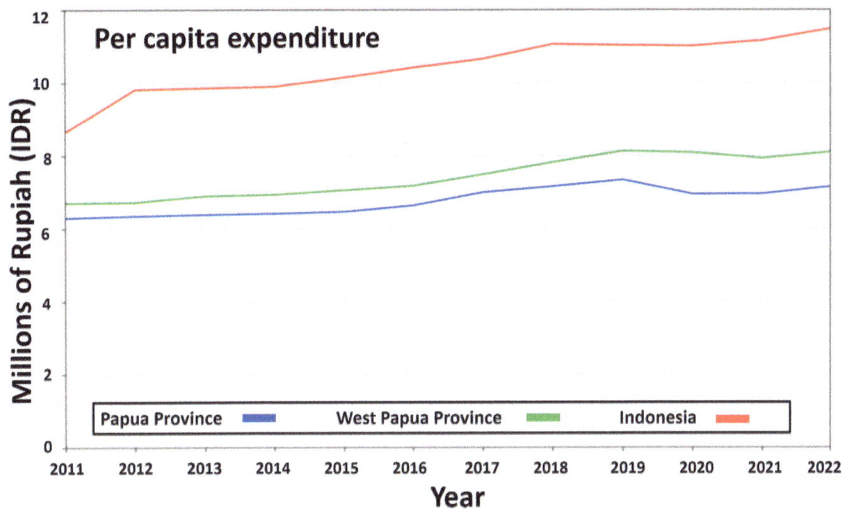

Fig. 12.2 Per Capita Expenditure in Million Rupiah (IDR). *Source* Ghifari (2023)

reveals that its primary motivation has been the prevention of national disintegration (Sullivan 2003). The policy of *pemekaran* was initially considered one of the alternative strategies to prevent independence—a demand proposed by a delegation of prominent Papuans known as "Team 100" during their meeting with President Habibie in 1999 (Sullivan 2003). This historical context has shaped Papuans belief in *pemekaran* today and will perhaps continue to do so in the years to come.

Another issue left as a legacy from Jokowi's government is that his diplomacy paid less attention to human rights abuses. Human rights conditions in Papua have remained dire under his administration. Although Indonesia has ratified and integrated eight fundamental international human rights treaties into its national legal framework (United Nations 2016), the government has not taken significant action to remedy breaches of human rights. Human rights organizations, both domestic and foreign, including the US Embassy, continue to draw attention to the appalling state of human rights conditions in their reports. In its 2023 country report on Human Rights Practices, for example, the US Embassy listed the various human rights violations that the Indonesian government had committed; nonetheless, the government "did not systematically take credible steps to identify and punish officials who may have committed human rights abuses" (US Embassy and Consulates 2023). Other past human rights violations, such as the Bloody Paniai massacre in 2014 and the Wasior incidents in 2001 and 2003, serve as reminders of the relatively high levels of conflict experienced throughout past terms of governance under previous administrations. While Jokowi's administration has led to relative decreases in human rights violations, the ongoing trauma and loss experienced by many have also not yet been addressed since the Jokowi government came into office in 2014 (James 2023). This is crucial because Pacific countries assess Indonesia's treatment of Papuans and, of course, this assessment affects how they welcome Indonesia into the Pacific and wider international community. Furthermore, foreign journalists and academics continue to cast doubt on Indonesia's political commitment to enhancing human rights conditions and defending the rights of Papuans. In retaliation, Indonesia has restricted their access and prohibited them from entering Indonesia.

12.6 Conclusion

As this chapter has shown, Papua's internationalization has shifted from a pre-reformasi era traditional focus on casting doubt over *Pepera* as part of a pursuit of political independence (*Merdeka*) toward a contemporary focus on a wide range of Indigenous quality of life issues. This form of contemporary internationalization has gained momentum in the post-reformasi period and may eventually garner sympathy and support for wider political independence. Addressing the political dynamics and internationalization of Papua still presents several complex challenges.

Jokowi adopted a down-to-earth pragmatic style of governance relative to all other post-reformasi presidents. Under his leadership, the internationalization of Papua was largely contained. Jokowi was less involved in Indonesian foreign affairs, but his vice president, minister of foreign affairs, and diplomats actively defended Indonesia's territorial integrity in international and regional fora, mainly at UNGA, MSG, and PIF. At UNGA, Indonesian diplomacy was centered on countering narratives and advocating for adherence to international norms of state sovereignty and non-interference. At regional fora, Indonesian diplomacy was directed toward containing Papuan freedom movement and its political aspirations. Furthermore, Indonesia bolstered its bilateral ties with neighboring Pacific nations, particularly PNG, with the major objectives being, among other things, the suppression of Papuan freedom fighters and support for Indonesia's attempts to preserve its territorial integrity.

Throughout his presidency, Jokowi has made several significant accomplishments. His record of achievement has been confined to the areas of infrastructure, social and economic development, the division of Papua into six provinces, the amendment of the Special Autonomy Law 2001, and his outreach and engagement through regular trips to Papua. Both the domestic and international community have praised his accomplishments relative to his predecessors. However, criticism from both state and non-state actors has persisted due to the inability of Indonesia to resolve human rights abuses and protracted conflict. The accomplishment will undoubtedly be seen as a significant component of his legacy. Prabowo, his successor, will still have to navigate the unfinished and unresolved issues. With that in mind, Prabowo needs to be a foreign policy president, unlike Jokowi, who has been less visible on the global stage. Prabowo should also adopt a constructive policy approach that engages not only with the Papuan community but also with multilateral institutions and intergovernmental organizations. In practice, Prabowo needs to move beyond a centralized top-down approach toward a consultative approach which places greater emphasis on strategic partnership with Papuan community. Infrastructure development in Papua and humanitarian approaches will help the Prabowo government focus on resolving past human rights violations and initiate the reconciliation process to foster peace in Papua. The relevant stakeholders, including Papuan customary leaders, religious leaders, and indigenous communities, should also be more involved along the way with the government to build a better future for Papua under the Prabowo government.

References

Ahmed, N. 2015. Land Rights: Counter-mapping West Papua. *Continent* 4(4): 12–19.

Amnesty International. (2018). *Don't Bother Just Lest Him Die: Killing with Impunity in Papua*. Amnesty.org. https://www.amnesty.org/en/wp-content/uploads/2021/05/ASA218198 2018ENGLISH.pdf.

Anderson, K. 2015. Colonialism and 'Cold Genocide': The Case of West Papua. *Genocide Studies and Prevention* 9 (2): 9–25.

BBC Indonesia. 2019, October 28. *Kunjungan kerja Jokowi ke Papua, Tokoh Papua: 'Jangan sampai sia-sia'*. BBC Indonesia. https://www.bbc.com/indonesia/indonesia-50200482

Beazley, J. 2024, February 14. West Papua Rebels Release Video of Kidnapped Former Jetstar pilot. *The Guardian*. https://www.theguardian.com/world/2024/feb/14/kidnapped-new-zealand-pilot-phillip-mehrtens-release-west-papua-rebels.

BPMI of Secretariat Presidential. 2022, September 1. *Presiden Jokowi: Pemekaran Wilayah di Papua untuk Pemerataan Pembangunan*. Cabinet Secretariat of Indonesia. https://men pan.go.id/site/berita-terkini/dari-istana/presiden-jokowi-pemekaran-wilayah-di-papua-untuk-pemerataan-pembangunan#:~:text=Presiden%20Joko%20Widodo%20menegaskan%20bahw a,upaya%20untuk%20menciptakan%20pemerataan%20pembangunan.

Brundige, E., W. King, P. Vahali, S. Vladeck, and X. Yuan. 2004. Indonesian human rights abuses in West Papua: Application of the law of genocide to the history of Indonesian control. *New Haven: Allard K. Lowenstein Internationaal Human Rights Clinic, Yale Law School*. https://cul ture2.org/archive/westpapua.net/docs/papers/paper12/yale-wphr.pdf

Cabinet Secretariat of Indonesia. 2023, March 21. *President Jokowi Addresses Papua as one of Development Priority Areas*. https://setkab.go.id/en/president-jokowi-addresses-papua-as-one-of-development-priority-areas/.

Chauvel, R. 2004. *Constructing Papuan nationalism: History, Ethnicity, and Adaptation*. East-West Center Washington. http://hdl.handle.net/10125/3515.

Chauvel, R. 2007. Refuge, Displacement and Dispossession: Responses to Indonesian Rule and Conflict in Papua. In *Conflict, Violence and Displacement in Indonesia*, ed. E-L. Hedman, 147–172. Cornell Southeast Asia Program Publications.

Connelly, A.L. 2014. *Indonesian Foreign Policy Under President Jokowi*. Lowy Institute for International Policy.

Connelly, A.L. 2015. Sovereignty and the Sea: President Joko Widodo's Foreign Policy Challenges. *Contemporary Southeast Asia* 37 (1): 1–28. https://www.jstor.org/stable/24916512.

Dirgantara, A., and D. Prabowo. 2022, August 29. *Mendagri: Pemekaran di Papua Punya Tujuan Mulia*. Kompas. https://nasional.kompas.com/read/2022/08/29/22524621/mendagri-pemekaran-di-papua-punya-tujuan-mulia.

Djumala, D. 2019, August 31. Diplomacy Becoming More Down-to-Earth. *The Jakarta Post*. https://www.thejakartapost.com/academia/2019/08/31/diplomacy-becoming-more-down-to-earth.html.

Drooglever, P. 2009. *An Act of Free Choice: Decolonization and the Right to Self-Determination in West Papua*. Oneworld.

Elisabeth, A. 2006. Dimensi internasional kasus Papua. *Jurnal Penelitian Politik* 3 (1): 43–65.

Farneubun, P., and J.R.W. Korwa. 2020. *The Competing Perceptions of Jakarta and Papua towards Special Autonomy*. Department of Pacific Affairs ANU, In Brief.

Fatima, Z. 2024, September 21. *Pilot Hostage Freed by Papua Rebels 'Very happy' to go home*. BBC News. https://www.bbc.co.uk/news/articles/c3wp8ndnv9qo.

Fry, G., and S. Tarte. (2015). The New 'Pacific Diplomacy': An Introduction. In *The New Pacific Diplomacy*, eds. Greg dan, Fry and Tarte, Sandra, eds. pp. 3–19. Australian National University Press.

Ghifari, D. 2023, October 25. Jokowi's 9-Year Reign Does Not Amount to Much for Papua. *The Jakarta Post*. https://www.thejakartapost.com/business/2023/10/25/jokowis-9-year-reign-does-not-amount-to-much-for-papua.html.

Hakim, R.N. and D. Meiliana. 2019, May 21. *Suara Jokowi Vs Prabowo per Provinsi di Pilpres 2014 dan 2019*. Kompas. https://nasional.kompas.com/read/2019/05/21/20014871/suara-jokowi-vs-prabowo-per-provinsi-di-pilpres-2014-dan-2019?page=all.

Human Rights Watch. 2023. *Indonesia: Events of 2022*. https://www.hrw.org/world-report/2023/country-chapters/indonesia.

IPWP. (2024, January 22). *Brussels Declaration on West Papua*. https://www.ipwp.org/ipwp-news/brussels-declaration-on-west-papua/.

James, R. 2023. Indonesia Confronts the Past, While Sidestepping the Present. *The Diplomat*. https://thediplomat.com/2023/02/indonesia-confronts-the-past-while-sidestepping-the-present/.

Joku, H. 2024, February 29. *Papua New Guinea, Indonesia Ratify Defense Deal to Expand Security Cooperation*. Radio New Zealand. https://www.rnz.co.nz/international/pacific-news/510486/papua-new-guinea-indonesia-ratify-defense-deal-to-expand-security-cooperation.

Kabuni, M. 2021, January 14. *Grassroots support for West Papua in PNG. Asia & the Pacific Policy Society*. https://www.policyforum.net/grassroots-support-for-west-papua-in-png/#:~:text=Even%20though%20PNG's%20prime%20minister,drawing%20about%202%2C000%20thousand%20people.

Korwa, J. R., N. Metherall, B. Rumabar, J. H. Mampioper, and T. Ranathunga. 2021. Peri-Urbanisation in Papua: A Participatory and Geospatial Impact Assessment of Peri-Urban Development and Transmigration in Port Numbay. *Asia & the Pacific Policy Studies* 8 (1): 129–150.

Lantang, F., and E.M.B. Tambunan. 2020. The Internationalization of "West Papua" Issue and Its Impact on Indonesia's Policy to the South Pacific Region. *Journal of ASEAN Studies* 8(1): 41–59. https://doi.org/10.21512/jas.v8i1.6447

Lawson, S. 2016. West Papua, Indonesia and the Melanesian Spearhead Group: Competing Logics in Regional and International Politics. *Australian Journal of International Affairs* 70 (5): 506–524. https://doi.org/10.1080/10357718.2015.1119231.

Leadbeater, M. 2008. Media blind spot over West Papua. *Pacific Journalism Review*, 14 (1), 168–175. https://search.informit.org/doi/abs/10.3316/informit.171547487397278

Mandibondibo, P., S. Wanggai, J. Korwa, Y. Lada, G. P. Utami, B. Rumabar, and D. Hardiansyah. 2022. State Borders as Center of Economic Growth: Case Study of the East Arso District in the Indonesia-Papua New Guinea Border. *Papua Journal of Diplomacy and International Relations* 2 (1): 35–58.

McGibbon, R. 2006. Pitfalls of Papua. *Lowy Institute for International Policy*, Sydney. https://www.lowyinstitute.org/sites/default/files/pubfiles/McGibbon,_Pitfalls_of_Papua_1.pdf

Melanesian Spearhead Group. 2009. *Annual Report*. MSG Secretariat. https://www.msgsec.info/wp-content/uploads/annualreports/2009-Annual-Report.pdf.

Metherall, N., D.R.D. Fretes, F. Mandibondibo, and T. Caucau. 2022. Assessing the Development Impact of the Sota Border Post Connecting Indonesia and Papua New Guinea. *Papua Journal of Diplomacy and International Relations* 2(2): 95–122. https://doi.org/10.31957/pjdir.v2i2.2209.

Ministry of Foreign Affairs of The Republic of Indonesia. 2022, December 8. *Indonesia Redoubles its Commitment to Enhance Partnership with the Pacific.* https://kemlu.go.id/portal/en/read/4240/berita/indonesia-redoubles-its-commitment-to-enhance-partnership-with-the-pacific#:~:text=%E2%80%9CWhen%20Indonesia%20talks%20about%20Pacific,of%20communities%20in%20this%20region.%E2%80%9D.

Ministry of Foreign Affairs of The Republic of Indonesia. 2023a. *Indonesia Sends Humanitarian Aid to Vanuatu.* https://kemlu.go.id/portal/en/read/4730/berita/indonesia-sends-humanitarian-aid-to-vanuatu#:~:text=Indonesia's%20diplomacy%20in%20the%20Pacific,New%20Zealand%2C%20in%20July%202019.

Ministry of Foreign Affairs of The Republic of Indonesia. 2023b. President Jokowi Meets PM Marape, Expresses Appreciation for PNG Support for Indonesia's Sovereignty and Integrity. https://kemlu.go.id/portal/en/read/5214/berita/president-jokowi-meets-pm-marape-expresses-appreciation-for-png-support-for-indonesias-sovereignty-and-integrity.

Ministry of Foreign Affairs of The Republic of Indonesia. 2024. In *Press Statement H. E. Retno L. P. Marsudi at the 4th Meeting of Indonesia-Papua New Guinea Joint Ministerial Commission.* https://kemlu.go.id/portal/en/read/5906/siaran_pers/press-statement-h-e-retno-l-p-marsudi-at-the-4th-meeting-of-indonesia-papua-new-guinea-joint-ministerial-commission.

Parlina, I. 2015, May 12. PNG PM praises Jokowi's Papuan Visit. *The Jakarta Post.* https://www.thejakartapost.com/news/2015/05/12/png-pm-praises-jokowi-s-papuan-visit.html.

PNG National Information Centre. 2024, February 27. Papua New Guinea Ratifies Outstanding PNG-Indonesia Defence Cooperation Agreement. https://info.gov.pg/papua-new-guinea-ratifies-outstanding-png-indonesia-defence-cooperation-agreement/.

Saltford, J. 2003. *The United Nations and the Indonesian Takeover of West Papua, 1962–1969: The Anatomy of Betrayal.* Routledge.

Silas, G. 2024, July 22. 10 Tahun Kunjungan Presiden Jokowi ke Papua, Berikut Datanya. CeposOnline. https://www.ceposonline.com/papua/1994890653/10-tahun-kunjungan-presiden-jokowi-ke-papua-berikut-datanya.

Sudarto, C. E. 2018. Jokowi's Development Policy for Papua. *German-Southeast Asian Center of Excellence for Public Policy and Good Governance (CPG).* https://cpg-online.de/jokowis-development-policy-for-papua/.

Sullivan, L. 2003. *Challenges to Special Autonomy in the Province of Papua, Republic of Indonesia [Discussion Paper].* Research School of Pacific and Asian Studies. https://core.ac.uk/download/pdf/156617367.pdf.

Taliawo, R.G. 2022. The Indonesianization of West Papua: Development of Indonesia's Attitudes and Policies towards West Papua and the Dynamics of the Papua Freedom Movement. *Journal of Humanities and Social Sciences Studies* 4(2): 71–83. https://doi.org/10.32996/jhsss.2022.4.2.10.

Tangkelangan, M. 2024, April 21. KSP: Di Era Jokowi Papua Mendapat Perhatian Lebih. Radio Republik Indonesia. https://www.rri.co.id/nasional/646388/ksp-di-era-jokowi-papua-mendapat-perhatian-lebih.

Tebay, N. 2005. *West Papua: The Struggle for Peace with Justice.* Ciir.

U.S. Embassy and Consulates in Indonesia. 2023. Country Reports on Human Rights Practices: Indonesia. https://id.usembassy.gov/our-relationship/official-reports/2023-country-reports-on-human-rights-practices-indonesia/.

United Nations. 2016. 1st Right of Reply by Indonesia, at the 71st Session of the General Assembly of the UN (New York, 20–26 September 2016). WebTV UN. https://webtv.un.org/en/asset/k1q/k1qu6qe1v4.

United Nations. 2018. Muhammad Jusuf Kalla, Vice-President of the Republic of Indonesia addresses the general debate of the 73rd Session of the General Assembly of the UN (New York, 25 September-1 October 2018). WebTV UN https://webtv.un.org/en/asset/k12/k12v8fzw94.

United Nations. 2019. Muhammad Jusuf Kalla, Vice-President of the Republic of Indonesia Addresses the General Debate of the 74th Session of the General Assembly of the UN (New York, 24-30 September 2019). WebTV UN https://webtv.un.org/en/asset/k1n/k1nl9hisro.
United Nations. 2020a. Jokowi Widodo, President of the Republic of Indonesia, Addresses the General Debate of the 75th Session of the General Assembly of the UN (New York, 22–29 September 2020). WebTV UN. https://webtv.un.org/en/asset/k1y/k1yw8yltnr.
United Nations. 2020b. 1st Right of Reply by Indonesia, at the 75th Session of the General Assembly of the UN (New York, 22-29 September 2020). WebTV UN https://webtv.un.org/en/asset/k1p/k1p3uhllwu.
United Nations. 2021. Jokowi Widodo, President of the Republic of Indonesia, Addresses the General Debate of the 76th Session of the General Assembly of the UN (New York, 21–27 September 2020). WebTV UN. https://webtv.un.org/en/asset/k1y/k1ycv90ej5.
United Nations. 2022. Retno Lestari Priansari Marsudi, Minister for Foreign Affairs of the Republic of Indonesia, Addresses the General Debate of the 77th Session of the General Assembly of the UN (New York, 20–26 September 2022). https://webtv.un.org/en/asset/k18/k18o5f9nea.
United Nations. 2023. Retno Lestari Priansari Marsudi, Minister for Foreign Affairs of the Republic of Indonesia, Addresses the General Debate of the 78th Session of the General Assembly of the UN (New York, 19–23 and 26 September 2023). https://webtv.un.org/en/asset/k10/k10azuh7pp.
Viartasiwi, N. 2018. The Politics of History in West Papua-Indonesia conflict. *Asian Journal of Political Science* 26(1): 141–159. https://www.tandfonline.com/doi/abs/10.1080/02185377.2018.1445535.
Wangge, H. R., and S. Lawson. 2023. The West Papua Issue in Pacific Regional Politics: Explaining Indonesia's Foreign Policy Failure. *The Pacific Review* 36 (1): 61–89. https://doi.org/10.1080/09512748.2021.1931417.
Webster, D. 2013. Self-Determination Abandoned: The Road to the New York Agreement on West New Guinea (Papua), 1960–62. *Indonesia* (95): 9–24. https://www.jstor.org/stable/10.5728/indonesia.95.0009.
Weya, Y. 2020, October 24. Ini Data Jumlah OAP di Papua yang Diumumkan Gubernur. Suara Papua. https://suarapapua.com/2020/10/24/ini-data-jumlah-oap-di-papua-yang-diumumkan-gubernur/.
Widjojo, M. 2006. Nationalist and separatist discourses in cyclical violence in Papua. *Asian Journal of Social Science*, 34 (3), 410–430. https://brill.com/view/journals/ajss/34/3/article-p410_5.xml
Wulandari, R. 2024, February 19. Pengakuan Hak-Hak Adat dalam Kebijakan Otonomi Khusus Papua: Tantangan dalam Implementasinya. https://setkab.go.id/pengakuan-hak-hak-adat-dalam-kebijakan-otonomi-khusus-papua-tantangan-dalam-implementasinya/#:~:text=Otonomi%20khusus%20terhadap%20Papua%20pertama,Kesatuan%20Republik%20Indonesia%20(NKRI).

Johni R.V. Korwa is a permanent lecturer at the Department of International Relations, Cenderawasih University, Papua, Indonesia. He obtained his master's degree from the Flinders University of South Australia. Currently, he is pursuing his PhD at the Department of Politics and International Relations, University of Southampton, the United Kingdom. He has a particular interest in border studies, transnational society, and international relations. Among his most recent articles published in the past two years are: (1) Indonesia's First Spaceport Plan in Biak Island: A View from International Relations, (2) Constructivist Analysis of the Establishment of the AUKUS Security Pact and its Implications for Regional Stability in the Indo-Pacific, (3) Indonesia's 2024 Simultaneous Elections from a Papuan Perspective.

Nicholas D. Metherall is currently completing his PhD in hydrology, forestry, GIS and remote sensing through the first Joint Cotutelle Program between the University of the South Pacific (USP) and the Australian National University (ANU). Nicholas led a team to win the grand prize for the first Space for Planet Earth Challenge. Previously, he worked in the Australian Government, regional intergovernmental organisations, NGOs and universities including the ANU, USP, Universitas Cenderawasih (UNCEN) in Jayapura and Universitas Nusa Cendana (UNDANA) in Kupang. Nicholas' most recent publications and contributions include: (1) A rapid spatial sampling design for ridge-to-reef water quality monitoring in Pacific SIDS: identifying pH hotspots along the Ba River in Fiji, (2) Environmental monitoring and the geospatial sprawl of historic mining legacies in Australia, and (3) GEF in the Pacific Islands: pathways to incorporate monitoring, ecosystem theory, and stewardship, (4) A field and remote sensing assessment of rates and drivers of tree cover loss in island catchments: variation in global model accuracy, (5) The impact of tropical cyclones Pam, Harold, Winston and Yasa on tree cover loss in Vanuatu and Fiji., (6) Assessing Papua New Guinea's Bid for ASEAN Membership: A Constructivist View.

Petrus K Farneubun is a permanent lecturer at the Department of International Relations, Cenderawasih University, Papua, Indonesia. He obtained his master's degree from the Australian National University (ANU). He completed his PhD in International Relations at the University of Groningen, the Netherlands, writing a dissertation on Indonesia-China relations under President Yudhoyono. His research interests are Indonesia's Foreign Policy, Theories of International Relations, Peace and Conflict, and Ethics of War. His most recent publications include: (1) Indonesia-China Strategic Partnership: Role of Vision, Bureaucratic Actors and Domestic Political Change, (2) China's Rise and its Implication for Australian Foreign Policy, (3) CoronaVirus (COVID-19) in Papua Province: An Increase in Case Numbers and the Challenges Ahead.

Made Selly Dwi Suryanti is a permanent lecturer at the Department of International Relations, Cenderawasih University, Papua, Indonesia. She is currently pursuing her Ph.D in the Department of Politics and Government, Gadjah Mada University, Yogyakarta. Her research interests are humanitarian issues, peace and conflict studies, and critical perspective in international relations. Her most recent publications in the past two years include: (1) Online gender-based violence in Indonesian context: the shadow pandemic study (academic journal), (2) Expanding the University Social Service for Empowering Sustainable Tourism Based on Sustainability Agenda (academic journal), (3) Book: Kebijakan Penanganan Orang Asing di Indonesia (Policy in Handling Foreigners in Indonesia).

Chapter 13
Between Great Powers in the South China Sea: Indonesia, the United States, and China

Amarulla Octavian

Abstract In responding to the US–China relations turbulence, Indonesia must skillfully "maneuver" in accordance with the Independent and Active policy. The goal is none other than to ensure Indonesia's national interests, namely to safeguard state sovereignty, territorial integrity, and sovereign rights in the Natuna Sea in accordance with the applicable national and international laws. At the same time, Indonesia must also contribute to maintaining stability and order in the Southeast Asia region from the possibility of war between great powers. This chapter suggests that Indonesia should take on a mediating role between the United States and China in addressing South China Sea issues. Through its diplomatic efforts, Indonesia can help reinforce ASEAN unity, enabling the organization to act collectively in discouraging either China or the US from escalating tensions into open conflict.

Keywords Indonesia · China · The United States · Power · Military

13.1 Introduction

In the last 10 years, many international events have changed the world's strategic environment landscape. This series of events resulted in a significant turmoil on international relations dynamics. Global terrorism threat has indeed decreased drastically due to the fall of ISIS, which was followed by the rise of Taliban to the government seat in Afghanistan. After Russia's attack on Ukraine, NATO members got more solid and undertook large-scale rearming to anticipate the possible scenario of Russian attack on Europe. Russia also experienced an economic crisis due to the fleeing Western investors and the increasing burden of war expenditures. Without potential allies, Russia faces difficulty to decisively conquer Ukraine, who still receives continuing support from the West. Countries known as Russia's "close friends" such as India and China seemed cautious in responding to their relations with Russia.

A. Octavian (✉)
The National Research and Innovation Agency of Republic Indonesia, Central Jakarta, Indonesia

I. Kartini and Kanisa (eds.), *Joko Widodo's Foreign Policy and Diplomatic Legacy*,
https://doi.org/10.1007/978-981-95-2085-5_13

The world economy, which was already fragmented due to the Covid-19 pandemic and trade war, is becoming way more fragmented and experiencing slow growth. The West and China are now finding it increasingly difficult to obtain essential rare earth materials (REMs) for economic growth and maintaining the progress of their electronics and automotive industries (Baskaran 2024). Slowing economic growth usually marks the beginning of interstate conflicts. Economic slowdown also triggers internal security instability and the emergence of radical/extremist political movements.

The increasingly complex global economic, defense, and security conditions also lands major impacts on Indonesia. Amid the weak world economy as the result of the Covid-19 pandemic and the Russo-Ukrainian War, Indonesia must still struggle to maintain its economy in the midst of disputes and conflicts between the US and China. Currently, Indonesia is highly regarded in the international political and economic arena, especially after it entered the G20 membership. Indonesia is also gaining important position in the international arena due to its capability to supply large amounts of nickel and bauxite which are important for the global electronics and automotive industries. The suspense of REMs supplies from Russia and China due to Western sanctions and the US–China trade war have created a shortage of supplies of these materials globally.

Amid the turmoil of US–China interests at the global level, Indonesia also faces a real threat to its sovereignty and sovereign rights in the North Natuna Sea as a result of China's unilateral claim to most part of the South China Sea (SCS). China's claim has been opposed at the Southeast Asia level, especially from Vietnam and the Philippines who have direct experience to the expansion, presence, and intrusion of Chinese military and paramilitary forces in their exclusive economic zones (EEZs).

In response to the increasing concern on China's economic and political influence development along with its military expansion in the Southeast Asia and SCS regions, US has stepped in as a major power who offers balancing for Southeast Asian countries that are in dispute with China. US is willing to provide not only political and economic assistance but also military assistance. This offer is a test to ASEAN's solidity.

SCS has become a turbulent region which at any time can turn into a battlefield for the conflicting parties. This condition will certainly affect the stability of Indonesia's defense, security, and economy, whose territory includes and is adjacent to the SCS. On the one hand, taking side with one of the conflicting powers is not an option because Indonesia adheres to the Independent and Active foreign policy principle that emphasizes neutrality.

In responding to the US–China relations turbulence, Indonesia must skillfully "maneuver" in accordance with the Independent and Active policy. The goal is none other than to ensure Indonesia's national interests, namely to safeguard state sovereignty, territorial integrity, and sovereign rights in the Natuna Sea in accordance with the applicable national and international laws. At the same time, Indonesia must also contribute to maintaining stability and order in the Southeast Asia region from the possibility of war between great powers.

13.2 US–China Relations from the Twentieth Century to the Twenty-First Century at a Glance

From the twentieth century to the twenty-first century, US and China have experienced many clashes of interests. China has experienced three changes in leadership systems, each with different tones and preferences towards the US. After the collapse of the Qing Dynasty leadership that ruled China from 1644 to 1911, China had transformed from a kingdom system to a nationalist republic state in 1912 and subsequently became a communist state since 1949. During the Qing Dynasty period, US–China diplomatic relations were more dominated by trade and religious affairs (Pruden 2009).

China was very closed to foreigners during the Qing Dynasty, where it set limit on the trading rights to only six months and restricted the distribution to particular locations. The situation changed when the Qing Dynasty was defeated by Western powers in the First Opium War (1839–1842) and the Second Opium War (1856–1860). China was "forced" to open itself to the presence of foreigners, especially Westerners. As a result of China's defeat in the First Opium War, US gained trading access in five Chinese ports, extraterritorial rights, and protection of missionaries' interests through Wangxia Treaty in 1844.

When gold was discovered in California during the mid-nineteenth century, tens of thousands of Chinese immigrants came to US. However, in 1882, an immigration law that prohibited Chinese immigrants from entering the country was passed. This law prohibited Chinese ethnic from obtaining US citizenship. US–China relations became increasingly strained when the Boxer Rebellion broke out. The movement's goal was to expel foreigners from China. To protect its interests and citizens, US sent its marines to suppress the rebellion along with other Western powers (and Japan). The suppression of Boxer Rebellion was sealed with the Boxer Protocol in 1901, which required the Qing Dynasty to pay a huge indemnity that ultimately led to its bankruptcy and downfall.

In 1912 China became a republic state under the leadership of Dr. Sun Yat-sen. China established Nanjing as its capital in 1927 after successfully controlling southern China using its military power, which was followed by the "purge" of communists in Shanghai by Chiang Kai-shek. In 1928, US became the first country to recognize China's sovereignty. After Japan's victory in the 1904–1905 Russo-Japanese War, US viewed Japan as a new power that could threaten its hegemony in the Pacific. US–China relations became closer after the outbreak of Sino-Japanese War in 1937. Until 1941, US provided material and political assistance to China to withstand Japanese attacks. After the Japanese attack on Pearl Harbor, US formed an alliance with China and fought together in the Pacific.

After World War II ended, a civil war broke out in China between the nationalist group led by Chiang Kai-shek and the communist group led by Mao Zedong. Battle between the two groups lasted for about four years, which was won by the communist forces that established People's Republic of China (PRC) in October 1949. The defeated Chinese nationalist group fled to Taiwan. US–China Relations was strained when Mao Zedong stated that he wanted to get close to the Soviet Union. US–China relations became increasingly colder, especially when the Korean War (1950–1953) broke out, where the two countries fought for about three years on the Korean Peninsula. US fought under United Nations (UN) mandate to expel North Korean troops from South Korean territories. When US troops succeeded in pushing back North Korean troops to the Chinese border, China deployed large numbers of troops to help North Korea. Both the US-aided South Korea and the China-aided North Korea were in a stalemate, which finally ended with an armistice agreement.

US–China relations were relatively hostile since the communist leadership in China until 1975. One of the peaks was during the First Taiwan Strait Crisis in 1954. After US naval blockade of Taiwan was ended in 1953, Taiwan deployed military forces to Quemoy and Matsu Islands in August 1954, and China responded with artillery fire. Taiwan finally formed an alliance with US. Afterward, US threatened to drop nuclear bombs on China in 1955 to compel China's will to negotiate. The conflict recurred when China bombed Quemoy and Matsu Islands again in 1958 (the Second Taiwan Strait Crisis). This time US responded by sending weapons and military equipment to Taiwan. Nine years later, China successfully conducted its first nuclear bomb test, which dismantled US' option to use the nuclear threats to China.

Since the ascendance of communist leadership in China, US had no formal relations with China and has often experienced clashes of interests. However, efforts to normalize US–China relations have been carried out slowly through non-official channels such as through "Ping-Pong Diplomacy" in 1971. The normalization of US–China relations was also driven by the collapse of friendship between China and the Soviet Union due to differences in communist ideology worldview, and defense policies.

US–China relations got warmer when President Nixon visited China and met with Mao Zedong in February 1972. It must be admitted that the normalization of US–China relations was more heavily driven by the urgency to counter the Soviet Union threat than a genuine motive. Normalization of relations with China consequently decreased the number of US enemies and increased its business opportunities. For China, normalization of relations with US could give a better balance of power against the Soviet Union.

In January 1979, US officially recognized the sovereignty of PRC, which was followed by formal diplomatic relations establishment between the two countries. US also recognized the One-China Policy, which considered Taiwan as China's territory, thus ending state-level relations with Taiwan. The Taiwan Relations Act, passed in April 1979, stated that US did not recognize Taiwan as a country, but still acknowledged its existence as an international political entity. The law also stipulated that US would continue to maintain close cultural, economic, and defense relations with the "People of Taiwan." Taiwan Relations Act also contains explicit clause that the unification of Taiwan with the PRC must be carried out through a peaceful approach. Shall there be any unification measure that is deemed not in adherence with the peace principles according to the US, then US would resort to intervention, including by the use of military force. Thus, Taiwan Relations Act implies that Taiwan is under US protectorate.

After the normalization of relations between US and China, there have been many changes, especially in the economic sector. According to Council of Foreign Relations (2023), in the period of 1980–2004, the trade value between the two increased significantly from US$ 5 billion to US$ 231 billion. During that period, China had transformed from a simple industrial state into an advanced industrial state. Some events have brought new seeds of hostility between US and China in almost every decade after the normalization of their relations. However, the number of events that exacerbated US–China relations have become more frequent during the period leading up to and after the collapse of Soviet Union (Table 13.1).

13.3 Litmus Test for US and China Interests

The ups and downs of US–China relations indicate that their clash of interests has prolonged since the end of World War II. The result of Plus-Minus Scoring analysis shows that there seems to be more disputes than cooperation between them. The ups and downs of US–China relations are driven by huge difference on their views regarding several interests. The first is ideological difference. Whereas the US adheres to liberal-democratic, China adopts to communist-authoritarian ideology. This ideological difference is the source of disagreements on each government's policies in treating its citizens.

China wants to defend its sovereignty and interests from US intervention to the fullest extent possible. An important sovereignty issue for China is the issue of Taiwan unification. China's nuclear weapons development is not only to neutralize US and Russian nuclear threats but also a part of its efforts to unify Taiwan. China's large military buildup is also suspected as an effort to unify Taiwan by force. The 2001 aircraft collision that took place in Hainan due to dangerous maneuvers by Chinese aircraft against US spy planes also demonstrated China's tough stance when it comes to sovereignty.

China considers the period spanning from the end of the Qing Dynasty to its independence from Japan in 1945 as "The Century of Humiliation" due to its inability to safeguard its sovereignty from external parties. Therefore, China (the Republic of China or the PRC) is determined to restore its past glory, or at least reclaim its territories that fell into foreign hands or regain its territory according to its own historical calculus. This thinking explains why China claims the SCS waters as its own and becomes ambitious to gain sovereignty over those territories. The same line of thinking also applies in how China mentions the city of Vladivostok which actually belongs to Russia. In a map published by the Ministry of Natural Resources of the PRC in August 2023, the city of Vladivostok in Russia is mentioned as Haishenwai. This map shows China's "subtle" intention to reclaim its past territory.

China's current ambition is not only limited to retaining the territory that it considers "lost," but also to gain as much influence as possible throughout the world. Through the One Belt One Road program, which subsequently renamed the

Table 13.1 Series of events that deteriorated US–China relations during the period leading up to and after the collapse of Soviet Union

Year	Event	Description
1989	Tiananmen Tragedy	Protests demanding democratization and corruption eradication in Tiananmen were stopped by the Chinese government using military force. US responded by halting military equipment sales to China and freezing bilateral relations
1995–1996	Taiwan Strait Crisis	China conducted military exercises and launched ballistic missiles in Taiwan Strait as a form of protest against democratization in Taiwan and the spread of Taiwanese people's movement to disintegrate from China
1999	The bombing of PRC Embassy in Belgrade	US warplanes under NATO command mistakenly targeted and bombed the Chinese ambassador's office
2001	Collisions between Chinese and US fighter jets in SCS	The collision occurred as a result of a dangerous maneuver by a Chinese jet that tried to intercept a US spy plane near Hainan
2006	Chinese submarine appearance incident near Kittyhawk aircraft carrier	A Chinese submarine managed to get close to a US and allied military training site undetected until the submarine surfaced, which demonstrated China's dangerous underwater capabilities
2007	China's significant defense spending increased by 18%, which reached US$ 44.94 billion (Indonesia's defense spending was only US$ 2 billion in 2006)	US viewed China's increased defense spending as disruptive to the region's balance of power and a form of China's intent to unify Taiwan by force
2009	SCS militarization by China through military bases construction	US considers this move as detrimental to the freedom of navigation principle and violating maritime law regarding the militarization of high seas
2012	The beginning of US–China trade war	The trade war was driven by the widening US trade deficit and allegations of unfair business practices by China in the form of REMs export restrictions
2015	US demanded to stop militarization of disputed areas in SCS	China claimed that islands reclamation in the SCS is for civilian purposes, but the facts show that there are military facilities and installations on the islands
2018	US begun to impose tariffs on Chinese products (manifestation of the trade war)	US imposed tariffs worth $50 billion of imports from China over allegations of technology and intellectual property theft, and Chinese measures to protect its domestic industries. Over time, the number and products subject to tariffs grew larger and became more diverse, especially those related to automotive and electronics
2018	US and Chinese destroyers almost collide in SCS CNN (2018)	A US destroyer sailing in the SCS (international zone) was confronted with a dangerous maneuver by a Chinese destroyer, which nearly resulted in a collision

(continued)

Table 13.1 (continued)

Year	Event	Description
2014–2019	There have been reports of massive theft of sensitive US government and corporate data by Chinese hackers	Theft of US sensitive information related to government secrets, trade secrets, and blueprints of US high-tech equipment
2021	Accusations of Chinese government's genocide against Uighur people	US accused China of genocide against Uighur people in the Xinjiang region who were about to disintegrate from China
2021	US formed AUKUS	AUKUS was formed to strengthen Australia's combat capabilities, arm it with nuclear submarines, and encircle China
2022	Refinement of the Quad, which was formed in 2007	The Quad's function was strengthened from just responding to the threat of natural disasters, to include defense-security cooperation which appears to be aimed at encircling China
2023	Chinese hot air balloons entered US territory illegally	A Chinese air balloon equipped with surveillance equipment entered US airspace and was shot down

Belt and Road Initiative, China is trying to spread its influence through large-scale infrastructure and development programs in various parts of the world.

On the other hand, US places great importance on maintaining world dominance, especially in matters of military power as its foreign policy cornerstone. US has interest in ensuring freedom of navigation, both at sea and in the air. Freedom of navigation is closely related to its survival as a state. US has the world's strongest navy to safeguard its global trade and facilitate its military power deployment for political and economic intervention. US product sales and investment markets are also mostly located across the ocean. By ensuring freedom of navigation, US can protect its interests in the world optimally. Thus, US places high premium on freedom of navigation in the oceans around China in its relations with China due to its economic survival concern and the maintenance of political, trade, and business relations with its allies in Asia, such as Japan, South Korea, Taiwan, and Singapore.

Economic freedom is closely related to US' efforts in maintaining its status as the world's strongest economy. The US' push-and-pull policy on Taiwan is also a measure to secure its position and economic influence in the Western Pacific. This political move is aimed at safeguarding US economic interests in Taiwan due to its large investment in the electronics and semi-conductor industry sectors. US also has the ambition to maintain its dominance as the world's strongest military and puts its best efforts to safeguard its superiority in various parts of the world. It will try to prevent the emergence of a new "super power," especially one that can disrupt its military hegemony in the world. US responds China's continuous defense spending increase and SCS militarization by resorting to economic sanctions and show of force. From the Cold War to the present, US is still conducting strategic encirclement measures against China.

13.4 China and US' Interests in SCS

13.4.1 China's Interests in SCS

SCS is a 3,500,000 km^2 body of waters that spans from the top of the North Natuna Sea in the south to the Taiwan Strait in the north. It also borders Vietnam's mainland in the west, and Philippines' Islands in the east. It is a sea trade route worth US$ 3.4 trillion annually (Dolven et al. 2022). According to the US Energy Information Agency (2013), the SCS contains 11 billion barrels of oil reserves and 190 trillion cubic feet of natural gas reserves. This data has not even included the natural beauty of the sea as tourism potentials along with its capture fisheries and aquaculture industries. SCS is also a gateway connecting West Pacific region with South Asia to sail to Africa and Europe, and vice versa. It also borders and connects large population countries, such as Indonesia, to large consumer markets, source of cheap labors, and attractive investment locations as indicated by the relatively stable economic growth figures in Southeast Asia. Currently, the role of SCS is getting increasingly important because it is a global nexus to gain access to REMs like tin, nickel, and bauxite, along with energy resources such as natural gas and coal, most of which are produced in Indonesia. Whoever controls the SCS not only has the opportunity to gain abundant natural marine resources (energy and food), but also access to the vast wealth of eastern world.

Both China and US have interests on SCS, despite the different nature and characteristics of each country's interests. For China, SCS contains the natural resources that it requires for its survival and ambitions. SCS possesses a very large fishing area, which is a huge food source for its population of more than 1.5 billion people. In recent years, China has begun to undergo food import from abroad because it has started to experience deficits on several food ingredients (Ferreira 2022). SCS does not only function as short-term food sources but also as a medium for food imports to China.

SCS has very large oil and gas reserves. Most of China's current energy needs has to be imported, especially from the Middle East and Africa. The far distribution distance does not only incur expensive logistics costs for oil and gas procurement but is also vulnerable to disruption from the threat of piracy, terrorism, or sea blockades by other countries' military forces.

However, the dominance of oil and gas projects in SCS will make it difficult for China's rivals, namely Taiwan, South Korea, and Japan, to gain closely-accessible oil and gas supplies. Access blockade to oil and gas in the SCS region will reduce industrial productivity, which will decrease state income, and ultimately result in defense spending reduction. This condition can give China an upper hand amid the current arms race between the four countries (China, South Korea, Taiwan, and Japan). The same condition also applies if China succeeds in monopolizing access and is able to block its rivals' access to SCS.

China's success in blocking access to food and energy resources in the SCS can also weaken the ability of Southeast Asian countries to negotiate. The economic progress of Southeast Asian countries is greatly dependent on SCS stability because the sea is not only important for economic growth, but also for political stability. The centrality of Southeast Asian countries within ASEAN is also at risk because each member state tends to prioritize its own interests when dealing with China.

China does not view the SCS merely as a location with abundant natural resources and very strategic position value. SCS is also important for China's determination and quest to restore its past glory. The magnitude of China's influence over SCS dynamics can at least signal its success in escaping from the century of humiliation stigma.

The SCS is also related to China's interests in other parts of the world. In the last decade, China has had relatively strong political and economic relations with countries in South Asia and Africa. Currently, China also has port access for military bases abroad such as in Sri Lanka (Hambantota), Pakistan (Gwadar), and Djibouti. These chains of bases are very important as China's logistics chain, both for business and defense purposes. China's energy needs are mostly fulfilled through oil and gas purchase from the Strait of Hormuz and West Africa. In the early 2000s, piracy was a serious threat in the Gulf of Aden, which also emerged in the Gulf of Guinea afterward.

In recent years, hostilities between China and India have become increasingly intense due to a border dispute in the Himalayas. China–India relations have become increasingly heated as China has been actively arming India's rival Pakistan with advanced military equipment, even to the point of developing fighter jets and submarines jointly (Eurasian Times 2024). Since open conflict between China and India can occur at any time, it becomes very important for China to prepare itself to anticipate conflict escalation, one of which is to have permanent base access in South Asia and Africa. The goal is to provide China flexibility in taking immediate response using military force in the events of piracy threat or sea blockade.

For China, having a base near the Strait of Hormuz and the Gulf of Aden is not only important to balance its power against India but also against the US who has permanent presence in the Middle East and East Africa region because of the region's high vulnerability to conflict and importance in world energy production. In the event of a conflict involving the US, China can still ensure open access to sea lanes through the concrete presence of its naval fleet. In addition, having a base near strategic chokepoints such as the Strait of Hormuz and the Gulf of Aden enables China to collect intelligence data such as radio messages or electronic signatures from warships or aircraft that traverse the areas. The presence of China's military bases in South Asia and Africa is intended none other than to safeguard Chinese trading fleet from piracy threat and counter military power confrontations from both India and US.

China's growing interests in the South and West Asia, as well as Europe and Africa, indicate the importance of China's access to the SCS. Chinese ships must first sail the SCS to reach areas in South Asia and Middle East, as well as Africa and Europe. If China loses access to the SCS, overseas bases will be cut off from the chain of command and logistics, thus hampering its operational effectiveness.

China can indeed take a detour to avoid the SCS, such as by passing through Sulu Sea and Makassar Strait to the Indian Ocean through South Java Sea; or through the Arctic route which is quite accessible during the summer. However, the Sulu Sea route is more dangerous due to piracy and kidnapping threats by armed groups in Southern Philippines (The Diplomat 2020). The Sulu Sea route is also farther and risks the potential for maritime closure or blockade due to the currently heated relations between China and the Philippines.

The Arctic route could be opted to reach Africa, Middle East, and South Asia. However, this route is very distant, which can increase logistics costs. In addition, the Arctic route is also too desolated, making it relatively difficult to replenish logistics, and is prone to the risk of iceberg crash incidents. The Arctic route is also quite difficult for China because it does not have access to any naval bases there. Furthermore, China does not completely trust neither Russia nor the Nordic countries. Therefore, taking the route through the Arctic, especially for military purposes is not a wise policy due to the risk of potential blockade.

China's freedom in SCS is also important to eliminate the "Malacca Dilemma," namely the serious threat of Malacca Strait's closure to the Chinese economy. As a result of these conditions, China will be isolated from the outside world and suffer destruction from the inside due to its high dependence on imported commodities, especially energy and food. China's blue water navy development along with the acquisition of aircraft carriers, nuclear submarines, nuclear-powered submarines, large amphibious landing ships, and the PLA Rocket Force establishment to integrate its nuclear power, is none other than to "break" the strategic encirclement scenario by the US and its allies.

13.4.2 US' Interests in SCS

The SCS also holds critical function for US' interests, which are not exactly similar to China's interests. Natural resources (energy and food) in SCS are very important and strategic resources. According to Energy Information Administration (2013), US possesses 48.3 billion barrels of oil reserves (or four times of the SCS) and 691 trillion cubic feet of gas reserves (or almost four times of the SCS). US' geographic position and condition that is surrounded by open sea and located in the middle of the Pacific and Atlantic oceans also allows it to obtain energy supplies from other parts of the world, both from South America, Southeast Asia, Africa, and the Middle East.

The same condition also applies to food resources. US is as a self-sufficient country in terms of agriculture and livestock. US neighbors, namely Canada and Mexico, are also large food producers who export their products to the US. If necessary, US can also supply its needs from other countries such as Australia and New Zealand. Therefore, US has very small or even no urgency to obtain food sources from SCS.

US' interests on natural resources in SCS are more about ensuring that China is incapable to use these resources freely, while preserving Southeast Asian countries' ability to manage them (U.S. Embassy & Consulates in Indonesia 2020). Besides not having any urgency on oil, gas, and food resources availability issues, US also wants Southeast Asian countries to have strong and stable economy and achieve a certain level of independence to counter China's actions in the region.

US is not really aiming for natural resources in the SCS but is very interested in its strategic position. SCS connects the Indian Ocean and the Pacific Ocean, or what is referred to as the central position in sea defense concept.[1] Freedom of access from its strategic position can guarantee the linkage of US fleet command in both ocean. By controlling the SCS, US is not only able to guarantee the security of its naval fleet but also to reduce Chinese naval fleet's effectiveness.

SCS also plays an important role in strategic encirclement policy of the US and its allies. The policy was previously implemented to encircle communist powers, namely the Soviet Union, China, North Korea, and Vietnam. Strategic encirclement is still implemented after the Soviet Union collapsed and communist ideology eroded. The aim is not to contain the spread of communist ideology, but to limit the maneuver space of other powers that it considers as threats. Currently, US has been successful in implementing strategic encirclement in Northeast Asia, mainly by building bases and stationing military personnel in Japan and South Korea. To respond the threats in East Asia region, US has access to bases or defense cooperation with Taiwan.

US is also deploying military forces in Darwin, Australia. US used to have access to two military bases in the Philippines (Clark Air Base and Subic Naval Base) after World War II, but the cooperation was stopped after the collapse of the Soviet Union. In the last 10 years, it is relatively difficult for US to regain access to bases in the Southeast Asia region, especially in the Philippines, due to the strong influence of China during Duterte administration (U.S. Department of Defense 2023). Nevertheless, Philippines has granted access to nine of its bases for the US military since 2023. US also has access to six military bases in Papua New Guinea.

US' interests in SCS are closely linked to advanced materials industry. Southeast Asia is a region that produces REMs for electronics and automotive goods production, which are the mainstay of US exports. Southeast Asia remains a relatively neutral region amid the turbulence of US–China trade war and the divided world opinion as a result of the Russo-Ukrainian War. The current circumstances make US unable to freely procure REMs from China and Russia. US can still procure REMs from Africa, but it is less efficient because REMs producing countries in Africa are still being in civil war, threatened by terrorism, or having poor diplomatic relations with

[1] To gain more comprehension on central position and concentration concepts, see Mahan, A.T. (1898). *The Influence of Sea Power Upon History, 1660-1783*. John Wilson and Son University Press.

the US. This condition requires US to place important priority on establishing good relations with Southeast Asian countries and gaining access to the SCS.

Securing access or at least neutrality in SCS is also important in maintaining the sustainability of US business in the region. The value of US trade that passes through SCS reaches US$1 trillion and still has continuing growth potentials in the future (U.S. Embassy & Consulates in Indonesia 2020). Following the return of US investment and tourists to Southeast Asia after the Covid-19 pandemic (Forward Keys 2022), security and response capability in SCS also becomes a priority for the US government in order to protect its citizens. Securing US' access to SCS is also related to the economic interests of its allies in the West Pacific. Important US allies, namely Japan, South Korea, and Taiwan, are economically dependent on the natural resources and freedom of navigation in SCS.

13.5 Indonesia's Interests in the South China Sea

The South China Sea (SCS) dispute becomes increasingly intensified when China's claim is followed by an aggressive political approach that combines military and paramilitary power. China also uses economic threats, such as embargoes or investment suspensions, as a way to influence the decision-making of ASEAN and its member states. China even goes so far as to divide ASEAN's cohesiveness to prevent internationalization of the SCS dispute (VOA News 2012). China is a super power, which remains difficult for all ASEAN member states to fight at once, both politically, economically, and militarily. It is inevitable for ASEAN to seek a balancing power who is willing to assist ASEAN member states in improving their bargaining power over China in the SCS dispute.

Indonesia's interests in international politics are always founded upon the Independent and Active policy, namely having as many friends as possible, avoiding hostility to the fullest extent possible, and putting its best efforts to realize world peace and order. Indonesia has complex interests in the SCS dispute, ranging from safeguarding sovereignty and territorial integrity to ensuring food and energy security. All these interests must be manifested within the Independent and Active policy framework in the cover of Indo-Pacific Outlook. Indonesia's interest in safeguarding its sovereignty in the North Natuna Sea is not merely driven by its desire to be free from other nations' domination, but also for the sake of its survivability.

Indonesia will have more than 300 million population in the near future, which indicates its increasing needs for greater and more stable supplies of food and energy. Indonesia must also be able to create as many jobs as possible because in the next few years it will experience a demographic bonus, where the total productive age population in the country far exceeds the total non-productive age population. If not managed effectively, this demographic bonus can become a boomerang that causes new problems like massive unemployment, criminal, and surge in government spending burden for subsidizing the poor.

To overcome these challenges, North Natuna Sea's potentials can offer solutions and play an important role in preserving the sustainability of national economic stability. In energy sector, Natuna East Block has 46 trillion cubic feet of gas reserves (The Ministry of Energy and Mineral Resources of the Republic of Indonesia 2022). Indonesia's economy has been under pressure after the Covid-19 pandemic and Russo-Ukrainian War due to the increasing need for energy imports which stems from its limited national oil and gas productivity and depleted reserves.

In food sector, Natuna Sea has 767,126 tons of annual capture fisheries potentials (The Ministry of Marine Affairs and Fisheries of the Republic of Indonesia 2017). Fisheries commodities are not only beneficial to strengthen national food security but also to increase community income and tax revenues. Fisheries potentials in Natuna Sea also plays an important role in opening employment opportunities for coastal communities. Any political and military threats from other countries that disrupt community and entrepreneurs' freedom and security in managing the resources in Natuna Sea will not only increase unemployment and poverty but also decrease public trust in the government and encourage criminal acts at sea, such as armed piracy.

Most of the relations and transactions between Indonesia and its strategic trading partners in the West Pacific region are conducted through SCS. The list of countries of origin and destination for Indonesia's imports and exports through SCS is shown as follows (Table 13.2):

Indonesia's total trade transactions that passed through SCS from 2019 to 2023 is at least valued at US$1.1 trillion. The data shows that Southeast Asia's economy also depends on the open access of SCS as a safe trade route. If this access is closed, it will trigger systemic economic paralysis in Southeast Asia, which can lead to regional political instability. Weakening political stability might encourage the return of classic problems in Southeast Asia, namely insurgent movements, separatists, and terrorism.

Table 13.2 Indonesia's export and import value in the period of 2019–2023 in US$

Country	Export Value	Import Value
China	$244.28 billion	$270.68 billion
Japan	$93.18 billion	$74.66 billion
South Korea	$45.83 billion	$46.93 billion
Taiwan	$30.47 billion	$20.04 billion
Malaysia	$56.76 billion	$47.39 billion
Thailand	$33.83 billion	$46.34 billion
Philippines	$45.20 billion	$5.67 billion
Vietnam	$32.96 billion	$21.30 billion
Total	$582.51 billion	$533.01 billion
AS (Benchmark)	$113.69 billion	$51.96 billion

Source Central Statistics Agency of the Republic of Indonesia (2024)

Indonesia's dependence on the Indochina Peninsula's economy is quite significant, especially in food procurement sector. In 2023, Indonesia imported up to three million tons of rice and is planned to increase the import to 3.6 million tons in 2024. Most of the rice supply comes from Thailand and Vietnam (Databoks 2024). Thus, Indonesia is at risk of suffering food insecurity if the access to SCS is closed.

Indonesia also has defense interests in the SCS, an outermost region that functions as a buffer zone to withstand enemy attacks as far as possible from its major islands. Natuna Sea has many island chains which can be resorted as strategic defense assets, such as forward bases, listening posts, enemy naval formation breaker, or permanent defense nodes equipped with anti-ship and anti-air missiles. Natuna Sea is also an important medium of connectivity for Indonesia and other Southeast Asian countries, especially for logistics routes and economic linkage in the event of war.

13.5.1 Common Interests of Indonesia and China in SCS

Despite the conflicting views between Indonesia and China on SCS matters, both countries still have some common interests that can dampen the existing conflict. In the last 20 years, Indonesia–China trade has increased rapidly from US$4.8 billion in 2000 to US$49 billion in 2011 (Sabaruddin 2014), and US$133.65 billion in 2022. When Western countries imposed economic sanctions and trade wars, Indonesia continued to conduct normal trade relations with China. After the Covid-19 pandemic, both countries' exports and imports recovered rapidly.

Indonesia is capable of supplying the required goods for China's industrial progress, such as REMs, palm oil, and coal. For China, whose economy is highly dependent on exports, Indonesia is a very strategic trading partner due to its large consumer market and purchasing power. The logistics network for China to access the Indonesian market is also less complicated compared to markets in the South America, the Middle East, Africa, and Europe which are very far and have to come across many US military bases. These advantages certainly become China's considerations in embracing Indonesia as a strategic trading partner.

Indonesia also considers trade relations with China as an important matter because Indonesian people favor Chinese products which are cheap but able to improve the life quality of its users. The purchasing power of Chinese people is also relatively strong and their demand for Indonesian products is increasingly diverse, providing a large market opportunity to advance Indonesian economy. Chinese people also enjoy traveling to Indonesia, which helps to advance the local economy. The close trade and bilateral relations between Indonesia and China put both parties in a difficult position in pursuing their interests on the maritime dispute on SCS. This condition may also explain why China "disturbs" the Philippines and Vietnam more frequently than Indonesia, which stems from the lesser bargaining power of the Philippines and Vietnam over the Chinese economy.

China also certainly considers Indonesia's geographical position, which has several key routes to the Indian Ocean, namely through the Sunda Strait, Karimata Strait, and Malacca Strait. Both China and Indonesia desire to manage energy resource potentials in SCS as soon as possible. Energy resources exploration in the SCS, such as oil and gas, is a shared priority since both countries are net energy importers in large quantities.

The same phenomenon also applies to fisheries management in the SCS, which is an important business for coastal communities. Furthermore, fish is also a main source of protein and a major source of foreign exchange for Indonesia and China. Several studies have shown that fish stocks in the waters near China, Vietnam, and the Philippines are depleting, which prompts fishermen from those countries to sail south to conduct fishing activities in the Natuna Sea. This condition causes a lot of illegal, unregulated, and unreported (IUU) fishing issues in Natuna Sea. There needs to be collaboration between Indonesian and Chinese fishermen on the fishing mechanism in Natuna Sea.

These conditions show that China and Indonesia have the same interests on SCS, namely to ensure that the region remains in a safe and open access condition to facilitate trade between the two countries.

13.5.2 Common Interests of Indonesia and US in SCS

Unlike Indonesia–China relations that is largely based on economic and business interests, Indonesia–US relations is based on more diverse common interests. Indonesia and US have relatively similar ideologies, especially in terms of democratic governance, freedom of the press, and civilian supremacy over the military. Indonesia actively promotes democratic values in Southeast Asia and mediates conflicts by prioritizing peaceful resolution paths. Indonesia's foreign policy is relatively compatible with US foreign policy, which also promotes democratic values and freedom. Indonesia and US are religious countries, despite the difference of majority religions that their people follow.

Regarding the SCS, Indonesia and US have same interests in economic terms. However, the similarity is not in terms of how the SCS becomes a link between the two countries' trade or natural resource exploration interests, but as a link between Indonesia and Southeast Asian region's trade with US allies in West Pacific. US has significant interests in the economic stability of Japan, South Korea, and Taiwan, which also depend on the security and freedom of access to SCS for trade. Economic instability of US allies can weaken each ally's military capability, which will incapacitate them from contributing effectively to US defense policy in the West Pacific region. For Indonesia, security and freedom of access to the SCS also play an important role for its trade interests with these US allies.

Southeast Asia's economic stability is also greatly affected by security stability in SCS. Southeast Asia's economic stability does not only contribute to economic growth of the US and Indonesia but also affect the domestic security stability of each country. Maintaining the economic stability of Southeast Asia through the security stability of SCS will consequently protect both Indonesia and US from the threat of terrorism.

Indonesia and US desire for all parties to follow the South China Sea Code of Conduct, a jointly-designed rule to ensure fair utilization of SCS as a common space and to avoid conflict escalation. The push by Indonesia and US to demilitarize SCS is based on the desire to uphold international law and maintain regional security stability.

Indonesia and US have the same interest in realizing a neutral SCS that is free from excessive militarization. However, Indonesia and US have different views in the way to achieve this goal, which appear somewhat contradictory. Indonesia places more emphasis on encouraging legal compliance to the disputing countries in the SCS and establishing mutually agreed rules in the form of code of conduct. Meanwhile, US takes a more aggressive approach in addition to promoting compliance with international law, such as strengthening military forces disposition, requesting access to bases in the region, providing military assistance and grants, and selling defense equipment to the countries in dispute (especially to the Philippines and Taiwan). For Indonesia, US' aggressive approach is a dilemma because support for these countries can indeed help Southeast Asian countries to become more effective in countering Chinese provocations, but on the other hand, it is feared that this power could also turn into a threat against fellow ASEAN members in the future.

13.6 Conclusion

The trajectory of political, economic, and military disputes between US and China tend to move towards open conflict. From merely accusing each other of human rights violations and political hypocrisy, dispute between the two countries has developed into an economic conflict which manifested in the form of trade war, economic sanctions, and even embargo. Military disputes, which has escalated since the collapse of Soviet Union, are also getting heated.

Many experts believe that it is only a matter of time before the SCS conflict turns into an open and prolonged war. Several indicators always show higher likelihood on the probability of war. An open war between US and China in the SCS would be very detrimental, especially for Southeast Asian countries who are very dependent on the security stability in SCS. Indonesia will not only experience slowing economic growth and massive unemployment but will also have to suffer the burden caused by waves of refugee arrivals.

Revisiting the past records of direct engagement between China and US during the Korean War, there is sufficient reason for the likelihood of open war breaking out between the two countries in SCS. The worst scenario is when Indonesia's territory in the Natuna Sea becomes a battlefield or occupied by the conflicting parties and ends in annexation. Loss of control over the Natuna Sea area will negatively impact Indonesia's sovereignty and economy. There are two options to deal with these dynamics. First, by undertaking the best efforts to prevent open war between US and China from breaking out in the SCS. Second, by strengthening Indonesian military capabilities to the extent that it is totally capable of preventing the possibility of Natuna Sea territorial annexation in the event of war between the two countries.

Politically, Indonesia must be able to become a mediator between US and China on SCS disputes. Indonesia's diplomacy could strengthen ASEAN's solidity to become a joint force in preventing China's and/or US' intention to engage in open war. Indonesia Armed Forces can invite all ASEAN military forces to jointly develop tactics and strategies to assist each other in military operations. Indonesia also needs to improve ASEAN naval exercises.

Relations with China must be maintained despite of Indonesia's desire to demilitarize the SCS. In an open war scenario in the SCS, Indonesia must be prepared for the possibility of embargo from China. China has previously conducted embargo to Japan and US in the form of exports and imports reduction of fish and REMs products (Conalghi 2023 and CNBC 2010). To overcome this possibility, Indonesia must be prepared to look for alternative markets.

Although Indonesia and US have more commonality in terms of interests and opportunities for cooperation in SCS, especially in matters of safeguarding sovereignty and freedom of navigation, Indonesian government must be able to demonstrate a neutral foreign policy as much as possible. Indonesia may refer to anti-access and prolonged war defense strategy with main idea of convincing that attacking Indonesia is a very hard and expensive military operation. Defense equipment procurement should be targeted at the types of capability that will not evoke suspicion from other countries, but still adaptive for future battles, such as advanced anti-missiles, anti-drone, and the capability to conduct naval blockades using small and medium sized ships. Indonesia can also strengthen the function and capability of Indonesia Coast Guard (ICG) as a legal instrument in defending its sovereignty and territorial claims in accordance with international legal rules. ICG must be reinforced to be capable of maritime law enforcement up to the EEZ, while also being ready for general mobilization at any time.

References

Baskaran, G. 2024. What China's Ban on Rare Earths Processing Technology Export Means (Commentary). *Center for Strategic & International Studies.*

Central Statistics Agency of the Republic of Indonesia. 2024. Data Ekspor Impor Nasional. Retrieved on July 6, 2024 from https://www.bps.go.id/id/exim.

CNBC. 2010. China Bans Rare Earth Exports to Japan Amid Tension. Retrieved on July 6, 2024 from https://www.cnbc.com/id/39318826.

CNN. 2018. Chinese Warship in Unsafe Encounter with US Destroyer, Amid Rising US-China Tensions. Retrieved on July 6, 2024 from https://edition.cnn.com/2018/10/01/politics/china-us-warship-unsafe-encounter/index.html.

Colnaghi, W.B. 2023. Japanese Seafood Ban Signals China's Shady Virtues. East Asia Forum. https://doi.org/10.59425/eabc.1698098436.

Council of Foreign Relations. 2023. U.S.-China Relations 1949–2023. Retrieved on May 14, 2023 at 20.39 GMT+7 from https://www.cfr.org/timeline/us-china-relations.

Databoks. 2024. Indonesia Impor Beras 3 Juta Ton pada 2023, Terbesar dalam Lima Tahun. Retrieved on July 6, 2024 from https://databoks.katadata.co.id/datapublish/2024/01/16/indonesia-impor-beras-3-juta-ton-pada-2023-terbesar-dalam-lima-tahun.

Dolven, B., C. Caitlin, and O. Ronald. 2022. China Primer: South China Sea Disputes. *Congressional Research Service*. Retrieved from https://crsreports.congress.gov/product/pdf/IF/IF10607.

Eurasian Times. 2024. China Arms Pakistan with AIP-Enabled Submarine; Keel Laying Ceremony For 2nd Hangor-Class Sub Held In Karachi. Retrieved on July 6, 2024 from https://www.eurasiantimes.com/china-arms-pakistan-with-deadly-submarine-keel/.

Ferreira, G. 2022. Does China Have Enough Food to Go to War? *Military Review*.

Forward Keys. 2022. US Visitors Lead Revival of Struggling Southeast Asia. Retrieved on July 6, 2024 from https://forwardkeys.com/us-visitors-lead-revival-of-struggling-southeast-asia/.

Pruden, G.B. 2009. American Protestant Missions in Nineteenth-Century China. *Education About Asia: Asian Intercultural Contacts*, Volume 14:2. Retrieved from https://www.asianstudies.org/wp-content/uploads/american-protestant-missions-in-nineteenth-century-china.pdf.

Sabaruddin, S.S. 2014. Dampak Liberalisasi Perdagangan Ri-China Terhadap Perubahan Perdagangan dan Kesejahteraan Masyarakat Indonesia: Sebuah Pendekatan Ekuilibrium Parsial (Smart Model) dan Pemanfaatan Sistem Neraca Sosial Ekonomi 2008. Jurnal Ekonomi dan Bisnis, Volume XVII No. 3, Desember 2014.

The Diplomat. 2020. Dangerous Waters: Maritime Crime in the Sulu Sea. Retrieved on July 6, 2024 from https://thediplomat.com/2020/06/dangerous-waters-maritime-crime-in-the-sulu-sea/.

The Ministry of Energy and Mineral Resources of the Republic of Indonesia. 2022. Pemerintah Targetkan Umumkan Pemenang Lelang East Natuna Juni 2023. Retrieved on July 6, 2024 from https://www.migas.esdm.go.id/post/pemerintah-targetkan-pemenang-lelang-east-natuna-diumumkan-juni-2023.

The Ministry of Marine Affairs and Fisheries of the Republic of Indonesia. 2017. Keputusan Menteri Kementerian Kelautan dan Perikanan RI No. 50 Tahun 2017 Tentang Estimasi Potensi, Jumlah Tangkapan yang Diperbolehkan, dan Tingkat Pemanfaatan Sumber Daya Ikan di Wilayah Pengelolaan Perikanan Negara Republik Indonesia. Retrieved from https://peraturan.bpk.go.id/Download/152201/Kepmen%20KKP%20Nomor%2050%20Tahun%202017.pdf.

U.S. Department of Defense. 2023. Philippines, U.S. Announce Locations of Four New EDCA Sites. Retrieved on July 6, 2024 from https://www.defense.gov/News/Releases/Release/Article/3349257/philippines-us-announce-locations-of-four-new-edca-sites/.

U.S. Embassy & Consulates in Indonesia. 2020. The South China Sea, Southeast Asia's Patrimony, and Everybody Own Backyard. Retrieved on June 26, 2024 from https://id.usembassy.gov/the-south-china-sea-southeast-asias-patrimony-and-everybodys-own-backyard/.

US Energy Information Administration. 2013. Contested Areas of South China Sea Likely Have Few Conventional Oil and Gas Resources. *Today in Energy*. EIA. https://www.eia.gov/todayinenergy/detail.php?id=10651.

VOA News. 2012. South China Sea Dispute Flares at ASEAN Summit. Retrieved on July 6, 2024 from https://www.voanews.com/a/south-china-sea-dispute-flares-at-asean-summit/1548880.html.

Amarulla Octavian is the Vice Chairman of The National Research And Innovation Agency of Republic Indonesia. He achieved his Bachelor of Engineering from Indonesian Naval High Technology College (1998–2001), Master of Science from Université Paris 2, Pantheon-Assas, France (2005–2006) and Doctor of Sociology from The University of Indonesia (2010–2021). He served as a military naval officer and commanded various operations during 1988–2020. He is also The Rector of the Republic of Indonesia Defence University from 2020–2023.

Chapter 14
Seizing the Minilateral Moment in the Indo-Pacific: Indonesia's Response to the Transformation of the Quad

R. R. Emilia Yustiningrum, Ganewati Wuryandari, Siswanto, and Dewi Fortuna Anwar

Abstract The Indo-Pacific region is experiencing dramatic geopolitical transformations, marked by the rise of China, intensifying great power rivalry between the United States and China, and the existence of numerous flash points for open conflicts due to geopolitics, geostrategic, and geoeconomic differences. Amidst these developments, the region also offers opportunities for cooperation, especially in the domain of non-traditional security challenges (NTS). As a contested region, Indo-Pacific also sees the rise of multiple forms of minilateralism, such as the Quadrilateral Security Dialogue (Quad), which has changed significantly from its early focus as a security grouping aimed against the perceived threat from China, by developing a much broader public goods agenda to deal with NTS challenges. This article examines how Indonesia's foreign policy has responded to the growth of minilateralism in the Indo-Pacific region under the administration of Joko Widodo (2014–2024), especially looking at the government's changing attitude towards the Quad after its transformation. This paper uses a qualitative approach, gathering and analyzing data from primary and secondary sources. The paper utilizes a theoretical framework from Non-Traditional Security (NTS) issues as the enabler for regional cooperation. By analyzing domestic and international factors pertaining to Indonesia's foreign policy, this paper argues that the transformation of the Quad has helped to change the perceptions of the Joko Widodo's government about the Quad from skeptical into becoming more open minded, which open opportunities to foster regional cooperation, stability, and resilience. The paper evaluates the Quad's ability to support Indonesia's national interests and a more equitable Indo-Pacific order by analyzing the opportunities and challenges that come with its evolution.

Keywords Minilateralism · Indo-Pacific · Indonesia · Foreign policy · Quad · Non-Traditional Security (NTS) issues

---

R. R. E. Yustiningrum (✉) · G. Wuryandari · Siswanto · D. F. Anwar
Research Center for Politics, National Research and Innovation Agency (BRIN), Jakarta Selatan, Indonesia

I. Kartini and Kanisa (eds.), *Joko Widodo's Foreign Policy and Diplomatic Legacy*,
https://doi.org/10.1007/978-981-95-2085-5_14

14.1 Introduction

Discussions about the Indo-Pacific region in the past decade have been dominated by geopolitical, geostrategic and geoeconomic issues, notably the rise of China, the US–China rivalry, China's aggressive behavior in the South China Sea and tension across the Taiwan Strait. China's rise as an economic superpower has transformed its relations with the outside world, replacing ideological conflict with new opportunities for economic cooperation and mutual benefits between China and other countries. Nevertheless, China's use of economic statecraft to widen its sphere of influence, its growing military might, as well as its increasingly assertive behavior to assert its claims in disputed maritime and land border areas have also raised concerns among the neighboring countries about Beijing's hegemonic ambition. The United States openly sees China as its strategic competitor and has shored up its alliances in the region to constrain, if not to contain China, including through the formation of new minilateral arrangements such as the Quadrilateral Security Dialogue (the Quad), comprising the United States, Australia, Japan—the latter two military allies of the US—and India, a non-aligned country, whose relations with Beijing have all deteriorated in the past decade and AUKUS, a trilateral security pact between Australia, the UK, and the US.

Yet it must also be remembered that the Indo-Pacific region also faces many non-traditional security challenges (NTS) including from natural disasters, pandemics, and climate change that are often transnational in nature and also necessitating cooperation between countries to overcome. Earthquakes, tsunamis, typhoons, flooding, and landslides are frequent occurrences in the region, causing destruction of properties and large losses of lives. The Indian Ocean tsunami of 26 December 2004 near Indonesia's Aceh province which also devastated coastal areas in Thailand, Sri Lanka, and Tamil Nadu, killed over 200,000 peoples, mostly in Aceh. Military assets are usually the first responders to the site of natural disasters, engaged in various activities such as search and rescue, providing floating clinics on war ships, transporting supplies by air and sea or evacuating victims and survivors to safer areas. It is also important to note that countries most vulnerable to devastating natural disasters are in many cases also not well-equipped to deal with such disasters on their own, thus needing greater external existence. The existence of more regional capabilities that can respond quickly and come to the assistance of countries in need are undoubtedly important in the Indo-Pacific region.

Indonesia's changing response to the emergence of various forms of minilateralism in the Indo-Pacific region, specifically to the Quad, comes from the Quad's internal transformation as well as the principles that non-traditional security issues are one of the foundations of regional collaborations. Whereas traditional security challenges and threats tend to divide countries between friends and foes, non-traditional security issues can bring countries together as humans face common challenges that call for cooperation and more collective approaches. Envall (2019) explains that Indonesia has largely been absent from the debates about the Quad due to its free and active foreign policy doctrine and non-alignment. Indonesia does not wish to take sides in an open conflict between the US and China, wants to ensure ASEAN centrality and Southeast Asia's strategic autonomy. Furthermore, Envall (2019) elaborates that Indonesia's long-term absence from the debates since the creation of the Quad also partly derived from the Quad's hiatus as well as underdeveloped status until recently. The earlier perception that the Quad was engaged in a policy of containing China had led to mixed perceptions of the grouping in Southeast Asia. Concerns arose amongst ASEAN countries that the activities of the Quad may heighten regional tension as well as undermine the centrality of ASEAN that seeks to engage all states and develop a regional architecture that is inclusive.

Other academics studying the Quad's transformation have noted that, in addition to meeting actual needs for more effective responses to a variety of non-traditional issues in the region, the expansion of the Quad agenda maybe interpreted in part as an effort to allay regional skepticisms about the Quad, which have been expressed by a number of countries. The Quad's transformation has taken place in several forms. Initially, the Quad was an impromptu grouping responding to the Indian Ocean tsunami of 2004 known as the Tsunami Core Group (Buchan and Rimland 2020; Chalivet 2022; Chanlett-Avery 2018; Chen 2022, November 21). However, the Quad was in hiatus for more than a decade afterward (Buchan and Rimland 2020). In 2017, led by the US under President Donald Trump (2017–2021) the Quad was revitalized into a grouping to promote a free, open, prosperous, and inclusive Indo-Pacific region to serve the interest of all powers worldwide (Chalivet 2022). Given the trade war against China launched by President Trump during his presidency, and the rising tension between China and the other three Quad members—Australia, Japan, and India—the Quad was widely seen as being aimed against China. However, in 2019 the Quad elevated the senior official-level dialogue into a ministerial-level dialogue to discuss not only traditional security issues but also non-traditional issues including counterterrorism, humanitarian assistance and disaster relief (HADR), maritime security, development finance, and cybersecurity (Wei 2022). When COVID-19 hit the world in early 2020, the Quad responded by creating vaccine working groups (Mehra 2022, June 7; Wei 2022) and delivering public goods to the Indo-Pacific countries on global health care, climate policy, infrastructure, disaster response, and education (Duggal 2022; Seah and Koga 2023, May 24). It can be said that COVID-19 pandemic triggered a fundamental shift of the Quad to focus more on NTS challenges and succeeded in using vaccine diplomacy to engage with the Global South.

By the time Indonesia took the chairmanship of ASEAN 2023, it can be seen that the government's views of minilateral initiatives, including the Quad, has begun to change. During an interview with the New Strait Times, President Joko Widodo (Jokowi) stated that ASEAN is the only regional organization that offers various forms of diplomacy as well as principles of collaboration, cooperation, and active involvement (NST 2023, May 8). Furthermore, Jokowi also mentioned that Indonesia needed to view the Quad and AUKUS as partners, not as competitors, of ASEAN in ensuring the maintenance of a stable and peaceful region (Basorie 2023, May 29; Chaudhury 2019, November 5; Connors 2023, May 9; NST 2023, May 8). In an interview with the Australian Financial Review before his official visit to Australia, Jokowi argued that groupings like the Quad and AUKUS must support efforts to build peace and stability in the region, so the spirit must be that of engagement and not containment (Symonds 2023, July 5). Additionally, Derry Aman, the Indonesian Permanent Representative to ASEAN, stated that ASEAN had always been open to cooperation with any parties as well as the individual members of Quad and AUKUS are already dialogue partner with ASEAN (Debora 2023, May 22; Shofa 2023, May 16).

Given the lack of comprehensive studies exclusively on the transformation of the Quad, there is a pressing need for a more thorough study on this issue. This paper investigates how Indonesia's foreign policy under the administration of Joko Widodo (2014–2024) responded to the rise of minilateralism in the Indo-Pacific, particularly to the transformation of the Quad. The paper analyses the period from the formation of the Quad in 2007 to 2024, beginning with the creation of the Tsunami Core Group in the aftermath of the Indian Ocean Tsunami up to the distribution of public goods in post-COVID-19 pandemics. This timeline is important since it reveals the transformation of the Quad from a forum for security dialogue focusing on great power competition, notably in facing China's rise in the Indo-Pacific terrain, into a regional grouping that gives priority to collaboration on non-traditional security issues, particularly in post natural disaster and global pandemics managements.

The paper is qualitative in nature. It gathers data from primary resources such as from elite surveys, speeches of heads of governments and ministers, official reports of state agencies, governmental regulations, multilateral agreements and treaties as well as from secondary sources including books, journals, magazines, and newspapers. The paper utilizes content analysis to scrutinize the written documents. The paper is divided into six parts. The Introduction provides the background on the dynamic in the Indo-Pacific, particularly the emergence of minilateralism such as the Quad and its transformation and how the Indonesian government's changing responses. The Non-Traditional Security issue as an important foundation for regional collaboration is used as the analytical framework. The early history of the Quad provides a context on how the Quad's members have exchanged views on ways to contain China. The frequency of global pandemics and natural disasters highlight the urgency of NTS and the rationale behind the Quad's transformation. Referring to various surveys from both domestic and international institutions, Indonesia's foreign policy responses to the Quad under Jokowi gradually changed from being non-committal to skeptical, into more positive views. In the final analysis it can be argued that one of Jokowi's legacies vis-à-vis minilateralism in the Indo-Pacific is that Indonesia is now open to cooperating with the Quad functionally on various NTS issues.

14.2 Non-Traditional Security (NTS) as an Enabler for Regional Cooperation

The discourse on NTS threats started in the 1990s. NTS are primarily non-military in nature and share similar characteristics. Their spheres are transnational that are neither entirely domestics nor exclusively interstate. They can appear at short notice but spread quickly because of globalization and technological innovation. NTS challenges and threats cannot be eliminated completely but can be mitigated by utilizing coping mechanisms. NTS mitigation requires not only state-level solutions but also regional and global collaborations. The objects of security are not just confined to the State but also include the wider public at both the individual and community levels (Mely Caballero-Anthony 2005, October 26–28; Mely Caballero-Anthony 2010).

The Copenhagen School provides a valuable instrument to investigate non-military perils that have been categorized as NTS challenges. The referent objects are both individuals and groups as well as issue areas that are rapidly identified as affecting security in a broader sense. The actors of securitization include states, political entities, military, and society by enunciating the existence of threats to the referent objects. The securitization process can be carried out through speeches, policy initiatives, or by creating institutions in order to influence specific audiences on the existence of threats (Buzan et al. 1998; Mely Caballero-Anthony 2005, October 26–28).

NTS issues have four characteristics that can contribute to regional collaboration. First, the transnational state of the NTS challenges presents opportunities for countries to search for commonality of purposes (Zimmerman 2014, p. 153). Second, NTS issues have no political impediments and are less likely to cause a similar degree of sensitivity compared to the traditional defense and security matters (Ball 2012, p. 22; Tow and Taylor 2009, p. 9; Zimmerman 2014, p. 154). Third, NTS issues provide favorable circumstances for developing regional cooperation because addressing natural disasters and pandemic threats are not only significant in themselves but also critical to nurture assurance and promote community building in the region (Tow and Taylor 2009, p. 10; Zimmerman 2014, p. 154). Lastly, the shared vulnerability to NTS issues call for the engagement of the coastal states of the Pacific and Indian Oceans to create multi-layered governance arrangements including multiple new players and various governance machineries (Hameiri and Jones 2011, p. 3; Zimmerman 2014, p. 154).

The creation of the ASEAN Regional Forum (ARF) in 1994 indicated a beginning of NTS collaboration in the region of Indo-Pacific. ARF gave priority to responding to contemporary security problems in the Indo-Pacific region in post-Cold War period (Khong and Nesadurai 2007, p. 33; Zimmerman 2014, p. 154). NTS challenges have become an integral element of the ARF and have largely become its jurisdiction (Haacke 2009; Zimmerman 2014, p. 155). In the period of 2000s, the ARF program has rapidly expanded into actions dedicated to NTS collaboration and between August 2013 up to December 2014 nearly most of the ARF' meetings focused on NTS problems including transnational crime, natural disaster managements, environmental degradation, and cyber security (Zimmerman 2014, p. 155). Within these issues, the ARF has also formed the on-going Inter-sessional Support Group (ISGs) or Inter-sessional Meetings (ISMs), and thus placing NTS issues as outstanding parts of the ARF program (Khong and Nesadurai 2007; Zimmerman 2014, p. 155). The ISMs covered the NTS issues on CBM and preventive diplomacy, counterterrorism and transnational crimes, HADR, disarmament, non-proliferation, and maritime security that have been the practical concern for ASEAN as well (Arase 2010, p. 815). Within the ISMs framework, there was a growing chain of working groups, agreements, policy coordination efforts, and relationships (Arase 2010, p. 815).

Furthermore, the ARF has not only developed a first track process that led to agreements at the ministerial meetings but also a second track one involving academic specialists as well as civil servants and military officers in an unofficial role (Arase 2010, p. 815). The second track process provides new ideas and policy inputs to the governments (Arase 2010, p. 815). The track two process has developed into various active global level meetings, producing policy advice, publishing papers as well as sending reports to funding agencies (Arase 2010, p. 815; Ball et al. 2006; Simon 2002).

Besides the ARF, ASEAN has also organized another platform that also carries out collaboration on the NTS challenges. The creation of ASEAN Defense Ministers' Meeting (ADMM) in 2006 was aimed at advancing regional and defence cooperation into a higher level (Zimmerman 2014, p. 156). It moved forward from confidence building measures into palpable defense and security collaboration within the framework of ASEAN (ADMM 2007, p. 2). In 2009, ADMM transformed itself into an NTS-based security dialogue (Zimmerman 2014, p. 157). ADMM's first concept paper, "ASEAN Military Assets and Capacities in Humanitarian Assistance and Disaster Relief," outlined the primary arrangements for regional military responses in the post natural disaster periods in the region (ADMM 2009b; Zimmerman 2014, p. 157). It was followed by "ASEAN Defence Establishments and Civil Society Organizations (CSOs) Cooperation on Non-Traditional Security" to endorse the collaboration between ASEAN's defence agencies and civil society groups and adopted into the ADMM's Three-Year Work Program (ADMM 2009a, p. 157; Zimmerman 2014).

14.3 Early History of the Quad

The root of Quad was based on the tsunami 2004. As one of the largest natural disasters in world record, the tsunami brought foreign militaries to deliver humanitarian assistance and disaster relief to the affected countries. The US Navy commenced Operation Unified Assistance by deploying USS Abraham Lincoln to Aceh, Indonesia, and it was succeeded by the navies of Japan, Australia, and India, to send troops and humanitarian action. These ad hoc actions started naturally and led to the birth of the Tsunami Core Group. This was the genesis for a broader collaboration in maritime affairs in the Indo-Pacific region (Buchan and Rimland 2020; Chanlett-Avery 2018).

The ideological elements of the group of four emerged when Shinzo Abe, who first served as Japanese Prime Minister from September 2006 to September 2007, proposed the establishment of an Arch of Freedom and Prosperity in the Asia-Pacific. In December 2006, Indian PM Mammohan Singh visited Tokyo. During his visit, both Japan and Indian declared a joint statement on their willingness to start a security dialogue with likeminded states in the region. Furthermore, US Vice President Dick Cheney expressed his nation's interest in a minilateral security grouping involving the four countries including Australia, India, Japan, and US, in February 2007, during his official visit to Australia and his meeting with Australian PM John Howard. Shinzo Abe's official visits to India in 2006 and to the US in April 2007 have come to be seen as the genesis of the formation of the Quad (Buchan and Rimland 2020; MFAJ 2006, December 15, 2006, November 30).

During the ARF session in Manila, the Philippines, in May 2007, governmental representatives of India, the US, Japan, and Australia talked informally which constituted the first official gathering of the Quad. The talk discussed the shared interests among member countries that led to the initiation of the Malabar joint military exercise between India and the US. However, the Quad's narrower purposes and objectives led to certain extreme speculation that it most likely turns into a military pact, an Asian NATO, primarily against the perceived China threats. As a consequence, India, Australia, and the US were hesitant to formalize the regional grouping and started to disassociate themselves from any minilateral security dialogue (Buchan and Rimland 2020).

Shinzo Abe's resignation inclined the Quad of its architect. Labour Party leader Kevin Rudd's election as Australian prime minister at the end of 2007 also led Canberra to disengage from the Quad, citing that it was not aligned with Australia's strategic interests and thus Australia would not participate in the Quad's meeting in January 2008. The Quad 1.0 was officially in hiatus from the beginning of 2008 to the early 2017 (Buchan and Rimland 2020).

The worsening geopolitical dynamics marked by the rise of China and the deterioration of Beijing's relations with Washington, Tokyo, Canberra, and New Delhi paved the way for the revivalism of the Quad as a much more active minilateral security arrangement. Shinzo Abe, who again served as PM of Japan from 2012 to 2020 published an article in a US think tank website in 2012 on "Asia's Democratic Security Diamond" as a response to China's assertive actions in the disputed areas in South China Sea and East China Sea (Wei 2022, p. 289). US President Donald Trump (2017–2021) pursued a more vigorous anti-China policy, including a trade war against China. Australia under the Prime Minister Scott Morrison from the Liberal Party (2018–2022) also became more critical of China in response to the latter's economic pressures against Australia. India under PM Narendra Modi also became alarmed at China's actions in the Indian Ocean which the former viewed as being against New Delhi's interests. While attending the East Asia Summit (EAS) in Manila, the Philippines, in November 2017, top leaders from the US, Australia,

Japan, and India again met as a group of four, marking the emergence of the Quad 2.0. Abe came with the proposal for regional order to constrain China's strategic dominance in the Indo-Pacific. The Manila meeting was considered as the rebirth of the Quad (Buchan and Rimland 2020; Chanlett-Avery 2018).

The Quad member states vehemently denied that the group of four was an anti-China movement. For instance, Indian Foreign Minister S. Jaishankar stated at the Raisina Dialogue in April 2021 that the Quad was not an Asian NATO (Chalivet 2022).

14.4 The Urgency of NTS Challenges and the Transformation of the Quad

As already discussed earlier, NTS challenges provide opportunities for enhancing, broadening, and developing modes of mutual partnership in the Indo-Pacific (Zimmerman 2014, p. 1). During the period of 2004–2016, the demand for UN annual international report aid for natural disasters worldwide expanded sharply from 3.7 billion to 20.1 billion USD (IFRC 2016). In the Indo-Pacific, the 2004 tsunami remained as the worst natural disaster in human history, followed by Myanmar's Cyclone Nargis in 2008, Japan's tsunami and earthquake in 2011, and the Philippines' tropical storm Haiyan in 2013.

A 2022 Asian Development Bank (ADB) report shows that states most impacted by natural hazards receive less funding for adaptation to change. Over the last two decades, 83 percent of natural calamities such as storms, heat waves, and floods, have taken the lives of 410,000 people and impacted 1.7 billion more (ADB 2022). The ADB report mentioned the huge stimulus deals that are currently being distributed worldwide in response to the pandemic (ADB 2022). In order to deal with the socio-economic impact of COVID-19, the ADB among others provided food and other goods to 92 million people, as well as cash assistance and vouchers to 5.9 million (ADB 2022).

The Indo-Pacific is the region where the majority of people displacement due to natural disasters occur. At least 225.3 million internal displacement happened in the region in 2010–2021 (ADB 2022). Natural calamities displace millions of people annually. Tropical storms were the major cause of displacements in the Pacific in 2010–2021, with 93,000 displaced in Vanuatu, Fiji, Solomon Islands, and Tonga (ADB 2022). After states struggled to deal with the spread of pandemic, measures to constrain COVID-19 hinder humanitarian intervention and foreign aid workers (ADB 2022).

In January 2021, Joe Biden replaced Donald Trump as president of the United States and carried out important changes in American foreign policy. Although President Biden did not reverse the US trade war against China, the US under President Biden has placed greater importance on multilateralism as well as working with allies and partners. While Trump revitalized the Quad in 2017 primarily as a grouping to constrain China, Biden has given greater emphasis to mitigating natural calamities and pandemics.

In a virtual summit on March 2021, the Quad leaders stated the need for the forum to pay more attention to humanitarian assistance and disaster relief (TWH 2021, March 12b). At the meeting, the Quad Vaccine Partnership agreed to accelerate the end of pandemic. The Quad leaders devoted themselves to taking imperative joint actions to extend the production of vaccines in 2021. The group of four leaders also agreed to collaborate to enhance and help the Indo-Pacific nations with vaccination by utilizing significant international framework such as WHO and COVAX (TWH 2021, March 12a).

The day after the virtual summit on 13 March 2021, an article jointly written by Joe Biden, Narendra Modi, Scott Morrison and Yoshihida Suga was published in the Washington Post. The leaders of the Quad member states highlighted the humanitarian crisis caused by the 2004 tsunami that propelled the four democratic countries to coordinate humanitarian assistance and disaster relief for people in need. The article also mentioned the challenges of climate change and pandemics, as well as the commitment of the Quad to cooperate with other countries in dealing with these transnational challenges (USECA 2021, March 15).

The Quad leaders' attention to humanitarian threats is increasingly reflected in the Quad Summit in May 2022 in Tokyo which produced the Quad vision. The Quad explored new initiatives, including frameworks for disaster management by establishing the Quad Partnership on Humanitarian Assistance and Disaster Relief in the Indo-Pacific (Chen 2022, November 21, p. 2). This is an effort to expand the scope of the Quad cooperation and open opportunities for ASEAN and Southeast Asian countries to join in human security cooperation.

Four months later, on 23 September 2022 in New York, the foreign ministers of Japan, India, and Australia and the Secretary of State of the United States signed the guidelines of Quad partnership on humanitarian assistance and disaster relief (TWH 2024, September 21). The aim of the guidelines is to create a foundation for the Quad member states to enhance coordination and accelerate operational strategy in post natural disasters. By utilizing the guidelines, the Quad countries will have a meeting twice a year to exchange lessons and current operational and carry out at least once plan-based table-top exercise.

The Quad's foreign ministerial meeting on 20 May 2023 made several points expressing supports for ASEAN. These include the Quad's support for ASEAN centrality and unity, being in line with ASEAN principles and priorities, support for the ASEAN Outlook on the Indo-Pacific, and Indonesia's Chairmanship on 2023 (TWH 2023, May 20). At the New York meeting, the transformation of the Quad Vaccine Partnership into the broader Quad Health Security Partnership was also announced. Through this partnership, the Quad will strengthen coordination and collaboration in supporting health security in the Indo-Pacific. Furthermore, the Quad also plans a series of activities to build regional capacity to detect and respond quickly to disease outbreaks that have the potential to become epidemics and pandemics in the forms of development of health workers, disease oversight, building system on electronic health information, and coordination of outbreak responses including Quad Pandemic Preparedness Exercise (TWH 2023, May 20).

The Quad partners also agreed to help each other produce vaccines for the global community. The Quad partners will work collaboratively to expand production of safe and effective COVID-19 vaccines at facilities in India, prioritizing increased vaccine capacity authorized by the Strict Regulatory Authority (SRA). The US government, through the US International Development Finance Corporation (DFC), will work with Indian company, Biological E. Ltd., to set up the fund for enhancing vaccine production up to 1 billion doses maximum by 2022. Japan, through the Japan International Cooperation Agency (JICA), will provide a soft Yen loan to the Government of India to expand production of the COVID-19 vaccine for export (TWH 2021, September 24). The Quad partners will ensure expanded manufacturing will be exported for global benefit, which will be achieved through multilateral initiatives such as COVAX, which provides life-saving vaccines to low-income countries and for countries in need. The Quad partners will also work together to strengthen vaccination by leveraging existing health security and development programs in the Indo-Pacific.

On the side-line of the G7 summit in Hiroshima on 15 May 2023 the leaders of Australia, India, Japan, and the United States met again to affirm their commitment to support the Indo-Pacific region that is free, open, inclusive, and resilient (USDS 2023, March 3). On that occasion, the Quad also conveyed its vision, which contained a statement that, in carrying out its activities, the Quad would act together to find solutions for the benefit of the entire region. In response to the impact of the COVID-19 pandemic, the Quad is working together to expand safe, affordable, and effective vaccine production and equitable access. Additionally, the Quad will address major challenges in the Indo-Pacific in terms of health security, technological change, climate change, and strategic challenges facing the region (TWH 2023, May 20).

At the Hiroshima meeting, the Quad emphasized its forward-looking Indo-Pacific orientation based on the following principles: 1) The Quad would like to invest in the prosperity and stability of the Indo-Pacific and commit to delivering benefits by providing options that build resilience, open communication and economic growth; 2) The Quad maintains and strengthens stability in the Indo-Pacific and manages competition responsibly, in accordance with international law, including the UN Charter; 3) The Quad recognizes the centrality, and leadership of regional institutions, including ASEAN, the Pacific Islands Forum, and the Association of Indian Ocean Rim Nations, and will work with them to advance shared interests; 4) The Quad works transparently and in open dialogue to implement a practical agenda that delivers sustainable economic and social value, and is responsive to regional and global partners (TWH 2023, May 20).

14.5 Joko Widodo's Foreign Policy on the Changing Perception and Engagement with the Quad: From Skepticism to Appreciation

14.5.1 Regional Perspective

From the very beginning, it was clear that the establishment of the Quad attracted mixed responses from countries in the Indo-Pacific region. Some countries supported it, some openly opposed it while others displayed ambivalence towards the Quad, which has widely been viewed as a response to increased Chinese economic and military power (Singh 2022, January 14). The Chinese government saw the Quad as an anti-China grouping and issued a formal diplomatic protest, calling it an "Asian NATO" (Wei 2022, p. 297). Others, such as South Korea, New Zealand, and Vietnam, are currently three important partners of the Quad, referred to as the Quad Plus. Among ASEAN countries, Vietnam and the Philippines are the most supportive of the Quad, as both face a looming threat from China in the disputed waters of the South China Sea. Nevertheless, not all ASEAN countries openly state their views of the Quad, neither expressing support nor renouncing the security grouping (Laksmana 2020, pp. 106–107; Wei 2022, p. 298). Besides concerns about the side-lining of ASEAN centrality, some Southeast Asian countries' skepticism stems from the fear that their region is getting caught in the superpower competition and may be forced to choose sides (Seah and Koga 2023, May 24, p. 1).

Indonesia's response to the Quad offers an interesting example of the evolution of perceptions regarding the Quad, from skepticism to a more positive perspective. Despite Indonesia's strategic position in the center of geopolitical competition in the Indo-Pacific and having friendly relations with all Quad member states, Indonesia was initially among the Quad sceptics and the minilateral security grouping was not seen as significant for Indonesian foreign policy. Indonesia was concerned that the grouping would compromise Jakarta's traditional non-alignment and sideline ASEAN centrality in the region (Kilic 2023, January 30). Since its establishment in 2007, the Quad was never referred to in Indonesia's foreign policy statements. For example, between 2008 and 2023, the term "Quad" was never mentioned in the Indonesian Foreign Minister's Annual Press Statement (PPTM). The annual press statement at the beginning of each year is an important review of Indonesia's diplomatic accomplishments as well as the foreign policy priorities for the coming years (DU 2024, January 9).

Till President Joko Widodo's positive pronouncement on the Quad in an interview in May 2023, the Indonesian government never referred to the Quad at all, due to the Quad's unpopularity with domestic audiences (Darmawan 2021, March 16). A 2018 survey carried out by Huong Le Thu from ASPI (Australian Strategic Policy Institute) (2018) involving respondents from the ten ASEAN countries found out that Indonesians were among the most ambivalent towards the Quad. Other surveys about Indonesians' support for cooperation with the Quad (Muhibat and Dzakwan

2018, December 7; Thu 2018), revealed that more than 50 percent of Indonesians surveyed did not regard cooperating with the Quad as desirable at the moment but probably desirable in the future (Muhibat and Dzakwan 2018, December 7; Thu 2018). Meanwhile, the survey conducted by the Lowy Institute in 2021 also showed the contradictory attitude within which the majority (66 percent) answered "do not know" when asked if they would prefer or decline Indonesia's membership in the Quad (Bland et al. 2021). This response might relate to other findings that very few Indonesians have heard of the Quad (7 percent) (Bland et al. 2021).

Domestic circumstances significantly affect Indonesia's free and active foreign policy. Due to its historical reluctance to take sides in great power conflicts, Indonesia initially felt uneasy about the Quad because of the earlier perception that it was aimed at containing China. An informal security framework such as the Quad does not have a strong appeal for Indonesia, which has steadfastly maintained its non-aligned position (Thu 2018). Any deviation from its foreign policy principles may cause domestic uproar, something the Indonesian government has strived to avoid (Sulaiman 2020, October 28). Le Thu's study (2018) revealed that the Quad had certain perception issues, such as the widespread belief that it was at odds with ASEAN and China. Furthermore, the ASPI survey also discovered that 54 percent ASEAN respondents perceived the Quad as being an "anti-China bulwark" (Thu 2018). This finding implies that this is a widespread belief in Indonesia as well.

As a result, in 2019 Indonesia launched its own concept of the Indo-Pacific, later known as the ASEAN Outlook on the Indo-Pacific, which emphasizes ASEAN Centrality, inclusiveness, transparency, openness, and rule of law as the underlying principles for promoting cooperation in the Indo-Pacific. This was done to counter the formation of perceived anti-China coalitions, such as the Quad, which Indonesia and other Southeast Asian states believed would further erode trust and escalate strategic rivalries in the region (Nabbs-Keller 2020).

14.5.2 Changed Attitudes

Indonesia's outlook appears to have lately changed despite its continued caution (Sulaiman 2020, October 28), as shown by its growing openness to the Quad. In the State of Southeast Asia Survey 2022, it revealed that the Indonesian respondents who agreed and strongly agreed that a stronger Quad will be a beneficial development for the region were 64.9 percent (Seah et al. 2022).This positive perception of the Quad went down to 51.3 percent in 2023 (Seah et al. 2023). The Research Center for Politics, National Research and Innovation Agency, Indonesia, also carried out a survey of elite's attitudes about various aspects related to the Indo-Pacific geopolitical dynamics in 2022. The survey results also indicate a positive shift in Indonesia's elite attitude towards the Quad. At the core of these survey findings is the view that the Quad plays a crucial stabilizing role in the region. The majority of respondents (64 percent) in the elites' survey on Indonesia in the US–China rivalry in the Indo-Pacific in 2022 agreed and strongly agreed that the Quad affects Indonesia's national interests by contributing to regional stability. The Indonesian elite in this survey see the Quad as, at best, a necessary strategic balancer to prevent the expanding hegemonic tendencies of China in the region. Participants in this survey included academic experts as well as representatives of state agencies and ministries having an interest in Indo-Pacific related issues in Indonesian foreign policy.

Indonesia's more positive response appears to be consistent with the general shift in Southeast Asian perspective on the Quad. A stronger Quad is seen as beneficial for the development of the region, by promoting practical cooperation, according to more than half of Southeast Asian respondents in ISEAS Yusof Ishak Survey 2023, with only 12 percent disagreeing (Seah et al. 2023; Yaacob et al. 2023). 69 percent of respondents who were asked what collaboration between ASEAN and the Quad would mean for the area said that it would be advantageous or complement ASEAN's own initiatives to promote cooperation in Southeast Asia and elsewhere (Seah et al. 2023).

The developments occurring within the Quad partly contributed to these shifts in perceptions of the Quad. The group has changed significantly since its first formation as a quadrilateral security dialogue in 2007, when its four members—Japan, the US, Australia, and India—focused primarily on opposing China (Wei 2022, p. 291). Its emphasis has been broadened particularly since 2021 to include promoting practical cooperation and providing regional "public goods." With this broadening scope focusing on human security, the Quad has re-framed its purpose from exclusively countering China to a partnership for sharing of capabilities for common good that Indo-Pacific countries are keenly interested in, including maritime, global health, education, climate, critical and emergency technology, cyber, and infrastructure (Seah and Koga 2023, May 24, p. 2). The spread of COVID-19 pandemic since early 2020 has changed the Quad's foreign ministerial meeting agenda into paying more attention to non-traditional security issues including disaster management and vaccine distribution (Chen 2022, November 21; USDS 2022, September 23). The Quad's humanitarian agenda, which includes vaccine collaboration, disaster relief, and other humanitarian assistance, is valued by some Southeast Asian nations (Wei 2022, p. 300). In fact, during the COVID-19 pandemic, Indonesia benefited from Australia's $100 million commitment to the Quad Vaccine Partnership with the United States, Japan, and India.

The Quad's ASEAN adaptation has had a significant impact on how the Quad was seen in Southeast Asia (Singh 2022, January 14). The words "free," "open," "peace," "stability," and "inclusive" were used in the Quad Leaders' Vision Statement on March 12, 2021. Moreover, the affirmation of the Quad's support for ASEAN's unity and centrality as well as the ASEAN Outlook on Indo-Pacific (TWH 2021, March 12a), delivered a signal that the two can exist on an equal footing. The Quad seems to be positioned as ASEAN's partner with a shared goal in the Indo-Pacific policy that focuses on delivering public goods in the area rather than defense cooperation. The changes the Quad made to align with ASEAN priorities helped to reduce the skepticisms felt towards the grouping, including in Indonesia. This was reflected in the recent favorable statement made by Joko Widodo (Jokowi), Indonesia's president, that has been regarded as a "softening stance" by Indonesia on the Quad (Connors

2023, May 9), by mentioning that the Quad and AUKUS should be perceived as partners and not competitors (Connors 2023, May 9; NST 2023, May 8) to create a stable and peaceful ASEAN region (Connors 2023, May 9).

Although Jokowi's comments are not yet viewed as an official Indonesian government policy (Connors 2023, May 9), his remark is strategically pivotal in terms of the relationship between Indonesia and the Quad. Indonesia's foreign ministry has so far been critical of AUKUS, a security pact between Australia, the United Kingdom, and the United States established on 15 September 2021, in which both the two nuclear weapon states—the US and the UK—would invest the nuclear-powered submarine to the non-nuclear weapon state—Australia (Doherty and Hurst 2023, March 14). Two days after the AUKUS establishment, Indonesia issued a diplomatic statement criticizing Australia's Nuclear-power Submarines Program on 17 September 2021 (MFARI 2021, September 17). Indonesia has also been active in proposing a revision of the Non-Proliferation Treaty to include nuclear materials in nuclear-powered submarines under the oversight of the International Atomic Energy Agency (IAEA) (Herscovitch and Priyandita 2021, November 8). Unlike its open criticism of AUKUS, as discussed earlier Indonesia has historically been officially silent on the Quad.

14.5.3 Broken Silence and More Can Be Done

Jokowi's remark on the Quad can be regarded as a partial change in Indonesia's thinking about the Quad. Although the implication of this development so far is still unclear, Indonesia's recent positive thinking about the Quad could be an entry point for cooperation with the Quad as a group. From Indonesia's perspective, the Quad's non-traditional security collaboration offers strategic benefits. The Quad resources can greatly enhance Indonesia's capacity to respond to emergencies and problems on the humanitarian front. Due to its geographic location in the Ring of Fire, Indonesia is a country vulnerable to disasters including earthquake, tsunami, flood, landslide, drought, extreme tide and abrasion, extreme weather, bush or forest fire, epidemic disease, and technology failure. Average economic losses from disaster in Indonesia is approximately US$ 2.2 billion annually (KPRI 2018, January 31-February 3). Through their collaboration, Indonesia may be able to develop a more complex system for disaster response and preparedness. The Quad countries are among the best at managing disasters when it comes to preparedness. India and Australia both have a tsunami alarm system for the Indian Ocean and the Pacific Ocean, whereas Japan has an earthquake early-warning system and the US has tools for mapping disaster susceptibility (Bose 2022, July 2; Rajagopalan 2020). As a result, there is plenty of room for knowledge exchange, technical expertise, best practices that are worthwhile learning, and poor practices that need to be avoided. The Quad countries can thus, besides provide the necessary technical expertise and training, engage in developing and coordinating better early warning systems for their partners in the region (Chen 2022, November 21). Indonesia and the Quad may also be able to increase their cooperation through COVID vaccine development efforts without offending China, making it more comfortable for Indonesia. China's opposition to nations working together on this vaccine would be hypocritical given that it has previously called for such collaboration (Rajagopalan 2020).

To foster closer collaboration with the Quad, Indonesia can also actively engage with the Quad through ASEAN-led mechanisms, including exchanges at the practical level on topics like disaster, climate change, connectivity, and infrastructure development. The partnership between Quad and ASEAN member states on the planning, preparation, and operations of natural disaster management is very likely (Chen 2022, November 21). The four Quad states have frequently taken part in cooperative drills with ASEAN member states, such as the ASEAN Regional Forum Disaster Relief Exercise. The joint exercises on natural disaster management remain a major platform to reach confidence building measure between Quad and ASEAN nations (Gong and Nanthini 2022). These joint efforts to develop and implement practical actions to deal with humanitarian hazards should be continued in the future since they would in turn benefit the Indo-Pacific as a whole.

14.6 Conclusion and Recommendation

The emergence of multiple forms of minilateralism in the Indo-Pacific such as the Quad has occurred in the last two decades. The emergence of minilateralism in the Indo-Pacific signify the need to form practical coalitions between likeminded countries to deal with specific issues that need immediate responses. In the Indo-Pacific the rise of minilateralism with its quicker decision-making process may be seen as a response to the less effective role of multilateralism in responding to emerging security challenges.

During the 10 years of the Joko Widodo's government, one could see the transformation of the Quad and the corresponding gradual changes in the Indonesian elites' towards the Quad. First, the Quad has transformed itself from being a forum security dialogues into a regional grouping focusing on non-traditional security issues including COVID-19 pandemic, vaccine, and HADR. The Quad's transformation has gradually changed the perception of the Indonesian people from skepticism into a more positive attitude as can be seen from various surveys organized by both Indonesian and international agencies. Finally, President Joko Widodo's statement that both the Quad and AUKUS should be regarded as partners, not competitors, reflected his pragmatic foreign policy in responding to the latest security dynamics in the Indo-Pacific region.

Indonesia's changing attitudes to the rise of minilateralism in the Indo-Pacific, specifically to the transformation of the Quad, is an important part of Joko Widodo's legacy. It is interesting to note that as defense minister during the second term of President Joko Widodo, Prabowo Subianto was more positive in his attitude towards AUKUS compared to the Indonesian Foreign Ministry. Minilateralism in the Indo-Pacific, notably the Quad has become more widely accepted as a complementary to the multilateralism promoted by ASEAN. Sworn in as president to replace Joko Widodo on 20th October 2024, Prabowo would likely also seize the opportunities offered by the Quad to work on various functional issues that can enhance Indonesia's capacity to deal with NTS challenges and HADR. However, based on the various surveys, Indonesia is still unwilling to join the Quad or Quad Plus.

With Indonesia's more positive view of the Quad, the latter should use this momentum to foster regional collaboration with a greater focus on functional and practical cooperation. NTS issues, particularly pandemics and HADR, should be given priority in future cooperation between Indonesia and the Quad states. Such future cooperation should prioritize concrete functional objectives and quantifiable goals on NTS issues.

As far as Indonesia is concerned, its preoccupation with domestic priorities should not lead to neglect of regional and international dynamics, given the myriad geopolitical, geoeconomic, and geostrategic challenges that exist, particularly in the Indo-Pacific region. Indonesia should not be averse to collaborating with the Quad in global pandemics and HADR, as doing so could prevent the former from benefiting from the resources and expertise that the Quad could offer.

Acknowledgements This study is supported by the Research Organization for Social Science and Humanity at the National Research and Innovation Agency (BRIN), Jakarta, Indonesia, under Grant Number KL.1.5/TRANSNASIONAL/IPSH.1.

References

ADB. 2022. *Disaster Displacement in Asia and the Pacific. A Business Case for Investment in Prevention and Solutions.* Geneva, Switzerland: Asian Development Bank (ADB) in Collaboration with Internal Displacement Monitoring Centre (IDMC).

ADMM. 2007. Protocol to the Concept Paper for the Establishment of the ASEAN Defence Ministers' Meeting (ADMM). Adopted at the Second ASEAN Defence Ministers' Meeting [Press release]. Retrieved from https://admm.asean.org/dmdocuments/2.%20Protocol%20to%20the%20Concept%20Paper%20for%20the%20Establishment%20of%20the%20ADMM.pdf.

ADMM. 2009a. ASEAN Defence Establishment and Civil Society Organizations (SCOs) Cooperation on Non-Traditional Security. Concept Paper [Press release]. Retrieved from https://admm.asean.org/dmdocuments/9.%20ANNEX%20F%20ADOPTED%20Concept%20Paper%20ASEAN%20Defence%20Establishments%20and%20CSO.pdf.

ADMM. (2009b). The Use of ASEAN Military Assets and Capacities in Humanitarian Assistance and Disaster Relief. Concept Paper [Press release]. Retrieved from https://admm.asean.org/dmdocuments/8.%20ANNEX%20D%20ADOPTED%20Concept%20Paper%20The%20Use%20of%20ASEAN%20Military%20Assets.pdf.

Arase, D. 2010. Non-Traditional Security in China-ASEAN Cooperation: The Institutionalization of Regional Security Cooperation and the Evolution of East Asian Regionalism. *Asian Survey* 50 (4): 808–833. https://doi.org/10.1525/as.2010.50.4.808.

Ball, D. 2012. Reflections on Defence Security in East Asia. RSIS Working Paper(237): 1–35.

Ball, D., A. Milner, and B. Taylor. 2006. Track 2 Security Dialogue in the Asia-Pacific: Reflections and Future Directions. *Asian Security* 2 (3): 174–188. https://doi.org/10.1080/14799850600920445.

Basorie, W. D. (2023, May 29). After Jokowi: What Indonesia's Next President Could Face in Foreign Affairs. The Interpreter. Retrieved from https://www.lowyinstitute.org/the-interpreter/after-jokowi-what-indonesia-s-next-president-could-face-foreign-affairs.
Bland, B., E. Laksmana, and N. Kassam. 2021. *Lowy Institute Indonesia Poll 2021. Charting Their Own Course. How Indonesians See the World*. Sidney, NSW: Lowy Institute.
Bose, S. 2022, July 2. Quad Cooperation on Disaster Management Prospects for Revitalizing the Partnership. Retrieved from https://www.nbr.org/publication/quad-cooperation-on-disaster-management-prospects-for-revitalizing-the-partnership/.
Buchan, P.G., and B. Rimland. 2020. Defining the Diamond. The Past, Present and Future of the Quadrilateral Security Dialogue. *CSIS Briefs*, 1–11.
Buzan, B., O. Wæver, and J. de Wilde. 1998. *Security: A New Framework for Analysis*. Boulder, CO: Lynne Rienner Publishing.
Caballero-Anthony, M. 2005, October 26–28. *Non-Traditional Security in Asia: The Many Faces of Securitisation*. Paper presented at the International Conference on Regionalisation and the Taming of Globalisation, University of Warwick, UK.
Caballero-Anthony, M. 2010. Non-Traditional Security Challenges, Regional Governance, and the ASEAN Political Security Community (APSC). *Asia Security Initiative Policy Series Working Paper* (7): 1–14.
Chalivet, A.l. 2022. India's Place in the Quad in Light of AUKUS. *Policy Brief*(18): 1–5.
Chanlett-Avery, E. 2018. Japan, the Indo-Pacific, and the "Quad". *Chicago Council on Global Affairs*, 2–7.
Chaudhury, D.R. 2019, November 5. PM Modi Proposes Indo Pacific Oceans Initiative. *The Economics Times*. Retrieved from https://economictimes.indiatimes.com/news/politics-and-nation/pm-modi-proposes-indo-pacific-oceans-initiative/articleshow/71915838.cms.
Chen, C. 2022, November 21. The Quad and HADR Operations: Prospects for Cooperation with Southeast Asia. *RSIS Commentary*(119): 1–4.
Connors, E. 2023, May 9. Jokowi Softens Stance on AUKUS, the Quad. *Financial Review*. Retrieved from https://www.afr.com/world/asia/jokowi-softens-stance-on-aukus-the-quad-20230509-p5d702.
Envall, Hans. D. P. 2019. The Quadrilateral Security Dialogue: Towards an Indo-Pacific Order? *RSIS Policy Report*, 1–11.
Darmawan, A.R. 2021, March 16. Public Opinion and Civil Society: Shaping Indonesia's South China Sea Policy? *New Mandala*. Retrieved from https://www.newmandala.org/public opinion and-civil-society-shaping-indonesias-south-china-sea-policy/.
Debora, C. 2023, May 22. Does Widodo want to embrace Quad and Aukus? *Dip.or.id*. Retrieved from https://dip.or.id/2023/05/22/widodo-wants-to-embrace-quad-and-aukus/.
Doherty, B., and D. Hurst D. 2023, March 14. What is the Aukus Submarine Deal and What Does it Mean?—The Key Facts. *The Guardian*. Retrieved from https://www.theguardian.com/world/2023/mar/14/what-is-the-aukus-submarine-deal-and-what-does-it-mean-the-key-facts.
DU. 2024, January 9. FH UNDIP Announced PPTM 2024 and Talkshow "Bincang Polugri" with the Indonesian Ministry of Foreign Affairs [Press release]. Retrieved from https://www.undip.ac.id/language/en/post/33108/fh-undip-announced-pptm-2024-and-talkshow-bincang-polugri-with-the-indonesian-ministry-of-foreign-affairs.html.
Duggal, M. 2022. Quad 4.0: To Securitize or Not to Securitize. *Focus Asia. Perspective and Analysis*, 1–16.
Gong, L., and S. Nanthini, S. 2022. Collective Response to Disasters in the Indo-Pacific. Opportunities and Challenges. *RSIS Policy Report*, 1–13.
Haacke, J. 2009. The ASEAN Regional Forum: From Dialogue to Practical Security Cooperation? *Cambridge Review of International Affairs* 22 (3): 427–449. https://doi.org/10.1080/09557570903104057.
Hameiri, S., and L. Jones. 2011. Non-Traditional Security and New Modes of Security Governance in Southeast Asia. In *Paper Presented at the 52nd International Studies Association (ISA)*

International Conference on Global Governance: Political Authority in Transition. Montreal, Canada.

Herscovitch, B., and G. Priyandita. 2021, November 8. Indonesia-Australia: Deeper Divide Lies Beneath AUKUS Submarine Rift. *The Interpreter.* Retrieved from https://www.lowyinstitute.org/the-interpreter/indonesia-australia-deeper-divide-lies-beneath-aukus-submarine-rift.

IFRC. 2016. *World Disasters Report. Resilience: Saving Lives Today, Investing for Tomorrow.* International Federation of Red Cross and Red Crescent Societies (IFRC).

Khong, Y. F., and H. E. S. Nesadurai. 2007. Hanging Together, Institutional Design, and Cooperation in Southeast Asia: AFTA and the ARF. In *Crafting Cooperation: Regional International Institutions in Comparative Perspective*, ed. A. Acharya and A. I. Johnson, 32–82. Cambridge, USA: Cambridge University Press.

Kilic, R. 2023, January 30. Inviting Indonesia to Become the Quad's Inaugural Strategic Dialogue Partner. *South Asian Voices.* Retrieved from https://southasianvoices.org/inviting-indonesia-to-become-the-quads-inaugural-strategic-dialogue-partner/.

KPRI. 2018, January 31-February 3. Disaster Management in Indonesia. In *Paper presented at the ASEAN Defence Ministers' Meeting (ADMM) Plus Early Working Group (EWG) on Humanitarian Assistance and Disaster Relief*. Hawaii, USA.

Laksmana, E. A. 2020. Whose Centrality? ASEAN and the Quad in the Indo-Pacific. *The Journal of Indo-Pacific Affairs* 3 (5): 106–117.

Mehra, J. 2022, June 7. An Evolving Agenda for the Quad. *Stimson.* Retrieved from https://www.stimson.org/2022/an-evolving-agenda-for-the-quad/.

MFAJ. 2006, December 15. Joint Statement Towards Japan-India Strategic and Global Partnership [Press release]. Retrieved from https://www.mofa.go.jp/region/asia-paci/india/pdfs/joint0612.pdf.

MFAJ. 2006, November 30. Speech by Mr. Taro Aso, Minister for Foreign Affairs on the Occasion of the Japan Institute of International Affairs Seminar "Arc of Freedom and Prosperity: Japan's Expanding Diplomatic Horizons" [Press release]. Retrieved from https://www.mofa.go.jp/announce/fm/aso/speech0611.html.

MFARI. (2021, September 17). Statement on Australia's Nuclear-powered Submarines Program [Press release]. Retrieved from https://kemlu.go.id/portal/en/read/2937/siaran_pers/statement-on-australias-nuclear-powered-submarines-program.

Muhibat, S.F., and M.H.A. Dzakwan. 2018, December 7. Indonesia and the Quad: Can't or Won't Decide? *The Strategist.* Retrieved from https://www.aspistrategist.org.au/indonesia-and-the-quad-cant-or-wont-decide/.

Nabbs-Keller, G. 2020. ASEAN Centrality and Indonesian Leadership in a Contested Indo-Pacific Order. *Security Challenges* 16 (3): 21–26.

NST. 2023, May 8. [Exclusive] Jokowi: Aim is to Make Asean Region Stable [NSTTV]. *New Straits Times.* Retrieved from https://www.nst.com.my/news/nation/2023/05/906846/exclusive-jokowi-aim-make-asean-region-stable-nsttv.

Rajagopalan, R.P. 2020. Towards A Quad-Plus Arrangement. Retrieved from https://policycommons.net/artifacts/1350527/towards-a-quad-plus-arrangement/1962688/.

Seah, S., and K. Koga. 2023, May 24. ASEAN and the Quad Inch Closer Together. *Foreign Policy.* Retrieved from https://foreignpolicy.com/2023/05/24/quad-asean-southeast-asia-china-geopolitics-indo-pacific/#cookie_message_anchor.

Seah, S., J. Lin, M. Martinus, S. Suvannaphakdy, and P.T.P. Thao. 2023. *The State of Southeast Asia 2023 Survey Report.* Singapore: ISEAS-Yusof Ishak Institute.

Seah, S., J. Lin, S. Suvannaphakdy, M. Martinus, P.T. Thao, F.N. Seth, and H.T. Ha. 2022. *The State of Southeast Asia 2022 Survey Report.* Singapore: ISEAS Yusof Ishak Institute.

Shofa, J.N. 2023, May 16. No Formal Dialogue between ASEAN and Aukus or Quad: Envoy. *Jakarta Globe.* Retrieved from https://jakartaglobe.id/news/no-formal-dialogue-between-asean-and-aukus-or-quad-envoy.

Simon, S. W. 2002. Evaluating Track II Approaches to Security Diplomacy in the Asia-Pacific: The CSCAP Experience. *The Pacific Review* 15 (2): 167–200. https://doi.org/10.1080/095127402 10131031.

Singh, G. 2022, January 14. The Quad adjusts to ASEAN. *The Jakarta Post*. Retrieved from https:// www.thejakartapost.com/opinion/2022/01/13/the-quad-adjusts-to-asean---.html.

Sulaiman, Y. 2020, October 28. Why Jakarta is Wary Over the Quad and Why Some Hope for a Trump Victory. *Asialink*. Retrieved from https://asialink.unimelb.edu.au/insights/why-jakarta-is-wary-over-the-quad-and-why-some-hope-for-a-trump-victory.

Symonds, P. 2023, July 5. Indonesian president visits Australia amid intensifying US-China tensions. *World Socialist Web Site*. Retrieved from https://www.wsws.org/en/articles/2023/07/06/thmn-j06.html.

Thu, H.L. 2018. *Southeast Asian perceptions of the Quadrilateral Security Dialogue: Survey Findings. ASPI Special Report*. Sydney, Australia: Australian Strategic Policy Institute (ASPI).

Tow, W., and B. Taylor. 2009. Emerging Regional Security Architecture: An Australian Perspective. In *ASEAN-India-Australia: Towards Closer Engagement in A New Asia*, eds. W. Tow and C.K. Wah, 3–21. Singapore: Institute of Southeast Asian Studies (ISEAS).

TWH. 2021, March 12a. Quad Leaders' Joint Statement: "The Spirit of the Quad" [Press release]. Retrieved from https://www.whitehouse.gov/briefing-room/statements-releases/2021/ 03/12/quad-leaders-joint-statement-the-spirit-of-the-quad/.

TWH. 2021, March 12b. Remarks by President Biden, Prime Minister Modi of India, Prime Minister Morrison of Australia, and Prime Minister Suga of Japan in the Virtual Quad Leaders Summit [Press release]. Retrieved from https://www.whitehouse.gov/briefing-room/speeches-remarks/2021/03/12/remarks-by-president-biden-prime-minister-modi-of-india-prime-min ister-morrison-of-australia-and-prime-minister-suga-of-japan-in-virtual-meeting-of-the-quad/.

TWH. 2021, September 24. Fact Sheet: Quad Leaders' Summit [Press release]. Retrieved from https://www.whitehouse.gov/briefing-room/statements-releases/2021/09/24/fact-sheet-quad-leaders-summit/.

TWH. 2023, May 20. Quad Leaders' Vision Statement—Enduring Partners for the Indo-Pacific [Press release]. Retrieved from https://www.whitehouse.gov/briefing-room/statements-releases/ 2023/05/20/quad-leaders-vision-statement-enduring-partners-for-the-indo-pacific/.

TWH. 2024, September 21. The Wilmington Declaration Joint Statement from the Leaders of Australia, India, Japan, and the United States [Press release]. Retrieved from https://www.whi tehouse.gov/briefing-room/statements-releases/2024/09/21/the-wilmington-declaration-joint-statement-from-the-leaders-of-australia-india-japan-and-the-united-states/.

USDS. 2022, September 23. Guidelines for Quad Partnership on Humanitarian Assistance and Disaster Relief (HADR) in the Indo-Pacific [Press release]. Retrieved from https://www.state.gov/guidelines-for-quad-partnership-on-humanitarian-assistance-and-disaster-relief-hadr-in-the-indo-pacific.

USDS. 2023, March 3. Joint Statement of the Quad Ministerial Meeting in New Delhi [Press release]. Retrieved from https://www.state.gov/joint-statement-of-the-quad-ministerial-meeting-in-new-delhi/.

USECA. 2021, March 15. Op-Ed: Our Four Nations are Committed to A Free, Open, Secure, and Prosperous Indo-Pacific Region (Washington Post) [Press release]. Retrieved from https://au. usembassy.gov/op-ed-our-four-nations-are-committed-to-a-free-open-secure-and-prosperous-indo-pacific-region-washington-post/.

Wei, Z. 2022. The Evolution of the 'QUAD': Driving Forces, Impacts, and Prospects. *China International Strategy Review* 4 (2): 288–304. https://doi.org/10.1007/s42533-022-00119-w.

Yaacob, A.R., G. Priyandita, and S. Laksmi. 2023. Southeast Asia's Security Landscape: Lessons for the ADF. *Australian Army Occasional Paper*(17): 1–54.

Zimmerman, E. 2014. Security Cooperation in the Indo-Pacific: Non-Traditional Security as A Catalyst. *Journal of Indian Ocean Region* 10 (2): 150–165. https://doi.org/10.1080/19480881. 2014.922325.

RR Emilia Yustiningrum is a senior researcher at the Research Center for Politics, National Research and Innovation Agency (BRIN), Indonesia. She holds a BA degree on international relations from Gadjah Mada University, Indonesia; MA degree on Humanitarian Action from the University of Groningen, The Netherlands; and a PhD from School Government and International Relations, Griffith University, Australia through Australia Award Scholarship (AAS). Her expertises are international politics, foreign policy, and disaster management. Her current publications are a book titled *Explaining Indonesia's Foreign Policy: The Role of the Military in Post Natural Disaster Management* (Palgrave Macmillan, 2024), examines Indonesia's post-Aceh tsunami disaster management through the lens of multifaceted foreign policy decision-making process involving a variety of bureaucratic actors. She also wrote a book chapter titled *Indonesia's engagement on the climate change negotiations: building national resilience* (Elsevier, 2023), investigates Indonesia's dynamic engagement on the climate change negotiations across administrations from the New Order era to the Reform era (1972–2021).

Ganewati Wuryandari is a researcher at the Research Center for Politics, National Research and Innovation Agency (BRIN) in Indonesia. She received her PhD in Asian Studies from the University of Western Australia in Perth in 2006 and a research professorship in the field of international relations in 2022. She has been involved in the ASEAN Regional Forum (ARF) since 2020 as an Expert and Eminent Persons (EEP) of the Republic of Indonesia and is now active as a Secretary General of the International Federation of Social Sciences Organization (IFSSO) and a member of the BRIN Social Humanities Ethics Clearance. Her areas of interest in academia are international relations, particularly as it relates to Indonesian foreign policy, Indo-Pacific, gender and security, and border issues. "Indonesia's engagement in Climate Change Negotiations: Building National Resilience" (in Uday Chatterjee and Rajib Shaw, et al., *Climate Change, Community Response, and Resilience*, Elsevier, 2023), *Indo-Pasifik dalam Politik Luar Negeri Indonesia: Dimensi Geopolitik, Geostrategi & Geoekonomi* (Andi, 2024), and *Politik Luar Negeri Indonesia Terhadap Tiongkok* (BRIN Press, 2024) are some of her most recent academic works.

Siswanto is a senior researcher at the Research Center for Politics, National Research and Innovation Agency (BRIN) since 2000. The author holds a Prof. Research degree in the field of International Relations. Their areas of expertise include the global politics of the United States and international relations. The author is an alumnus of the Doctoral Program in History at the Faculty of Cultural Sciences, University of Indonesia, graduating in 2008. The author's scholarly contributions include the following publications: *John F. Kennedy's Breakthrough in the West Irian Crisis* (Gading Inti Prima, 2015); *Decolonization Versus Status Quo: History of Papuan Diplomacy* (Gagas Nusantara, 2018) *Donald Trump's Leadership Controversy (Gagas Nusantara, 2018) Indonesia and West Irian Diplomacy 1949–1962: Utilizing the Cold War* (LIPI Press, 2020); The Professor's oration book *Research on the Development of a Rational Mediation Model Based on Changes in United States Policy and Mediation Action in the West Irian Dispute* (BRIN Publisher, 2021).

Dewi Fortuna Anwar is an Academician of the Social Science Commission of the Indonesian Academy of Sciences (AIPI), a Research Professor at the Research Center for Politics, National Research and Innovation Agency (BRIN), Chairman of the Board of Directors of The Habibie Center (THC) and Co-Founder of the Foreign Policy Community of Indonesia (FPCI). In 2001–2010 she was Deputy for Social Sciences and Humanities, Indonesian Institute of Sciences (LIPI). In 2010–2017 Dewi served as a Deputy Secretary to the Vice President of the Republic of Indonesia. Dewi was a Distinguished Visiting Professor at the S. Rajaratnam School of International Studies (RSIS), NTU, Singapore in 2017–2018, and a Distinguished Visiting Professor at SAIS, Johns Hopkins University in 2007. She has written widely on Indonesia's foreign policy, and ASEAN regional political and security issues. Dewi sits and has sat in a number of national and international advisory boards i.a. Centre for Humanitarian Dialogue (HD) since 2019, Stockholm International Peace Research Institute (SIPRI) in 2010–2020, and the UN Secretary General Advisory Board on Disarmament Matters in 2008–2012. She obtained her PhD from Monash University, Melbourne, while her M.A. and B.A. (Hons) were from SOAS, University of London.

Chapter 15
The Rise of Minilateralism, Middle Powers, and Indonesia's Diplomacy

Dian Wirengjurit

Abstract Rivalry between superpowers today is no longer based on ideology but on a more complex combination of military, economic, technological, and diplomatic capabilities. However, the global economic crisis in the 1990s, which was initially expected to rebuild multilateral collaboration, has created obstacles to a more effective cooperation. Minilateralism has now emerged as an alternative for solid and closer cooperation among countries in the sense of more rational, realistic, and effective ones. As a middle power (MP), Indonesia has begun to thoroughly examine the minilateralism model in its foreign policy. However, its implementation will not be smooth, due to the fact that a lot of middle power countries are dependent on the superpowers. In this global order, as a persistent non-aligned country, Indonesia is striving hard to employ minilateralism as a creative alignment in its future endeavors.

Keyword Indonesia · Miniteralism · Diplomacy · Middle Power

15.1 Introduction

The First Vice President Mohammad Hatta in 1948 had correctly reminded us that Indonesia "will be rowing between two reefs" in its future foreign policy. However, the current geopolitical setting has been evolving from bipolar (US–Soviet Union) during the Cold War (1947–1990) to unipolar post-Cold War (early to late 1990s) and to multipolar or tripolar ones (US, Russia, and China) in the "new Cold War" since the beginning of this millennium. The rivalry between the superpowers is no longer based on mere ideology, but has now turned into military, economic, technological, and diplomatic power. Consequently, Indonesia's foreign policy needs to be revamped in order to be able to grow between the three reefs (*Wirengjurit*, July 2020).

The last decade has also begun to further shift this trend, and now we are witnessing different challenges in international relations. The Covid-19 disease has exposed

D. Wirengjurit (✉)
Institute of Civilization Studies, Jakarta, Indonesia

I. Kartini and Kanisa (eds.), *Joko Widodo's Foreign Policy and Diplomatic Legacy*,
https://doi.org/10.1007/978-981-95-2085-5_15

human vulnerability and the helplessness of the current multilateral system in dealing with a global pandemic crisis. Besides, climate change has now also become an immediate challenge to human survival to mitigate. In the meanwhile, the world is also confronted by intensified military conflict both in the traditional region (Middle East) in the non-traditional ones (East Europe). The whole crisis has forced the world community to question how countries can work together to address problems that cannot be addressed by one country alone.

During President Joko Widodo's (Jokowi) presidency, Indonesia had to live up with these dynamics, particularly by the two ongoing military conflicts (the Israel-Hamas and the Russia–Ukraine War) which had no precedence in the post-Cold War. Moreover, the rapid development of artificial intelligence (AI) and the pressing problem of food and energy shortages in many parts of the world further complicate the various problems that leaders must address. Overall, the challenges facing the world appear daunting and seemingly insurmountable. The traditional multilateral cooperation system is starting to fall apart or breaking down (*Mladenov*, 2023), so a new or more creative approach is needed to answer these challenges.

15.2 New "Toy": Minilateralism

One response to these challenges has been the growing popularity of "minilateralism," a concept in international relations that involves small groups of countries collaborating to address issues of common interest and purpose. This concept emerged to answer the global order that has undergone significant developments in the last century with the end of the Cold War to confront a multipolar world with its multidimensional challenges. However, it turns out that the new millennium and the emergence of multipolarity have generated significant obstacles to effective collaboration.

Minilateralism is in fact not a new concept, having existed since the Concert of Europe in the early nineteenth century and is now making a comeback in the form of smaller socio-economic and politico-military corporations. The idea of bilateralism, minilateralism, and multilateralism collaborations have co-existed in global governance since 1945. Notably, the multilateral institutions that were created in the post-war era were negotiated through "disguised" minilateralism, pursued between the United States (US) and other Atlantic powers (*Tirkey*, 2021).

Minilateralism refers to informal and more targeted initiatives intended to address a "specific threat, contingency or security issue with fewer states (usually three or four) sharing the same interest for resolving it within a finite period of time." Certain features are frequently associated with minilaterals: they have a small number of participants, ad hoc in nature, and their outcomes and commitments are voluntary ones (Tirkey, 2021). However, with the vast dynamics of geopolitics and international relations, for sure this is not a rigid format.

The global economic crisis, which was originally expected to build multilateral collaboration, has instead turned into a "slack" and slow mechanism and is unable to drive effective cooperation. Climate change, nuclear (weapons) proliferation, terrorism, pandemic, trade protectionism, and many more, none of which can be resolved completely or contained collaboratively. Looking back at history, successful multilateral efforts only materialized in the 1990s.

In 1994, 123 countries agreed to establish the World Trade Organization (WTO) and new international trade rules, but since then other efforts to agree on world trade contentious issues have been practically unsuccessful. In 1995, 185 countries agreed to extend the Non-Proliferation Treaty (NPT), but since then not only has that multilateral initiative failed, but India, Pakistan, and North Korea, along with Israel have become de facto nuclear states. And in 1997, the Kyoto Protocol to reduce greenhouse gas emissions was ratified by 184 countries but has not been approved by the US and China, the world's two biggest air polluters.

The number of global problems that world leaders need to address continues to grow and diversify. There is indeed an increasingly urgent need for alternatives to the traditional multilateral system of cooperation. According to Dominika Urhova, minilateralism has the potential to open new doors for more effective and rapid decision-making. The most widely cited advantage of minilateralism is its flexibility and adaptability, which stems from the idea that collaboration within small groups itself. Member states can thus come together to address common challenges without having to reach consensus on other issues (*Urhova*, 2023).

The pattern is clear. Since the early 1990s, the need for multilateral collaboration has increased, but at the same time negotiations have failed; deadlines have been missed; financial commitments and promises have not been met; implementation has stalled; and collective action has fallen far short of what is offered and needed. This failure is due not only to a lack of collaborative commitment but also to a misguided obsession with multilateralism as a *panacea* for all the world's ills (*Naim*, 2009). Such an obsession continues to lead the way of many countries, especially the least developed and developing ones, the UN's members, down this path of minilateralism.

The hallmark of this concept is the emphasis on common interests and goals, rather than shared values or ideological alignment. As a result, countries can collaborate on important issues without having to concur on everything or have the same worldview. These countries were "excited and noisy" in seeking friends through minilateralism collaboration, including by forming "small" (mini) security/military pacts (*Kompas Editorial*, August 2023). The popularity of minilateralism continues to grow as the superpowers are grappling with major conflicts, such as the wars in Gaza and Ukraine, and the widening gap between the US and Russia, as well as China.

This minilateralism has several advantages (Rana 2024): *First*, small clusters are easier to manage than large clusters. *Second*, informal methods work well in such groups. *Third*, success depends on careful selection of partners, plus clarity of purpose. *Fourth*, this method may be suitable for small countries, provided they do not have large and dominant neighbors who may view this path as hostile. In the current circumstances, these advantages may be enjoyed by the MPs, including Indonesia.

Richard Haass, President of the Council on Foreign Relations, classifies four types of minilateralism (*Moret*, 2016). *First*, *"elite multilateralism"* describes groups such as the G20, which are made up of powers that play a leading or important role in a particular domain. *Second*, *"regional multilateralism"* refers to the proliferation of bilateral and regional trade pacts that have emerged as a result of the WTO impasse. *Third*, "*functional* multilateralism" refers to "coalitions of the willing and relevant" that can handle some of the governance of broader issues. *Fourth*, "*informal* multilateralism" describes the actions used by national governments to implement measures consistent with global norms, particularly in the financial realm in cases where multilateral agreements are blocked by national parliaments.

The more agile and flexible minilateral format allows countries to react more quickly to crises or opportunities without being bogged down by bureaucratic red tape. This flexibility also allows governments to build stronger, more intimate relationships based on shared interests, rather than being forced to work in large but less cohesive groups. Another advantage is that minilateralism is more agile and adaptable than traditional diplomatic channels, including global multilateral organizations (*Elliott*, 2024).

However, minilateralism has its own drawbacks, one of which is that the grouping can be exclusionary and exacerbate power imbalances, potentially prioritizing short-term interests over long-term goals. However, the benefits of minilateralism outweigh the drawbacks, and that minilateralism is a smarter and more targeted approach; that is, bringing together the fewest number of countries needed to have the greatest impact on a given problem. This is the "magic number" for minilateralism, according to Moises Naim (2009), which varies widely depending on the problem.

Furthermore, minilateralism also poses the danger of forum-shopping, which endangers important international organizations and reduces accountability in global governance. Minilateralism promotes voluntary, non-binding, and non-legally binding commitment targets. For countries increasingly showing a preference for more easily negotiated "soft law" mechanisms, minilateralism provides an attractive alternative to multilaterals (*Tirkey*, 2021), which was categorically construed as the cornerstone of international collaboration until the recent past.

15.3 The Player: Middle Powers (MPs)

Meanwhile, MPs are countries that are not superpowers or large countries but can exert influence and play an important role in international relations. The MPs phenomenon is also not new because the Italian philosopher and diplomat Giovanni Botero formulated it in the sixteenth century by dividing countries into three groups: *grandissime* (large), *mezzano* (medium), and *piccioli* (small). According to Botero, MP is a country that has enough power and authority to stand alone without requiring assistance from other countries (K. Rudd, 2006).

Although there is no agreed standard method to date, it is generally recognized that these have the ability to influence the attitudes and behavior of other countries.

These countries often have certain capabilities, such as strong economies, advanced technology, and diplomatic influence, and perhaps even adequate military, that allow them to "have a say" in global affairs (Elliott, 2024). They consist of Global North countries including Australia, Canada, and South Korea, and Global South countries including Argentina, Brazil, and Indonesia. Other countries "earn middle power status because they 'go hard' in global affairs," such as Norway.

Even though the grouping looks simple and easy (straightforward), there is still disagreement about how MPs should be defined and how they should act in global politics. There are two ways to define MPs: *first* is based on the military power, capabilities, and geostrategic position of a country; *second* is based on the leadership capabilities of a country (Baç 2015). In other words, these countries are considered liberal, oriented towards democracy, and have legitimate concerns in international politics, security, and economy.

The concept of MPs gained serious attention towards the end of World War II in relation to Canada and Australia. Both countries tried to find ways to increase their influence based on their contributions to the Allies. In 1944, Canadian Prime Minister Mackenzie King asserted that middle power countries should work together to secure their influence in the international community, proposing the concept of "*Middlepowerhood.*" In Australia, FM Herbert Evatt discussed the concept of MPs with the aim of securing his country's national interests in the new world order after WWII. Since then, the concept of MPs has become a hallmark of the foreign policies of both countries (*Dong-min Shin*, 2015). MPs are essential to the creation and maintenance of the world order, and they continue to support the establishment of necessary international institutions. In that sense, they act as stabilizers in the world system.

MPs can also be distinguished, according to Bac, specifically between traditional and emerging ones. From a historical perspective, traditional MPs are mostly Western European countries; while an important characteristic of emerging ones (e.g., South Africa, Malaysia, and Turkey) is that they are also major regional players who are most often liberal and democratic oriented states. MPs can also be distinguished, most notably between traditional and emerging ones.

The rise of MPs was emphasized by Amb. Dino Patti Djalal (2024), chairman of the Indonesian Foreign Policy. Djalal argues that "in a world of increasingly intense geopolitical competition, MPs make a difference by bridging gaps, providing solutions, and building regional architecture." Similarly, these groups of countries are often seen as bridge builders between larger powers, using their diplomatic skills to mediate conflicts and foster cooperation on international issues (*Jordaan*, 2003). Also, as the tectonic plates of the global balance of power shift, leaving ever-widening fault lines in the international order, some countries face both formidable challenges and unparalleled opportunities to build bridges. The countries in this position are MPs whose unilateral actions cannot project economic or military power on a global scale but are large enough to exert influence (*Wihardja and Verico*, 2022). It is fortuitous that the three G20 presidencies after Indonesia fall again on MP countries, namely India, Brazil, and South Africa.

The role played by these MPs stems in part from the perception of their legitimate concern for human security issues. These states can successfully effect change because of their diplomatic prowess and their ability to project credible positions, which allows them to act as moral and intellectual leaders (Baç 2015). MP states also usually have a highly institutionalized diplomatic apparatus and are able to disseminate their foreign policy ideas and objectives through the network of diplomatic missions they maintain in accredited countries.

In furtherance, MPs are essential for the creation and maintenance of the world order, and they continue to support the establishment of the necessary international institutions. In that sense, they act as stabilizers in the world system. According to international relations theory (Baç 2015), hegemonic powers are responsible for the creation of international institutions, but their maintenance and survival depend on the convergence of interests among other players; that is where the role of MPs is enhanced.

Currently, of the 193 countries in the world today, about two dozen qualify as MPs; a few are in the Global North but the majority are in the Global South. While all MPs in the North are mainly committed to military pacts, most of them in the South are nonaligned and tend to pursue limited strategic interests. Collectively, these MPs have been a major factor in the shift in global power. The world now has the highest number of MPs in history, and that number will grow as countries like Pakistan, Mongolia, and Ethiopia enter the category (*Djalal*, 2024).

But it is not just quantity that matters; there are qualitative differences too (*Djalal*, 2024). The MPs of the global south are stronger and richer than they were when the NAM was born six decades ago. By 2030, the estimated purchasing power parity (PPP) of Indonesia, Brazil, Mexico, Turkey, and South Africa will be greater than that of Germany, Italy, Spain, the Netherlands, and Sweden. As they grow stronger, these MPs are gaining in confidence and a sense of entitlement.

MPs in the global South and North can work together to strengthen multilateralism and enhance cooperation to address global agendas such as climate change, nuclear nonproliferation, food security, AI governance, and the Sustainable Development Goals (SDGs). By expanding the scope of cooperation and intensifying confidence-building efforts, MPs can reduce, rather than expand, the space for geopolitical competition and at the same time reduce the prospective dominance of any one great power. By consistently hedging and diversifying their strategic relationships, MP can make multipolarity less volatile and more stable. And perhaps as they steadily improve, they can eventually shape the more durable, inclusive, and equitable world order that has so far eluded humanity.

15.4 Minilateralism in Action

"I think in the 21st century, the world order will not be shaped by superpowers or big countries, but by the proliferation of MPs" Amb. Djalal said (2024). Moreover, there are more MPs across the world than ever before, and each has the size, ambition

and resources to play a bigger role. Djalal gave examples of the emerging alignment between MPs in the global south and those in the western hemisphere, such as Australia and India, adding: "*This is a very significant geopolitical development.*"

What is unique about this innovative practice of minilateralism is that it is not just regional, but also continent-wide and even more cross-continent collaboration, aiming to spur non-traditional cooperation. The emergence of various cross-continental minilateral forums/organizations today—including QUAD, AUKUS, BRICS, IPEF, RCEP and MIKTA—significantly marks the increasing role of MPs on the world stage, especially in the minilateral format.

So far MPs continue to perceive multilateralism as a viable vehicle, but they are increasingly separating themselves from larger pan-regional forums, in order to benefit from minilateral configurations led by major powers, and in an effort to achieve more focused and concrete strategic goals (Hynd and Wilkins 2024). Minilateralism is now gaining ground on all issues of importance to international cooperation, such as climate change, economic cooperation, trade, connectivity, financial regulation, security and transition of energy. The growth of regional minilateral organizations for international economic cooperation, such as the EU and ASEAN, and more diverse groupings, such as the G20, are now beginning to replace "global" multilateralism.

In the Middle East, minilateralism has gained popularity as for decades countries have struggled with the effects of conflict, instability, and foreign intervention. In this context, countries have increasingly sought to form partnerships and coalitions to help them address common challenges. For the Gulf states and superpowers, minilateralism is a strategic necessity that has significant implications for the region's security and prosperity. China, for example, has capitalized on this minilateralism by successfully forging a peace agreement between Saudi Arabia and Iran in March 2023.

The historic Abraham Accords of 2020 not only normalized relations between Israel and three major Arab states, but also opened the door to new forms of cooperation that were previously unimaginable (*Wirengjurit*, September 2020). The nascent Negev Forum, which brings the US together with Israel, the UAE, Egypt, Morocco, and Bahrain in a new regional cooperation framework, is one example. In addition, the UAE, announced commitments with India and France, to cooperate in defense, energy and technology. While in 2021, India, Israel, the UAE and the US announced a "*Partnership for the Future*" which later became known as the I2U2 format. While it took 15 years since the QUAD was formed for its leaders to meet in Japan in 2022, the first I2U2 summit, albeit virtual, took place less than a year after its formation.

In Asia and the Pacific, the optimistic vision of an "*Asia-Pacific Century*" embracing broad economic integration, institution-building, and multilateral cooperation, combined with a relatively smooth recovery from the 2008 Global Financial Crisis, had helped fuel a surge of interest in MPs in the region in the early 2010s. Many spoke of an "*MPs moment*" and celebrated the "*rise of MPs*" (Hynd and Wilkins 2024), while political leaders from countries as diverse as South Korea, Australia, and Indonesia freely associated themselves with the MP label.

However, minilateralism in this vast region has so far been more associated with defense/security issues as the rivalry between the three superpowers has intensified. They are the facilitators of minilateralism and interestingly, the idea has actually been more widely embraced by the US, which has had many strategic partners in various regions of the world and has been most active in seeking friends by forming new alliances. In this region, the US has formed the QUAD Pact which took 15 years until it was finally formalized in 2021, at the same year AUKUS (Australia, the UK, and the US) was formed. Although it has various goals, it cannot be denied that its main goal is to stem the growth of Chinese influence. Especially for India, QUAD is needed considering that China has a strategic partnership with Pakistan and the border dispute with China is a thorn in the flesh for New Delhi. Most recently, in East Asia, the US has formed a military alliance with Japan and South Korea which are "at odds" with each other overshadowed by old wounds (1910–1945) due to the colonization of the Land of Sakura, to confront the same enemy, North Korea (Wirengjurit 2023b).

Still in the region, MPs have maintained their role as "norm entrepreneurs," but in a way that focuses more on power and protecting their national interests. Most notably, Indo-Pacific MPs from India to Indonesia have supported the building and assertion of a rules-based order, which offers protection from the "might is right" alternative to pure power politics (Hynd and Wilkins 2024). Further, for Japan and India, this is not the case—they have some great power capabilities, but they lack the ability to define the security order in the way that the US or China do. Nevertheless, the MPs label remains useful because influential second-tier states such as Australia and Indonesia continue to identify themselves in this way. Both countries also pursue forms of "classic" MPs diplomacy adapted, modified, and updated through coalition building, multilateralism, and norm entrepreneurship.

In Southeast Asia, the US sees an opportunity to enter the region in the increasingly intense and open disputes in the South China Sea between China and the Philippines and Vietnam over the Spratly and Paracel Islands, respectively. Now, the US has not only set foot back in the Philippines, which it left at the end of the Cold War, but is also getting closer to Vietnam. Indonesia, Malaysia and the Philippines in 2016 formed the IMPC to combat piracy, sea robbery, violent extremism and terrorism, as well as air and sea patrols in the Sulu Sea. Meetings are held at the level of foreign ministers, defense ministers, and defense chiefs.

At the same time, MP initiatives range from the Asia-Pacific Community, envisioned by Australian Prime Minister Kevin Rudd, to the creation of a cross-regional MPs minilateral grouping known as MIKTA. Such initiatives raise hopes that MPs have the opportunity and the will to restructure the international system, following the approach set out by "emerging power" forums such as BRICS and India, Brazil and South Africa (IBSA) in the early 2000s. Simply put, the benefits of minilateralism outweigh the costs (*Mladenov*, 2023). The rise of MPs and rapid technological advances make it clear that minilateralism is a viable way forward for many countries.

There is at least one MP that stands out in every region of the world, which was not the case in the twentieth century. These MP countries share a growing desire and ability to shape the geopolitical dynamics and international relations in their region, and specifically via minilateralism. These MP countries feel compelled to maximize their role and sometimes must be ready to lead in the region, while still curbing the ambitions of external powers (*Djalal*, 2024). Thus, in Southeast Asia, for many years Indonesia works with fellow ASEAN members to ensure that external powers such as the US, China and Russia adhere to the rules and mechanisms designed by ASEAN.

Beyond their immediate region, there is also a trend of MPs uniting and becoming more assertive, creating and expanding their cooperation across the board in minilateralism, despite the competition around them. As Western countries begin to mitigate the risks posed by China and Russia, and as the space for dialogue between them shrinks, MPs in the Global South are forming unprecedented economic, diplomatic, and strategic ties with each other.

15.5 Indonesia's Diplomacy in Minilateralism

As a new generation of influence is emerging on the global stage, namely non-Western MP countries—such as Brazil, India, Indonesia, South Africa, and Turkey—they are now beginning to show their strength in global governance. Their influence spans a wide range of areas, from trade and finance, environment, human rights, and energy transition. The evolution of the G20 as an important forum for global economic discussions, to a forum for discussing various world issues, underscores this shift, highlighting the growing influence of these non-Western middle countries.

In Indonesia, domestic pressures have been more obvious towards addressing domestic challenges than expanding its global footprint. Historically, Indonesia's foreign policy, institutionalized during the Suharto era, has prioritized regional engagement. This raises questions about Indonesia's motivations and the role it is taking on the global stage. Although Indonesia's material capabilities are consistent with MP status, its foreign policy circles initially rejected the label, viewing it as something that is on the wane.

However, despite the relevance of the NAM which is often being questioned, Indonesia's foreign policy continues to uphold a non-alignment posture, which in so far it can navigate between the three superpowers. For example, up to date Indonesia managed to properly keep the distance among those big powers, by remaining away from QUAD or SCO (Shanghai Cooperation Organization) as well as from OECD and BRICS. On the other hand, India which was one of the founders, now seems to have been left behind playing "here and there," by joining both QUAD and SCO. With its posture, Indonesia may continue to be able to express itself freely in various international forums, without being trapped or held hostage or voluntarily pushed to take sides.

In the current geopolitical environment, as a leading MP, Indonesia seems unable to escape the desire to take advantage of minilateralism and empower it for more realistic and pragmatic purposes. For example, Indonesia now trades more with India than with the US and is building naval submarines and fighter jets with South Korea—a strategic venture that would be impossible with any Western country, Russia, or China. Also significant is that three middle-country countries in the southern hemisphere—Indonesia, India, and Brazil—have held successive presidencies of the G20 and work closely together in the group's "troika" mechanism for policy coordination.

Meanwhile, more countries in the southern hemisphere, including Ethiopia, Egypt, and Iran, are lining up to join the BRICS than those seeking to join the OECD. But there is also a new pragmatism that has seen them break old taboos: India, Indonesia, and Vietnam have signed strong strategic partnership agreements with the US, something that would have been unthinkable four decades ago due to their domestic political circumstances and historical background (Djalal, 2024).

Among the MPs, under the guise of a cross-regional grouping, Indonesia is currently actively promoting MIKTA, which consists of Mexico, Indonesia, South Korea, Turkey, and Australia. This grouping was established as a cross-regional grouping of G20 member countries in 2013 on the sidelines of the UNGA meeting in New York. The basic idea is—the five countries across the region work together, bridge the gap between developed and developing countries, and build consensus on issues relevant to all regions. As a country -that is geographically, culturally, linguistically, and religiously diverse-, Indonesia's commitment to democracy and working together in the multilateral system is therefore essential.

One of MIKTA's main strengths is its flexibility, where it provides an informal environment where current issues can be discussed, without the pressure to reach a consensus. It seeks to assume an adaptive form of multilateral cooperation, different from the traditional "bloc," thereby enhancing the ability to manage effective global governance in a world with rapid development. As a diverse grouping of five MPs in the G20, MIKTA immediately attracted the attention of observers and policymakers. They view MIKTA as "a potential significant advance in global governance" (*Cooper and Dal*, 2016), their contributions in the G20 are recognized.

The emergence of MIKTA marks the *third wave* of MPs diplomacy (*Cooper and Dal*, 2016). In the *first wave*, MPs focused on developing multilateralism through the UN and UN agencies; in the *second wave* they shifted to activism and leadership on specific issue areas such as human security. MIKTA is seen as the *third wave* of MP diplomacy, marking joint efforts outside the G7 and BRICS blocks to develop platforms for collective action.

Since all MIKTA members are also members of larger international organizations, such as the UN, WTO, and G20, issues relevant to these forums can be the subject of consultation at MIKTA. This can involve working on commitments made within these organizations or, sometimes, committing to additional goals or standards. MIKTA also provides an opportunity for these regional powers to engage in dialogue with the larger economic powers independently.

MIKTA members are also active contributors to major international forums and play a constructive role in the international environment, serving as a platform for cross-regional consultations that enhance mutual understanding, deepen bilateral relations, and help find common ground for cooperation. MIKTA is also active as a catalyst or facilitator in launching initiatives and implementing global governance reforms. One of programs that is currently being seriously explored is working on new and sustainable energy cooperation in the field of geothermal, where the potential for this natural resource is very large, as well as bio-energy (palm oil) and CO2 emission mitigation.

The UAE and Indonesia, and later joined by five other countries (India, Sri Lanka, Australia, Japan and Spain) formed the Mangrove Alliance (MAC) within the framework of climate change at COP27 in Egypt to enhance the rehabilitation and preservation of their ecosystems. MAC seeks to raise awareness about the role of mangroves as a nature-based climate change solution, and, through its members, will work towards expanding and rehabilitating mangrove forests globally. Indonesia is currently exploring the formation of a new mini-group to protect its vital natural resources (nickel, cobalt, and manganese) with an OPEC-like structure to oversee its trade.

15.6 Conclusion

In an increasingly multipolar world where traditional institutions are struggling to cope with new and complex challenges, the emergence of minilateralism presents a more creative and flexible approach to diplomacy and international relations. Minilateralism brings a breath of fresh air to the current landscape of international collaboration. With its relative ease and directness, it has the potential to improve relations between states and generate innovative solutions. Furthermore, as MPs take on the role of active agents in shaping the international order, as they seek to capitalize on the momentum brought by the grouping. However, most minilateral groupings are still in their infancy and have a number of challenges to overcome before they can be labeled as successful.

The magic number of minilateralism is not a magic solution; however, it is a much better bet at this time than multilateralism that is merely wishful thinking (Naim, 2009). Overall, MPs are rewriting the playbook of world affairs. Collaboration involving MPs—within the Global South, between the Global South and the North, between the Global South and the US, Russia and China—is reshaping the international order. Although the rise of MPs will certainly be a long process, they have provided an enabling environment to manage risks and expand opportunities in world affairs.

As a leading MP, Indonesia will carry on to employ it carefully, not just join in the "*hustle and bustle*" of minilateralism. Indeed, MPs and minilateralism have clearly become major players in geopolitics and international relations, both now and in the future, and Indonesia will always be in the midst of its dynamics. Indonesia's involvement in minilateralism remains a preliminary attempt in Jokowi's diplomatic endeavors that are focused on economic growth, social welfare and political independence. For Indonesia, which has vast and long experience in ASEAN, minilateralism continues to be attractive, including in the cross-regional framework because of its specific objectives, especially in bringing minilateralism as an alternative and creative alignment.

Despite the fact as an old phenomenon, since it reappears with a "new" packaging, for many countries, including Indonesia, this is the "new toy" supposed to be ridden in the global "playing ground." However, in the post-Jokowi era Indonesia must be able to rapidly adapt to the fast growing minilateralism with careful considerations: *First*, minilateralism needs to maintain and reaffirm the non-alignment posture, and to keep it from not involving military cooperation, or even becoming a military alliance. *Second*, the areas of interest must be selective and focused on the potential of proven (natural) resources and the backbone of the country's revenue, such as mining, agriculture, and marine industry, as well as pressing issues such as climate change and clean and renewable energy.

As the minilateral approach has become the preferred approach for many countries, MPs are taking up the role of active agents in shaping the international order, they seek to leverage the momentum that it brings to the table. Speaking at the Anwar Gargash Diplomatic Academy on September 6, 2022, Indian External Affairs Minister S. Jaishankar underlined minilateralism as "*a form of diplomacy that is here to stay, and is the way forward for many countries.*" For sure minilateralism will continue to progress and grow, and become a main vehicle for the MPs, including Indonesia, to drive ahead in its foreign policy and diplomacy.

References

Baç, M. M. 2015, October 13. Middle power | definition. *Role & Impact. Encyclopedia Britannica.*. https://www.britannica.com/topic/middle-power#ref1231330.

Cooper, A. F., and E. P. Dal. 2016. Positioning the third wave of middle power diplomacy: Institutional elevation, practice limitations. *International Journal* 71 (4): 516–528. https://www.jstor.org/stable/26414055.

Djalal, D. P. (2024, July 3). Middle powers are the new architects of the world order. Nikkei Asia. https://asia.nikkei.com/Opinion/Middle-powers-are-the-new-architects-of-the-world-order.

Elliott, D. 2024, September 10. What are middle powers and why do they matter?. World. *Economic Forum.*. https://www.weforum.org/agenda/2024/01/middle-powers-multilateralism-international-relations/.

Hynd, A. M. and Wilkins, T. (2024, April 4). Indo-Pacific Middle Powers: Rethinking Roles and Preferences. Pacific Forum. https://pacforum.org/publications/pacnet-43-indo-pacific-middle-powers-rethinking-roles-and-preferences/

Jordaan, E. 2003. The concept of a middle power in international relations: distinguishing between emerging and traditional middle powers. *Politikon* 30 (1): 165–181. https://doi.org/10.1080/0258934032000147282.

Karim, M. F. (2023, October 3). Indonesia's role in international relations: A perspective from the global south. E-International Relations. https://www.e-ir.info/2023/10/03/indonesias-role-in-international-relations-a-perspective-from-the-global-south/

Kompas. (2023, Agustus 19). Tajuk Rencana Riuh. Minilateralisme di Indo-Pasifik. *Mathrubhumi News.* (2022, September 1). EAM Jaishankar highlights key pillars of India's foreign policy, ties with UAE. https://english.mathrubhumi.com/news/world/eam-jaishankar-highlights-key-pillars-of-india-s-foreign-policy-ties-with-uae-1.7838018.

Mladenov, N. (2023, April 14). Minilateralism: A Concept That Is Changing the Changing World Order. The Washington Institute for Near East Policy. https://www.washingtoninstitute.org/policy-analysis/minilateralism-concept-changing-world-order

Moret, Erica. (June 2016) Effective minilateralism for the EU What, When and How. European Union Institute for Security Studies (EUISS). https://css.ethz.ch/content/dam/ethz/special-interest/gess/cis/center-for-securities-studies/resources/docs/Brief_17_Minilateralism.pdf

Naim, M. (2009, June 21). Minilateralism. *Foreign Policy.* https://foreignpolicy.com/2009/06/21/minilateralism/.

Rana, K. (2024, April 4). Minilateralism: A Trend in Regional Diplomacy - Diplo. *Diplo.* https://www.diplomacy.edu/blog/minilateralism-a-trend-in-regional-diplomacy

Rudd, K. (2006). Making Australia a force for good. *Labor eHerald Archived* 27 June 2007 at the Wayback Machine; https://en.wikipedia.org/wiki/Middle_power#cite_note-Rudd-3.

Shin, D. (2015, December 6). A Critical Review of the Concept of Middle Power. E-International Relations. https://www.e-ir.info/2015/12/04/a-critical-review-of-the-concept-of-middle-power/.

Tirkey, A. (2021, September 1). Minilateralism: Weighing the prospects for cooperation and governance. *ORF Observer Research Foundation* (ORF). https://www.orfonline.org/research/minilateralism-weighing-prospects-cooperation-governance

UAE Ministry of Climate Change & Environment. (2022, November 8). UAE, Indonesia Unveil Mangrove Alliance for Climate at COP-27. https://www.moccae.gov.ae/en/media-center/news/8/11/2022/uae-indonesia-unveil-mangrove-alliance-for-climate-at-cop27.aspx#page=1.

Urhova, D. (2023, June 6). The Shifting Balance of Power and the Rise of Minilateralism: The Indo-Pacific and Beyond. *China* Observer. https://chinaobservers.eu/the-shifting-balance-of-power-and-the-rise-of-minilateralism-the-indo-pacific-and-beyond/

Wihardja, M and Verico, K. (2022, November 21). Defending multilateralism as Indonesia passes the G20 baton. *East Asia Forum*, https://eastasiaforum.org/2022/11/21/defending-multilateralism-as-indonesia-passes-the-g20-baton/

Wirengjurit, D. Mendayung di Antara Tiga Karang. *Kompas*, 18 Juli 2020a.

Wirengjurit, D. Indonesia antara Quad dan SCO. *Kompas*, 5 Desember 2020b.

Wirengjurit, D. Indonesia dan BRICS. *Kompas*, 29 September 2022.

Wirengjurit, D. Bola Salju Menggelinding di Timur Tengah. *Kompas*. 29 Maret 2023a.

Wirengjurit, D. Minilateralisme, 'Mainan Baru' Geopolitik. *Kompas* 10 Oktober 2023b.

Wirengjurit, D. Maju Bersama OECD?. *Kompas*, 28 Desember 2023c.

Wirengjurit, D. The Power of MP and Minilateralism. *Middle Power Insights*, MPSN-FPCI, May 2024a.

Wirengjurit, D. Diplomasi: Indonesia dan Dunia. *Kompas*, 21 Februari 2024b.

Wirengjurit, D. Menuju Politik Luar Negeri Fokus. Kompas, 10 Mei 2024c.

Dian Wirengjurit completed a bachelor's degree in International Relations, University of Indonesia (1984) and a master's degree in International Relations, University of Wollongong, Australia (1990). He started working at the Ministry of Foreign Affairs of Indonesia in 1985 and ended his career in 2022. The former Deputy Chief of Mission of Indonesia Embassy in Moscow (2010–2012) and Indonesian Ambassador Extraordinary and Plenipotentiary to the Islamic Republic of Iran and Turkmenistan (2012–2016) spent most of his diplomatic career in multilateral post (UN New York 1990–1994; Geneva 1997–2001 and 2003–2007). Meanwhile, his multilateral experience and knowledge are outlined in his books "Zone of Peace and Nuclear Weapon-Free Zones: Definition, History and Development" (Updated, 2004), Geopolitics an International Relations: Capita Selecta (2023) and Iran: Nuclear, Sanctions, Military, and Diplomacy (2022). Currently he is a geopolitics and international relations analyst at the Institute of Civilization Studies in Jakarta.

Chapter 16
Indonesia's Leadership in Multilateral Forum: Reflecting from G20's 2022 Presidency and ASEAN'S 2023 Chairmanship

Humphrey Wangke, Khanisa, and Pandu Prayoga

Abstract In 2022 and 2023, Indonesia has the rare opportunity to lead two multilateral forums back-to-back, G20 and ASEAN. Both have their own significance, and Indonesia was striving to optimize its political and economic gain by setting an agenda that fits with its national interest. In G20 presidency 2022, Indonesia prioritized issues such as sustainable development, equitable economic recovery post-pandemic, global financial stability by using digital currency, and how to apply diplomatic approach amidst the Russia–Ukraine war. Then, during its ASEAN Chairmanship in 2023, Indonesia focused on enhancing regional cooperation, addressing pressing security concerns, and driving economic integration among member states. Although its roles received many criticisms such as pragmatic achievement and inward-looking foreign policy, Indonesia has demonstrated its commitment to multilateralism. Indonesia's leadership in both forums illustrate its ability to navigate complex international dynamics and advocate for the interests of developing nations on the global stage. The goal of this chapter is to examine how Indonesia has utilized its diplomatic capabilities and strategic vision to influence global and regional policies. The discussion of this chapter will be divided into three parts. First, Indonesia's approach to multilateralism, specifically Joko Widodo's signature approach which many recognize as being pragmatic and straightforward. Second and third, Indonesia's strategy in leading G20 and ASEAN. The chapter will be concluded with a critical outlook on how Indonesia would carry its foreign policy in a multilateral forum in the future.

Keywords Indonesia · Multilateralism · G20 · ASEAN

H. Wangke
Research Center for Politics, BRIN, Jakarta, Indonesia
e-mail: hump001@brin.go.id

Khanisa (✉)
Research Center for Politics, BRIN, Jakarta, Indonesia
e-mail: khan002@brin.go.id

P. Prayoga
Research Center for Politics, BRIN, Jakarta, Indonesia
e-mail: pandu.prayoga@brin.g.id

I. Kartini and Kanisa (eds.), *Joko Widodo's Foreign Policy and Diplomatic Legacy*,
https://doi.org/10.1007/978-981-95-2085-5_16

16.1 Introduction

With foreign policy embedded in its constitution, Indonesia has always put making good relations with other countries as one of its political priorities. Further, as issues and interests of states are intertwined between one another, it is crucial for Indonesia to ensure that relations with the foreign partners are done strategically. Therefore, Indonesia has been continuing to carry its foreign policy in various forms, bilaterally or multilaterally. Both have their own virtues. In bilateral arrangement, one-on-one relations enable Indonesia to have a more intimate discussion where specific issues can be discussed in more thorough negotiation. Whereas, multilateral forums will give opportunities to create a vast network and find more like-minded countries.

Indonesia's strategy in approaching multilateral forums can be found in two forms, joining or building. Indonesia's attempt to join the existing multilateral cooperation is a realization of its effort in proving its capability to contribute to international society. From a global organization like the United Nations to a more distinguished group of states like G20 that will later be discussed in this chapter, Indonesia has been actively taking part in them, even trusted to lead in the recent years.

Aside from joining, Indonesia has also built its legacy in multilateralism. Often started with a parallel interest, several of Indonesia's multilateralism projects have successfully voiced the concern of the member states to be heard globally. The Non-Aligned Movement whose goal was to maintain neutrality amidst two fighting polars in the cold war era is one of the first multilateral projects which Indonesia orchestrated. However, another example that still stands strong is ASEAN. Built within the time of uncertainty, Indonesia with four other Southeast Asian states was aware of the importance of regional stability in order for the region to move forward. This regional effort has created one of the most renowned regionalisms where states from outside the area are heavily interested to take part in the cooperation.

At this stage, although multilateralism seems to be one of Indonesia's preferred diplomatic strategies, we still need to consider the difference of diplomatic approach from one president to the other. In this context, under the ten years of President Joko Widodo's administration, Indonesia's foreign policy has been seemingly repositioned from being outward looking in the era of President Susilo Bambang Yudhoyono to being inward-looking during his time. This change also brings adjustment to how Indonesia approaches multilateralism. The transformation of Indonesian diplomacy in the era of President Joko Widodo places Indonesia in the problem of how to make national interests and multilateralism two complementary issues that benefit Indonesia. This kind of approach encourages the implementation of Indonesia's foreign policy to be in line with domestic demands.

Interestingly, the change in President Joko Widodo's diplomatic style occurred at a time when Indonesia's strategic environment was also changing. In 2015 ASEAN is heading towards the formation of the ASEAN Community which of course requires a more active role from each member. Meanwhile, in the Indo-Pacific region, Indonesia has to face competition between China and the United States to become the main country in the region. This change in the strategic environment will determine

Indonesia's international relevance, position, and role. Changes like this do not mean that Indonesia is no longer involved in multilateral diplomacy, but it is carried out at a more limited level compared to the era of President Susilo Bambang Yudhoyono.

Indonesia's distinguished diplomatic style at this era was challenged by the two events where Indonesia must lead two respectable groupings back-to-back, G20 in 2022 and ASEAN in 2023. Indeed, that was a great opportunity for Indonesia to achieve political and economic gain from these two groupings as well as externalizing its national interest. Therefore, this chapter will solely focus on how Indonesia's strategy on multilateralism was implemented in those two years. The question that will be the basis of the examination is, "how Indonesia has utilized its diplomatic capabilities and strategic vision to influence global and regional policies?" The authors argue that although President Joko Widodo is likely to have a more straightforward and pragmatic vision on foreign policy, Indonesia was still carrying its multilateral foreign policy by highlighting the importance of discussion and personal approach in smoothing its diplomatic strategy. This gesture gives a more open and relaxed ambiance in order to build trust within the diplomatic network.

The discussion of this chapter will be divided into three parts. First, Indonesia's approach to multilateralism, specifically Joko Widodo's signature approach which many recognize as being pragmatic and straightforward. Second and third, Indonesia's strategy in leading G20 and ASEAN. The chapter will be concluded with a critical outlook on how Indonesia would carry its foreign policy in a multilateral forum in the future.

16.2 Multilateralism and Indonesia's Foreign Policy

Multilateralism has various interpretations. Some view it as a coordinated relationship among three or more countries adhering to certain principles (Ruggie, 1992). Others define it as norms, rules, and principles agreed upon collectively that guide and regulate international relations, with reciprocity being a common basis for multilateral institutions (Hampson & Heinbecker, 2011). From these definitions, multilateralism can be understood as cooperation between countries based on mutual understanding and shared values by creating international institutions. According to the United Nations, multilateralism is a method of cooperation and organizing the international system where cooperation is governed by collectively developed rules that ensure sustainable and effective collaboration (United Nations, 2024).

When Joko Widodo assumed the presidency of Indonesia, the role of multilateral institutions began to wane alongside the decline of American unionism. According to Ruland, multilateral institutions started to diminish as they became arenas of power competition among the US and Western countries, involving struggles over access, decision-making, and normative order (Ruland, 2012). Julia and Keohane use the term "contested multilateralism" to describe the situation resulting from efforts by countries, multilateral organizations, and non-state actors to use both existing and newly formed multilateral institutions to challenge the rules, practices, or missions of these institutions (Morse & Keohane, 2014). Emerging multilateral powers, such as BRICS (Brazil, Russia, India, China, South Africa), BRI (Belt and Road Initiative), QUAD (United States, India, Japan, and Australia), AUKUS (Australia, the United Kingdom, and the United States), IPEF (Indo-Pacific Economic Forum), and developing countries in the Global South, have diminished the legitimacy of the multilateral architecture. These developments challenge the importance of the UN system and highlight the breakdown of international cooperation. The crisis in existing multilateral agreements has been ongoing for two decades and has become more acute.

Facing an unfavorable external environment, President Joko Widodo declared the end of his predecessor Susilo Bambang Yudhoyono's "One Thousand Friends, Zero Enemies" approach, which emphasized greater participation in international forums. Under President Joko Widodo, this policy was reinterpreted to mean engaging with all countries without compromising Indonesia's sovereignty or national interests. After attending his first multilateral meetings in November 2014, including the G20, ASEAN Summit, and APEC, President Joko Widodo stated that he would prioritize foreign relations that offer significant benefits to Indonesia, saying, "If it is not beneficial, I will not do it" (Qin, 2015). Analysis indicates that Indonesia's foreign policy under President Joko Widodo is more transactional, aiming to benefit Indonesia more directly (Lundry 2018).

Foreign Minister Retno Marsudi noted that President Joko Widodo's foreign policy is more results-oriented and people-focused. She emphasized that Indonesia's foreign policy addresses the urgent needs of Indonesia's national interests (Parameswaran, 2015). After two years in office, there was a positive domestic perception of Indonesia's foreign policy. Actions such as executing transnational drug traffickers, destroying illegal fishing vessels, and implementing the high-speed rail policy have satisfied the public by prioritizing short-term national interests (Rosyidin, 2015).

Despite strongly emphasizing national needs, President Joko Widodo did not abandon the spirit of multilateralism initiated by his predecessors. In a speech at the Brookings Institution in Washington, DC, President Joko Widodo stated that Indonesia would balance national interests with international interests (Parameswaran, 2015). In other words, while conducting his foreign policy, President Joko Widodo remains actively involved in global issues without diminishing Indonesia's interests (Rosyidin, 2017). Joko Widodo's increasingly active engagement in recent years shows a larger vision that shapes Indonesia's regional and global engagement under his administration. For example, during the COVID-19 pandemic in 2020, Indonesia faced international challenges while maintaining its focus on national interests.

In order to maintain a balance between national and international interests, President Joko Widodo pursued bilateral and multilateral diplomacy in response to the COVID-19 pandemic. Bilaterally, Foreign Minister Retno Marsudi visited several countries, including China, the United Kingdom, and Switzerland, to secure vaccine availability for the Indonesian people. By empowering Indonesian embassies to engage in direct diplomacy, Indonesia successfully gained access to vaccines for

20 percent of its population through the WHO's provisions. Multilaterally, President Joko Widodo participated in the 75th United Nations General Assembly in September 2020 for the first time. In his speech, President Joko Widodo emphasized the importance of collaboration and collective global leadership in handling the pandemic. Following up on this proposal, on December 20, 2020, Indonesia adopted the UN General Assembly resolution on "Global Health and Foreign Policy: Strengthening Health System Resilience through Affordable Healthcare for All."

Through this resolution, Indonesia strongly emphasized the importance of accessible and affordable healthcare services to strengthen global health system resilience. Indonesia's diplomacy did not stop at proposing the UN resolution; it also actively worked to ensure the availability of COVID-19 vaccines for all countries, not just for developed nations. Indonesia believes that bilateral and multilateral approaches will pave the way for all countries to address various issues, including the COVID-19 pandemic, through coordination and cooperation, thereby finding a middle ground for national interests. Indonesia sought to collaborate with global and regional organizations to mitigate the pandemic while maintaining its support for the WHO (Luerdi and Azhari 2022, p. 41).

President Joko Widodo's shift in leadership style in Indonesia's foreign policy has sparked opinions that Indonesia's foreign policy appears paradoxical (Rosyidin and Kusumawardhana, 2022). On the one hand, Indonesia strongly emphasizes a bilateral approach, but on the other hand, it pays attention to its involvement in various multilateral forums. Not only at the global level, such as the UN but at the regional level, President Joko Widodo has also demonstrated his leadership capacity in multilateral diplomacy. As a leader of ASEAN, Indonesia has expressed discomfort with the United States' approach to China, viewing it as an exclusive strategy aimed at isolating China. Indonesia perceives the QUAD, consisting of the United States, Japan, Australia, and India, as a potential coalition of external powers that does not include ASEAN.

In response, Indonesia developed the ASEAN Outlook on the Indo-Pacific (AOIP) as a more ASEAN-centered Indo-Pacific strategy aligned with principles of inclusivity, including towards China, and building consensus, with a focus on normative approaches in politics and diplomacy (Acharya, 2019). While Western strategies in the Indo-Pacific region highlight the competition between China and the United States, AOIP does not mention specific countries but emphasizes collaboration in strengthening maritime cooperation, connectivity, sustainable development goals, and cooperation in economic and other fields (ASEAN Secretariat, 2024).

Furthermore, as an Indonesian initiative, AOIP was developed as a driving force to bring competing major powers together to cooperate in the Indo-Pacific region. Since then, the key principles of AOIP are strengthening ASEAN centrality, openness, transparency, non-intervention, and respect for international law have become mainstream across various pillars of ASEAN cooperation. Regional stability is crucial for Indonesia, considering Southeast Asia is a global hotspot for tensions involving superpowers like the United States and China. Without dynamic security stability, it would be challenging for Indonesia and ASEAN countries to catch up economically. ASEAN centrality is understood in terms of the region's geographical position and as a reconstruction of perspectives that have hindered ASEAN countries from growing together. For Indonesia, ASEAN must be solid in its unity to grow collectively and complement one another.

Indonesia has also been actively communicating with other Asian countries in the Indo-Pacific region through platforms such as the East Asia Summit and the Indian Ocean Rim Association (IORA). While significant progress may still need to be visible from these forums, it is undeniable that President Joko Widodo has actively participated in these multilateral dialogues. In November 2022, Indonesia successfully hosted the G20 Summit in Bali, showcasing its growing regional leadership. With the summit's compelling theme, "Recover Together, Recover Stronger," President Joko Widodo once again emphasized the importance of equitable sustainable development. The Bali G20 Summit yielded concrete outcomes, including establishing a pandemic fund that has gathered commitments totaling up to 1.5 billion USD. As the first Southeast Asian country to host the G20 Summit, President Joko Widodo demonstrated his capability to address global issues in multilateral forums like the G20. When Indonesia served as G20 Chair in 2022 and ASEAN Chair in 2023, President Joko Widodo showcased his capacity as a multilateralist who strongly emphasizes inclusivity, collaboration, and resilience.

Despite shifts in emphasis and leadership style under President Joko Widodo's administration, Indonesia's commitment to regional stability, security, and cooperation has remained consistent. Through active participation in various multilateral meetings, Indonesia has effectively influenced the geopolitical dynamics in the Indo-Pacific and Southeast Asia regions. As seen in 2022 and 2023, Indonesia's foreign policy experienced a significant escalation in multilateralism. In November 2022, President Joko Widodo attended the ASEAN+3 Summit in Cambodia. During the meeting in Cambodia, Indonesia was appointed as the ASEAN Chair for 2023. Additionally, Indonesia hosted the G20 Summit in Bali on November 15–16, 2022.

President Joko Widodo's success in leading the G20 Summit in Bali in 2022 left a strong impression on the world, as he refused to expel Russia from the G20 or exclude President Vladimir Putin from the summit. The summit took place despite pressure from Western countries seeking to punish Russia for invading Ukraine. Undeniably, these two high-level meetings were highly significant and strategically valuable for Indonesia. From an international relations perspective, President Joko Widodo's active role elevated Indonesia's position and role as a key actor in regional and global international relations.

President Joko Widodo sought to address Indonesia's challenges by prioritizing a more limited multilateral approach compared to his predecessor, yet one that had substantial national and international impact. The limitation is whether the multilateral meeting will be beneficial to Indonesia. Such conditions arise because each president has their way of practicing Indonesia's free and active foreign policy. Indonesia's multilateral approach during President Joko Widodo's era represents an adaptation in responding to global challenges that affect the country. Foreign policy is not static but can evolve according to the national interests of the time.

16.3 G20's Presidency Amidst the Global Tension

When receiving Indonesia's turn of the G20 presidency from Italian Prime Minister Mario Draghi in November 2021, President Joko Widodo brought post-pandemic optimism by stating the theme of Indonesia's 2022 presidency will be "Recover Together, Recover Stronger." Indonesia's leadership would be committed to growth that is inclusive, people-centered, environmentally friendly, and sustainable (PresidenRI, 2021). Indonesia's envision a broad set of goals that might be hard to be fully achieved.

Indonesia's excitement to use G20 as a platform to gain multilateral benefit is understandable. G20 is one of the elite global groupings which comprises developed and developing countries. Its members are Argentina, Australia, Brazil, Canada, China, France, Germany, India, Indonesia, Italy, Japan, Mexico, Russia, Saudi Arabia, South Africa, South Korea, Turkey, the United Kingdom, and the United States. As one of the developing states within the organization, Indonesia is expected to voice the interest of its fellow developing states in facing the bigger powers (Kartini, 2021).

In terms of importance, the combined economies of the G20 members are equal to around 80 percent of GDP and 75 percent of exports of the world. The citizens of these states also represent 60 percent of the total world's population (OECD, no date). These remarkable numbers are great modalities for Indonesia to expand its target market as well as gain more investors. In this context, the G20 is an international role that suits President Widodo's focus on diplomacy which is expected to bring tangible economic results.

In a press release published by the Coordinating Ministry for Economic Affairs, it is stated that one of the focuses of Indonesia's G20 Presidency is to "ensure the world economy remains open, fair, mutually beneficial, and ensures that no one is left behind." This statement is a strong signal that Indonesia is aware of its position as a developing country and demands an equal opportunity to not be stirred by the interest of larger economies.

The existing developmental gap is something that can not be easily put away by friendly multilateral rhetoric. According to Beson and Bell (2009), historically it is indeed formed by the developed countries who were aware of the importance of their developing counterparts. However, Beeson and Bell explain that there is an interplay between efforts by developed states to maintain their hegemonic grasp of smaller economies and dynamics created by the smaller economic members within the institution.

In its 2022 presidency, it can be seen that President Joko Widodo is intended to maximize gain in the critical fields where Indonesia has been pushing forward. Al-Fadhat (2022) explains there are three fields mentioned by the president, global health, sustainability, and digital economy. However, Al-Fadhat explains that each of those three fields have their own challenges, for example, the problem of inequality within the global health industry and the complexities to achieve green economy vis-a-vis Indonesia dependency to conventional and resource based economy.

During its term as the G20 president, Indonesia faced an unexpected global political security disruption as Russia launched military action in Ukraine in February 2022. The conflict between Russia and Ukraine itself is not new since Russia already annexed Crimea, a border area between the two states, in 2014. What makes the 2022 conflict different is Ukraine's leniency to the western counterpart. As explained by Masters (2023), Ukraine holds cultural, political, and economic significance to Russia. Masters also explored that Russian action in Ukraine might be heavily triggered by Ukraine's move to be a member of NATO. Russia as the power that holds Eastern Europe hegemony feels angered by the US and westerns maneuver to expand their NATO umbrella.

This conflict brings a direct implication with the G20 dynamics in 2022. In this case, Russia and those who aligned with Ukraine, such as the US, the UK, France, and Germany are the member states of G20. Hostilities among them can bring failure to the Indonesian presidency's vision and goal. Heated situation happened at the G20 Foreign Minister Meeting in July 2022. The attendance of Russian Foreign Minister, Sergei Lavrov, was greeted with resentment from the Western counterpart (DW.com, 2022).

In an effort to manage this situation, President Joko Widodo embarked on official visits to Kyiv and Moscow at the end of June 2022. This effort, although cannot be regarded as a successful peace-broker maneuver, still resulted in three gains, securing Indonesian wheat supplies, creating the remarkable image as the G20 President and earning positive imagery of active statesmanship that can take global role on the eyes of the domestic audience (Dharmaputra, 2022).

As the summit approaches, Indonesia focuses on building collective commitment to address significant issues which will be featured in the summit outcome both in the Finance track and the Sherpa track. On the finance track, there are seven issues that have been discussed: exit policy to support recovery, addressing scarring effect to secure future growth, central bank digital currency, cross border payment, sustainable finance, digital financial inclusion, and international taxation (Nurhanisah, 2022). Meanwhile, on the Sherpa track, which focuses more on socio-economics issues, Indonesia brought the issue of global health architecture, standardization of digital proficiency, improvement of multilateral trade to benefit small businesses, as well as improvement of education (Wisnubroto, 2022).

The summit itself was held on 15–16 November 2022 in Bali. Alongside the magnificent setup and banquet, under its presidency Indonesia successfully produced the Bali Leaders' Declaration. This 19-page document highlights the importance of recovering from various impacts of the pandemic as well as other challenges for example food security, digital innovation, climate and energy crises, and other related issues.

With all the hurdles that looming beneath Indonesia's presidency, this result cannot be understated. Indonesia successfully carried out its diplomatic role and finished its turn to lead one of the prestigious multilateral organizations until it passed the torch to India for 2023. Virgianita (2022) stated that there are several notes that can be highlighted such as the dominant voice by the western counterpart, especially when it comes to the Ukraine–Russia conflict, and how Indonesia can ensure the implementation of the frameworks agreed on the declaration.

16.4 ASEAN'S Chairmanship in the Time of Regional Uncertainty

In 2023, Indonesia managed the ASEAN Chairmanship during a period marked by significant regional uncertainties, both political and economic, following its successful leadership of the G20 in 2022. This period was marked by geopolitical tension, namely the ongoing crisis in Myanmar, intensifying US–China rivalry, and global economic recovery challenges from COVID-19. Under the proposed theme "ASEAN Matters: Epicentrum of Growth," Indonesia aimed to strengthen ASEAN's role as a central force in addressing challenges, promoting economic growth, and enhancing regional security (Coordinating Ministry for Economic Affairs, 2023). Throughout its chairmanship, Indonesia successfully demonstrated leadership by focusing on key areas, addressing pressing issues, and driving the ASEAN agenda forward. Indonesia's leadership within ASEAN refers to the role played by the country in navigating the regional bloc's direction, especially during periods of geopolitical, the grouping unity, economic, health, and how ASEAN took a global role.

One of Indonesia's main priorities was to strengthen ASEAN's institutional capacity and ensure its role as a stabilizing force in the Indo-Pacific as Joko Widodo said at 43rd ASEAN Summit (Susilo, 2023). The Indo-Pacific region in 2023 was marked by increasing tensions, particularly involving major powers like the United States and China (Nabbs-Keller, 2020). Indonesia's leadership was pivotal in maintaining ASEAN's neutral stance, promoting peace and stability through dialogue and diplomacy in the middle of multi-alignment and engagement (Nagy, 2024).

Strengthening the ASEAN Outlook on the Indo-Pacific (AOIP) has been initiated by Indonesia to support this priority (Chambers & Bunyavejchewin, 2023). By emphasizing the need for ASEAN to engage with all stakeholders in the Indo-Pacific, Indonesia played a critical role in advancing regional cooperation and ensuring that ASEAN remained relevant and central in discussions on Indo-Pacific strategy, as opposed to allowing external powers to dominate the narrative. This openness and inclusiveness of ASEAN to dialogue partners was historically started from the beginning and was integrated in the Treaty of Amity and Cooperation (TAC) of 1976 (Drysdale et al., 2023). Through ASEAN, Indonesia played a non-aligned and active engagement strategy to release pressure from bigger powers as well as cooperate with them that can be seen also in ASEAN Regional Forum (Katsumata, 2014).

Under the 2023 ASEAN Chairmanship, one of the most significant internal challenges is the ongoing crisis in Myanmar. Alongside differing responses among ASEAN members, Indonesia worked tirelessly to maintain unity within ASEAN. The military coup in Myanmar in February 2021 led to widespread violence and political instability, testing ASEAN's long-standing principle of non-interference and consensus decision-making. The grouping commonly adheres to the principle of non-interference, which complicates direct interventions. However, Indonesia emphasized ASEAN's Five-Point Consensus to address the political crisis in Myanmar and pushed for ending to violence, dialogue among all parties, humanitarian assistance to the country, and the appointment of a special envoy. Indonesia was proactive in sending its special envoy to engage with stakeholders in Myanmar including opposition forces, proving its commitment to restore peace and stability in the region. This was part of Indonesia's effort to mediate the conflict, take advanced action than previous chair nations that avoided strong action. This diplomatic push was seen as a bold move to maintain ASEAN's credibility and unity. Until the end of the year, as President Jokowi addressed honestly that implementation of the Five-Point Consensus has made no significant progress (Office of Assistant to Deputy Cabinet Secretary for State Documents & Translation, 2023a).

Simultaneously, Indonesia had to navigate the differing national interests and foreign policy orientations of ASEAN members, which often leaned towards international pressures from the United Nations, the United States of America, and China. For example, while countries like Thailand and Cambodia showed hesitation to impose sanctions or take a hard stance against Myanmar's junta, others like Malaysia and Singapore advocate for a tougher approach. Since its membership entrance in 1997, Myanmar has presented challenges to ASEAN like human rights violence and military influence (Seah, 2023). Many critics say that Indonesia failed to meet this pressure (Jaehyon, 2023).

Indonesia held multiple diplomatic engagements to encourage Myanmar's military junta to cooperate with ASEAN's peace efforts (Gunawan, 2024). Although there were challenges, Indonesia's proactive stance ensured that the Myanmar issue remained on ASEAN's agenda, preventing the group from being seen as ineffective or irrelevant on the international stage. Indonesia's leadership was a testament to its commitment to ASEAN's unity, even in the face of internal challenges and different member states' responses.

Aside from Indonesia's effort to strengthen ASEAN's unity, in its 2023 ASEAN chairmanship, Indonesia aims to broaden the perspective on the Indo-Pacific region, moving beyond the security domain to focus on inclusive economic cooperation, prioritizing maritime, connectivity, SDGs, and economics under AOIP (Darwis & Wambrauw, 2023). By organizing the ASEAN Indo-Pacific Forum (AIPF) as part of the 43rd ASEAN Summit, Indonesia aims to enhance partnerships, boost regional supply chains, and support a fair energy transition, fostering a more peaceful and prosperous region (Office of Assistant to Deputy Cabinet Secretary for State Documents & Translation, 2023b).

Amid global economic uncertainties, including post-pandemic recovery challenges and global inflation pressures, Indonesia was trying to set ASEAN as a vital hub for economic growth. This effort highlights ASEAN's role as a dynamic center for regional and global economic potential. As a hub for economic growth, ASEAN serves as a focal point for trade, investment, and innovation, leveraging its integrated supply chains while at the same time expanding its consumer base to attract global business. In the 55th ASEAN Economic Ministers' (AEM) Meeting, ASEAN has focused on strengthening economic ties within the region by deepening intra-regional trade and enhancing connectivity among member states (ASEAN Secretariat, 2023a). Indonesia has sought to solidify partnerships with external stakeholders, creating avenues for foreign investment as well as technology transfer.

Indonesia has identified the digital economy as a pivotal driver of future growth in ASEAN, recognizing its potential to transform industries, boost innovation, and create opportunities across the region. To support this idea, Indonesia has prioritized fostering cooperation among ASEAN member states on critical issues like digital connectivity, e-commerce, and digital literacy (Yusgiantoro & Nasution, 2023). Strengthening digital connectivity is essential to bridging the digital divide within the region, ensuring that all member states have access to reliable digital infrastructure including for those who live in remote areas. E-commerce, a rapidly growing sector, has been a focal point for enhancing regional trade, enabling small and medium-sized enterprises (SMEs) to access broader markets and integrate into the global value chain.

Meanwhile, improving digital literacy has been a cornerstone of Indonesia's efforts, as it empowers individuals and communities to participate meaningfully in the digital economy while addressing significant challenges such as cybersecurity and misinformation, including hoaxes. Acknowledging the transformative power of digital skills, Indonesia has an intention to ensure that citizens across ASEAN are equipped with the knowledge and tools to navigate the digital landscape effectively, encouraging inclusivity and resilience. As cyber threats increased in complexity and frequency, Indonesia managed dialogue on regional cybersecurity cooperation, ensuring that ASEAN could better protect its digital infrastructure and maintain stability in the digital space (ASEAN Secretariat, 2023b). Through these initiatives, Indonesia aims to position ASEAN as a globally competitive digital hub, promoting inclusive and sustainable economic growth while preparing the region to thrive in an increasingly digitized global economy.

Furthermore, Indonesia has taken a leading initiative to boost ASEAN's resilience in public health, considering the critical needs for stronger regional collaboration to address emerging challenges, such as potential future pandemics threats. Drawing on lessons from the COVID-19 pandemic, Indonesia has advocated for comprehensive efforts to strengthen regional health systems, ensuring they are equipped to handle crises effectively. This includes enhancing the capacity of healthcare infrastructure, increasing the availability of medical resources, and realizing innovation in public health strategies. By prioritizing these areas, Indonesia has sought to create a robust foundation for ASEAN to collectively respond to health emergencies while safeguarding the well-being of its people. Pushed for stronger regional health systems, better preparedness for future pandemics, and improved vaccine distribution mechanisms to ensure that ASEAN members could recover together from the impact of COVID-19.

A key focus of Indonesia's leadership has been on improving regional preparedness for future pandemics. Recognizing that swift and coordinated responses are vital in mitigating the impact of health crises, Indonesia has emphasized the importance of building early warning systems, conducting joint simulations, and establishing mechanisms for rapid information sharing among ASEAN members. Additionally, Indonesia has pushed for the integration of pandemic response frameworks to align national strategies with regional objectives, creating a unified approach to tackling potential health threats. These efforts not only enhance ASEAN's capacity to respond to crises but also contribute to global health security, underscoring the region's commitment to proactive and collaborative solutions.

Another significant aspect of Indonesia's initiative has been improving vaccine distribution mechanisms to ensure equitable access across ASEAN. During the COVID-19 pandemic, disparities in vaccine access highlighted the urgent need for a fair and efficient distribution system. To address this, Indonesia has championed efforts to strengthen regional supply chains, expand local vaccines production capabilities, and promote partnerships with international organizations and private sectors. By doing so, Indonesia aims to ensure that all ASEAN member states, regardless of their economic status, can recover collectively from the pandemic's impact. These initiatives demonstrate Indonesia's commitment to fostering a resilient, inclusive, and united ASEAN, capable of overcoming public health challenges and building a healthier future for its people.

In the context of ASEAN's role on the global stage, Indonesia placed strong emphasis on elevating and advocating the important move for the region to move beyond being a passive observer in international affairs. Indonesia asserted that ASEAN should act as an active contributor to addressing global challenges, leveraging its strategic position and collective strength. By focusing on issues such as sustainable development, regional stability, and multilateral cooperation, Indonesia aimed to showcase ASEAN as a key player capable of influencing global discourse and providing solutions to solve international problems.

This vision was met with skepticism from scholars like Artner (2017) suggesting that Indonesia's leadership might deprioritize ASEAN. These critiques stemmed from observations that President Joko Widodo's administration had focused heavily on domestic development agendas, potentially leaving ASEAN without the strong leadership required to drive the grouping forward. Concerns were raised that this inward-looking approach could result in ASEAN losing momentum on critical regional and international initiatives, leaving it vulnerable to external pressures and diminishing its relevance on the global stage.

Despite these criticisms, Indonesia's chairmanship demonstrated a renewed commitment to strengthening ASEAN's position and ensuring its collective voice was heard internationally. Through diplomatic engagements and a focus on inclusive initiatives, Indonesia worked to dispel doubts about its leadership and reaffirmed its dedication to regional unity. By addressing both domestic priorities and ASEAN's broader goals, Indonesia sought to balance internal development with regional and global responsibilities, reinforcing its role as a cornerstone of ASEAN and its aspiration to contribute meaningfully to global problem-solving efforts.

In 2023, despite the challenging context of regional uncertainties, Indonesia successfully led ASEAN by fostering unity, promoting economic growth, and strengthening ASEAN's role in global and regional affairs. Indonesia's focus on diplomacy, sustainability, and digital transformation laid a solid foundation for the grouping's continued relevance and resilience in an increasingly complex world. Through its leadership, Indonesia reinforced the notion that ASEAN matters and that the region can act as an epicenter of growth and stability in the face of global challenges. These strategies not only underscored ASEAN's relevance in a shifting geopolitical landscape but also reinforced Indonesia's role as a stabilizing force within the bloc. Through a combination of pragmatic diplomacy and inclusive leadership, Indonesia demonstrated its commitment to upholding ASEAN's unity and collective resilience during a time of regional uncertainty.

16.5 Conclusion: Looking Back at Two Leaderships

Indonesia has successfully led two multilateral forums back-to-back. Both G20 and ASEAN are prestigious groups in their own way and Indonesia's role to carry its vision during the time of leadership must be appreciated. However, as this chapter's aim is to examine how Indonesia has utilized its diplomatic capabilities and strategic vision to influence global and regional policies, there are three lessons that might be useful for Indonesia to engage in multilateral diplomacy in the future.

First, in both G20 and ASEAN's leadership Indonesia proves the seriousness in maintaining multilateral engagement. As previously stated, President Joko Widodo prefers a balanced approach between bilateralism and multilateralism. On these occasions, we can see a mixed strategy as Indonesia tries to use bilateral ways in untying diplomatic knots when there are hurdles in the multilateral platform. By doing so, Indonesia is aiming to lower the tension when conflicting parties meet at the multilateral event.

Second, fighting for Indonesian domestic interest is possible on a multilateral platform. This was particularly visible in the effort to recover post-pandemic era. During G20 and ASEAN's leadership, the agenda to ensure better vaccine distribution and reforming cooperation on global health was on both times. This is parallel with President Joko Widodo's attempt to ensure a national level recovery that will be supported by a better system in facing similar challenges in the future.

Third, when it comes to output, multilateral forums tend to be more general or even vague points. In this case Indonesia must prioritize following up these points so that it can be more realistic. Output from G20 and ASEAN are both important and promising, however the implementation of them might be problematic as the leader of these multilateral forums change in the following years. Therefore, in holding a leadership position, formulation of realistic outputs and locking the commitment of stakeholders before the turn ended is important.

Reflecting on what Indonesia has done in these years, it is evident that President Joko Widodo has not abandoned the potential of multilateralism. However, it must be noted that multilateralism engagement must be maintained not only when Indonesia is the chair. It is a long-running diplomatic effort that needs strong commitment. Indonesia's continuous role in ASEAN is a good example of how it works with the organization and makes the organization as their cornerstone of Indonesia's foreign policy. Indonesia's role in the G20 might not be as substantial as in ASEAN, but Indonesia can play a role in leading the developing countries' voice in the organization.

References

Acharya, A. (2019). Why ASEAN'S Indo-Pasifiic matters, the strategist, https://wwwaspistrategist.org.au/why-ASEANs-indo-pasific-outlook-matters/.

Al-Fadhat, F. 2022. Indonesia's G20 presidency: neoliberal policy and authoritarian tendencies. *Australian Journal of International Affairs* 76 (6): 617–623. https://doi.org/10.1080/10357718.2022.2070598.

Artner, A. 2017. Role of Indonesia in the evolution of Asean. *The Journal of East Asian Affairs* 31 (1): 1–38. http://www.jstor.org/stable/44321271.

ASEAN Outlook on the Indo-Pasific. (n.d.). https://asean.org/asean-outlook-on-the-indo-pacific/7 Agenda Prioritas Finance Track Presidensi G20 Indonesia (n.d.). https://indonesiabaik.id/infografis/7-agenda-prioritas-finance-track-presidensi-g20-indonesia.

ASEAN Secretariat. (2023a). The 55th ASEAN Economic Ministers' (AEM) Meeting 19 August 2023, Semarang, Indonesia.

ASEAN Secretariat. (2023b, October 18). Chairman's Statement of the 8th ASEAN Ministerial Conference on Cybersecurity. https://Asean.Org/Chairmans-Statement-of-the-8th-Asean-Ministerial-Conference-on-Cybersecurity/.

Beberapa Catatan Penting Pertemuan Sherpa Track G20. (n.d.). https://indonesia.go.id/kategori/kabar-g20/4674/beberapa-catatan-penting-pertemuan-sherpa-track-g20?lang=1

Beeson, M., and S. Bell. 2009. The G-20 and international economic governance: Hegemony, collectivism, or both? *Global Governance* 15 (1): 67–86. http://www.jstor.org/stable/27800739.

Chambers, P., and P. Bunyavejchewin. 2023. Complex positioning: Thailand's role in the context of ASEAN "centrality" and Indo-Pacific security. In *ASEAN and regional actors in the Indo-Pacific*, ed. S. Sudo and C. Yamahata, 17–37. Springer.

Coordinating Ministry for Economic Affairs, R. of I. (2023, May 10). Presiden Jokowi Jelaskan Makna "ASEAN Matters: Epicentrum of Growth" dan Dukungan dari Seluruh Kepala Negara ASEAN. https://www.ekon.go.id/publikasi/detail/5131/presiden-jokowi-jelaskan-makna-asean-matters-epicentrum-of-growth-dan-dukungan-dari-seluruh-kepala-negara-asean.

Darwis, D., and M. S. Wambrauw. 2023. Indonesia's foreign policy in the Indo-Pacific region in the 2023 ASEAN chair. *International Journal of Multidisciplinary Research and Analysis* 06 (11):

Dharmaputra, R. (2022). Jokowi's visits to Russia and Ukraine are more about domestic gains than the global interest. The Conversation. https://theconversation.com/jokowis-visits-to-russia-and-ukraine-are-more-about-domestic-gains-than-the-global-interest-186217

Drysdale, P., D. Narjoko, and R. St Maria. 2023. ASEAN and the role of Asian regionalism in managing asymmetric power. In *Navigating prosperity and security in East Asia*, ed. S. Armstrong, T. Westland, and A. Triggs, 189–214. ANU Press.

Gunawan, T. 2024. Between principles and actions: ASEAN and Indonesia in dealing with the political crisis in Myanmar. *Jurnal Global & Strategis* 18 (1): 81–106. https://doi.org/10.20473/jgs.18.1.2024.81-106.

Hampson, F. O., and P. Heinbecker. 2011. The "new" multilateralism of the twenty-first century. *Global Governance: A Review of Multilateralism and International Organizations* 17 (3): 299–310. https://doi.org/10.1163/19426720-01703003.

Indonesia Terima Presidensi G20, Presiden Jokowi Undang Para Pemimpin Dunia ke Bali Tahun Depan. (1 November 2021). https://www.presidenri.go.id/siaran-pers/indonesia-terima-presidensi-g20-presiden-jokowi-undang-para-pemimpin-dunia-ke-bali-tahun-depan/

Jaehyon, L. 2023. The 2023 ASEAN, ASEAN+3, & East Asia summits: Focusing on ASEAN issues amid US-China indifference. *Contemporary Southeast Asia* 45 (3): 494–519.

Kartini, I. (29 November 2021). Arsitektur Kesehatan dan Ekonomi Global. republika.id. https://www.republika.id/posts/22640/arsitektur-kesehatan-dan-ekonomi-global

Katsumata, H. (2014). What Explains ASEAN's Leadership in East Asian Community Building? Pacific Affairs, 87(2), 247–264. doi:https://doi.org/10.5509/2014872247.

Luerdi, and S. Azhari. 2022. Upholding multilateralism: Indonesia's foreign policy in responding Covid-19 pandemic. *Jurnal Ilmiah Hubungan International, Universitas Parahyangan* 18:1. https://doi.org/10.26593/jihi.v18i1.5255.29-54.

Lundry, C. 2018. Assessing Indonesia's foreign policy under Jokowi. *Asia Policy* 13 (4): 30–35. https://doi.org/10.1353/asp.2018.0058.

Masters, J. (2023, February 14). Ukraine: Conflict at the Crossroads of Europe and Russia. Council on Foreign Relations. https://www.cfr.org/backgrounder/ukraine-conflict-crossroads-europe-and-russia

Morse, J.C, and Keohane, R.O. (2014). *Contested Multilateralism. the Review of* International Organization, 9(1), 358–412.

Nabbs-Keller, G. 2020. ASEAN centrality and Indonesian leadership in a contested Indo-Pacific order. *Security Challenges* 16 (3): 21–26.

Nagy, S. R. 2024. Southeast Asia's Indo-Pacific balance: FOIP and BRI as hedging strategies to secure strategic autonomy. In *Southeast Asia and the Indo-Pacific construct*, ed. I. Saxena and S. Nagy, 1–24. Nova Science Publisher.

OECD and G20 (n.d.). https://www.oecd.org/en/about/oecd-and-g20.html.

Office of Assistant to Deputy Cabinet Secretary for State Documents & Translation. (2023a, May 11). ASEAN Leaders Discuss Follow-up to Five-Point Consensus. https://setkab.go.id/en/asean-leaders-discuss-follow-up-to-five-point-consensus/.

Office of Assistant to Deputy Cabinet Secretary for State Documents & Translation. (2023b, September 5). President Jokowi Opens ASEAN-Indo-Pacific Forum. https://setkab.go.id/en/president-jokowi-opens-asean-indo-pacific-forum/.

Parameswaran, P. (2015). Jokowi Defends Indonesia's Foreign Policy During US Trip, The Diplomat. https://thediplomat.com/2015/10/jokowi-defends-indonesias-foreign-policy-during-us-trip/.

Qin, S. 2015. A retreat from multilateralism: Foreign policy restructuring under Jokowi. *Australian Institute of International Affairs*.. https://www.internationalaffairs.org.au/australianoutlook/a-retreat-from-multilateralism-foreign-policy-restructuring-under-jokowi/.

Rosyidin, M. (2015). Redesigning Indonesia's foreign policy in the 21st century. Niche diplomacy the way forward on the regional and global stage. *Strategic Review.* https://sr-indonesia.com/web-exclusives/view/redesigning-indonesia-s-foreign-policy-in-the-21st-century/

Rosyidin, M. 2017. Foreign policy in changing global politics: Indonesia's foreign policy and the quest for major power status in the asia countries. *South East Asia Research* 1-17. https://doi.org/10.1177/0967828X17706570.

Rosyidin, M. and Kusumawardhana, I. (2022). Participating for Profit: Bilateral-Multilateral Nexus in Indonesian Foreign Policy in the Era of Joko Widodo (2014–2019). Politik Indonesia: Indonesian Political Science Review, 7 (1), pp. 11–35.

Rosyidin, M., and S. C. H. Pattipeilohy. 2020. Regionalism under challenge: Ideas and Joko Widodo's foreign policy towards ASEAN, 2014-2019. *Journal of ASEAN Studies* 8 (2): 147–172. https://doi.org/10.21512/jas.v8i2.6596.

Ruggie, J. G. 1992. Multilateralism: The anatomy of an institution. *International Organization.* 46 (3): https://www.jstor.org/stable/2706989.

Rüland, J. 2012. The rise of "diminished multilateralism": East Asian and European forum shopping in global governance. *Asia Eur J* 9:255–270. https://doi.org/10.1007/s10308-012-0311-9.

Seah, S. 2023. Non-interference versus self-determination: ASEAN'S cold war normative framework. *Contemporary Southeast Asia* 45 (3): 494–519.
Susilo, N. (2023, September 7). Presiden Jokowi: "Kapal" ASEAN Harus Terus Berlayar. Https://Www.Kompas.Id/Baca/Internasional/2023/09/07/Presiden-Jokowi-Kapal-Asean-Harus-Terus-Berlayar.
United Nations. The Virtues of Mullateralism and Diplomacy, https://www.un.org/en/observances/multilateralism-for-peace-day.
Virgianita, A. (2022). Deklarasi G20 Bali 2022: Kesuksesan siapa? The Conversation. https://theconversation.com/deklarasi-g20-bali-2022-kesuksesan-siapa-194915.
Welle, D. (2022, July 8). Russia faces Ukraine heat as G20 foreign ministers meet. dw.com. https://www.dw.com/en/russia-faces-ukraine-heat-as-g20-foreign-ministers-meet/a-62402935.
Wicaksana, I. G. W. 2021. *How does Indonesia exercise agency in contested and complex regional environment, the pacific reviews* 35 (2): 297–318. https://doi.org/10.1080/09512748.2021.1998202.
Yusgiantoro, F. C., & Nasution, N. A. (2023, September 11). ASEAN digital connectivity under Indonesian chairmanship. Https://Www.Thejakartapost.Com/Opinion/2023/09/11/Asean-Digital-Connectivity-under-Indonesian-Chairmanship.Html.

Humphrey Wangke is a senior researcher in International Relations at the BRIN Political Research Center. He completed his environmental studies at UI in 2018 with the topic of Sustainable Development. At BRIN, he is actively involved in ASEAN studies in aspects of sustainable development, political-economy, and socio-culture. Can be contacted via hump001@brin.go.id

Khanisa is a researcher at the Research Center for Politics, BRIN. She received her bachelor's degree from the Department of International Relations (HI), Gadjah Mada University, in 2010 and completed her Master of Arts (International Relations) and Master of Diplomacy from the Australian National University in 2015. She focused her research on ASEAN and Southeast Asian issues. Khanisa can be contacted at khan002@brin.go.id

Pandu Prayoga is an international studies researcher at the BRIN Political Research Center. Pandu completed his undergraduate degree at Hasanuddin University Makassar (2011) and his master's degree at Coventry University, England (2019) in the field of International Relations. At BRIN, he is actively involved in research on ASEAN in terms of security, political-economic, and socio-cultural aspects. Pandu can be contacted at pandu.prayoga@brin.g.id

Chapter 17
Indonesian Diplomacy for Palestinian Peace and Independence: Challenges and Sustainability

Nostalgiawan Wahyudhi, Hamdan Basyar, Dhurorudin Mashad, Fakhry Ghafur, and Defbry Margiansyah

Abstract The foreign policy of the Republic of Indonesia towards the State of Palestine under the presidency of Joko Widodo (Jokowi) has received great attention. The Ministry of Foreign Affairs has been highly active in advocating for Palestine at various international forums. Consequently, Indonesia is currently at the fore front of countries that are both consistent and active in their efforts to secure Palestinian independence through the two-state solution on the international stage. However, this policy is markedly divergent from that of previous regimes, such as Soekarno, Soeharto and SBY, who were notably less vocal in their defense of Palestine. This paper examines why Jokowi's foreign policy towards Palestine is so prominent? What are the desired outcomes of this policy? What differentiates it from the foreign policy of Soekarno, Soeharto, and SBY? This paper finds that Indonesia's foreign policy towards Palestine is strongly influenced by the character of the regime in power and the existing domestic and international political context. Furthermore, it can be observed that Jokowi's policy places greater emphasis on political approach than that of Soeharto and SBY, yet it does not adopt a confrontational political approach as did that of Soekarno. In addition, it can be seen that Indonesia's foreign policy objectives towards Palestine are strongly influenced by the norms of international institutions, notably the UN, which advocates for a two-state solution.

Keywords Indonesia · Foreign Policy · Palestine · Middle East · Two state solution · Joko Widodo

---

N. Wahyudhi (✉) · H. Basyar · D. Mashad · F. Ghafur · D. Margiansyah
Research Center for Politics, Indonesia (BRIN), The National Research and Innovation Agency, Central Jakarta, Indonesia
e-mail: nostalgiawan.wahyudhi@brin.go.id

I. Kartini and Kanisa (eds.), *Joko Widodo's Foreign Policy and Diplomatic Legacy*,
https://doi.org/10.1007/978-981-95-2085-5_17

17.1 Introduction

During the two terms of President Joko Widodo's tenure, Indonesia's foreign policy towards Palestine underwent a notable shift in focus. During this period, Indonesia's defense of Palestine in various international forums was maximally utilized and felt very loose. This recalls the dynamic Indonesian foreign policy of the Soekarno reign. Nevertheless, a closer examination reveals a slight divergence from the previous governments, which appears to fade. It is interesting to study why Palestine has a prominence place in Indonesia's foreign policy under Jokowi? In addition to the Indonesian government's assertive stance on foreign policy towards Palestine under President Joko Widodo (Jokowi), it is imperative to examine the roots of Palestine in Indonesian foreign policy's DNA.

In essence, foreign policy represents an elaboration of Indonesia's interests, with the objective of achieving its political goals. This encompasses Indonesian foreign policy towards Palestine both material aspects that are rational in nature and non-material aspects which may at times appear irrational. However, beyond these fundamental aspects, foreign policy is consistently determined through rational decision-making process. It is intriguing to examine the specific objectives Indonesia seeks to achieve through its robust foreign policy on Palestine during the Jokowi era, which differs considerably from those of previous administrations. To this end, this paper will examine Indonesian foreign policy under Jokowi, analyzing its dynamics and distinguishing features that set it apart from the policies of the Soekarno, Soeharto and SBY eras.

This chapter presents an analysis of Indonesia's foreign policy using the Rational Actor Model (RAM), offering an insight into the country's strategic decisions. RAM is an indispensable framework for the study of foreign policy analysis (MacDonald 2003). The model posits that decision-makers are rational agents who are driven to seek the optimal outcome when faced with a range of options. The model is based on the tenets of expected utility theory, which is a concept that is worthy of further investigation. The model posits that individuals evaluate potential outcomes based on their preferences and make decisions that are aligned with achieving the highest level of satisfaction (Mintz and DeRouen Jr 2010; Alden 2017). Despite the existence of some critiques pertaining to this methodology, the relevance of RAM remains undiminished, particularly in the context of elucidating Indonesia's foreign policy.

In analyzing President Joko Widodo's foreign policy towards Palestine, RAM posits that states act as rational entities, making decisions to maximize their national interests (Setiawati 2024). The Jokowi administration has consistently and passionately advocated for Palestinian independence, reflecting Indonesia's strong historical and cultural ties to the Palestinian cause while skillfully navigating complex geopolitical dynamics (Gunn 2022). Indonesia has consistently advocated for Palestinian statehood in international forums, reflecting a calculated alignment with its national identity and regional leadership aspirations. By leveraging diplomatic channels and fostering relationships with other Muslim-majority nations, Jokowi aims to balance the act between domestic expectations and international pressures. The RAM suggests that decision-makers assess potential outcomes before acting; thus, Jokowi's government has engaged in multilateral diplomacy while maintaining strong ties with Western powers. This duality is evident in Indonesia's participation in organizations like the Organization of Islamic Cooperation (OIC) and its support for UN resolutions favoring Palestinian rights (Setiawati 2024).

This chapter is developed in several stages. The initial section examines the fluctuations of Indonesian foreign policy, which exhibits a common thread but is implemented with disparate characteristics by Sukarno, Suharto, and Susilo Bambang Yudhoyono (SBY). The second section provides an overview of Indonesia's foreign policy under President Jokowi, emphasizing the utilization of international institutions to advance Indonesia's foreign policy objectives through an active-reactive approach in international forums. The third section examines Indonesia's foreign policy achievements in advocating a two-state solution that takes into account the preferences of international institutions. The final stage will present a conclusion evaluating Indonesia's foreign policy towards Palestine under President Jokowi.

17.2 Pre-Jokowi's Reign: Indonesian up and Down Diplomacy on Palestine

Indonesia has demonstrated a consistent and unwavering commitment to a foreign policy that demonstrates unquestionable solidarity with the Palestinian struggle since the inception of this country. This was made possible due to Indonesia's struggle for independence from Dutch colonialism, which received significant international support, particularly from Palestine. Palestine was among the first countries to extend congratulations to Indonesia on its independence on September 6, 1944. Among the prominent figures who lent their backing was Syaikh Mohammed Amin al-Husaini, the Grand Mufti of Jerusalem. He conveyed his felicitations over Indonesia's independence via Radio Berlin, Germany (MOFA 2019a). His contributions were instrumental in securing recognition for Indonesia's independence among Arab nations (Rasyidi 1989), within the context of a broader anti-colonial movement (Reid 1974).[1] It helped Indonesia to gain broader international acceptance (Anderson 2006).

> Until the independence of the Palestinian nation is rightfully granted to its people, Indonesia will persist in challenging the oppressive presence of Israeli occupation. (Cabinet Secretariat 2016a).

This famous speech of Soekarno present strong commitment to Palestinian, thus reflect strong opposition to Israel and its colonialism. This political standpoint serves as the fundamental basis for Indonesia's foreign policy stance towards Israel to

[1] Husaini encouraged Arab countries to extend their support to Indonesia in its pursuit of independence. This led to a resolution whereby all member countries of the Arab League were advised to lend their backing to Indonesian sovereignty (Rasyidi 1989).

date. Despite a 1950 Ben Gurion's telegram to Soekarno-Hatta offering congratulations and expressing unwavering support for Indonesia's independence (Barton and Rubenstein 2005), this gesture did not alter the trajectory of Indonesia's foreign policy with regard to Palestine.

The robust Indonesian foreign policy towards Palestine during the Soekarno era was seemingly reinforced by two key factors. Firstly, Palestine was the first country to recognize Indonesia's independence. Secondly, the international context of the 1950s and 1960s, which was characterized by a complex interplay between anti-colonial and anti-imperialist sentiments, was strongly reflected in foreign policy stance adopted by Indonesia.

The anti-imperialist and anti-colonialist spirit constituted a significant element of Sukarno's policy towards Palestine, as evidenced by an analysis of the various justifications he provided for this stance. These also was characterized by a dynamic interplay between a spirit of confrontation, the rallying of solidarity, and the strategic utilization of international entities, which significantly amplified its diplomatic impact. First, He emphasized the ideology of nationalism on the call for pan-Islamic solidarity to save the third holiest place in Islam. This was in addition to his espousal of anti-colonialism and anti-imperialism (Sihbudi 1997). Second, Indonesia adopted an active yet confrontational foreign policy related to anti-imperialist and colonialist issues. The Indonesian government's support for Palestine was demonstrated through tangible actions, including the imposition of a diplomatic embargo on Israel and the rejection of Israel's participation in Indonesian-sponsored forums. This in a certain degree resulted in the emergence of numerous adversaries with the potential to pose a threat to the newly independent Indonesian state.

Third, Sukarno operationalized an anti-colonialism and anti-imperialism spirit with the objective of maximizing the potential for the mobilization of the solidarity of developing and newly independent countries. This solidarity is exemplified by Indonesia's defense of Palestine, which was still engaged in the struggle for independence including the pivotal rejection of Israel's presence at the 1955 Asia-Africa Conference and the 1962 Asian Games in Bandung (Hasyim 2023).[2] This served as the foundation for Indonesia's foreign policy in favor of Palestinian independence.

Fourth, this spirit of international solidarity was organized by Soekarno through the formation of international entities. The presence of organizations such as the Asia Africa Conference and Non-Aligned Movements (NAM) provides the main foundation for Indonesia's foreign policy (Suryadinata 2023, 172—185). The existence of organizations serves as the primary foundation for Indonesia's foreign policy, which is characterized by a neutral stance with regard to any particular bloc. These forums provide an opportunity for Indonesia and other developing countries to voice their support for the Palestinian cause in the international arena.

---

[2] As a result of this, the 1962 Asian Games in Indonesia were not recognized by International Olympiad Committee (IOC). Consequently, Soekarno initiated a counter-international sporting competition, namely Ganefo (Games of New Emerging Forces), which was held in 1963.

During the Soeharto regime, while still espousing the fundamental tenets and principles of Soekarno's foreign policy, he appeared less inclined to emphasize confrontational rejection as Soekarno did in his foreign policy (Sukma 1995). Soeharto did, however, continue to embrace the spirit of anti-colonialism and non-alignment in his foreign policy. Nevertheless, his approach of prioritizing diplomatic functions to express solidarity with the Palestinian cause seemed to align with public demands.

In his diplomatic engagement with Palestine, Soeharto more frequently established relations with the formal Palestinian political organization recognized by the United Nations, namely the Palestinian Liberation Organization (PLO). The close relationship between Soeharto and the PLO was evidenced by the diplomatic visits of Yasser Arafat, the PLO leader, to Jakarta in 1984, 1986, and 1992. During these visits, Soeharto extended substantial diplomatic support to Arafat, including a warm welcome on his first visit to Indonesia, but got significant national attention (Tempo 1993). Furthermore, Arafat requested that Indonesia continue to play an active role in garnering international support for the Palestinian cause, particularly through platforms such as the Non-Aligned Movement and the OIC.

According to Azra, in deference to the Indonesian government's policy, Yasser Arafat was meticulous in underscoring that the PLO was not a religious entity but rather a pluralist political organization. Soeharto pledged to facilitate the establishment of a PLO office in Indonesia. This decision was reinforced by Adam Malik's statement during his visit to several Arab countries in 1972 and announced that Indonesia had no objection to the PLO opening an office in Jakarta. However, this promise remained unfulfilled until the conclusion of his tenure. However, Indonesia facilitated the establishment of the Palestinian embassy in Jakarta and could only extend to the PLO the prospect of diplomatic relations at the UN (Azra 2006; Barton and Rubenstein 2005).

Soeharto pursued a similar course of diplomatic action, utilizing international entities that had been employed by Soekarno. At the Non-Aligned Summit in Jakarta in 1992, Soeharto underscored the necessity for NAM to advocate for a just and comprehensive resolution to the Palestinian issue and the Arab-Israeli conflict, in alignment with UN resolutions, particularly UNSC resolutions 242, 338, and 425. These resolutions called for Israel's withdrawal from Palestinian territories, including the al-Aqsa Mosque, the Golan Heights, and South Lebanon (Rachmadita 2023).

It is intriguing to observe. It seems probable that Soeharto was at ease with the idea of pursuing a selective foreign policy, which set him apart from Soekarno. This resulted in Indonesia's foreign policy towards Palestine being relatively passive. Rather than adopting a confrontational approach like Soekarno, Soeharto placed significant emphasis on ASEAN non-intervention values in conducting diplomacy with Palestine. Soeharto was hesitant to engage directly with Palestinian domestic affairs, including rejection of the PLO office in Jakarta. He typically utilized formal channels such as the UN-recognized PLO or the NAM, which offered a relatively secure avenue for Indonesian politics (Suryadinata 2023, 160).

In contrast, Soeharto also accorded priority to domestic affairs, particularly domestic political stability. This emphasis on domestic affairs shaped Soeharto's approach to foreign policy, which was characterized by a limited engagement with international issues (Yani 2009). The Palestinian issue, for instance, did not constitute a significant element of the Soeharto regime's foreign policy agenda. In general, Indonesian foreign policy did not prioritize the Middle East region (Suryadinata 1995). However, for domestic consumption, Soeharto upheld the political memory of support for Palestinian independence through the issuance of Indonesia-Palestine stamps in 1978, which appear to have been a means of maintaining domestic support.

Despite Indonesia's official support for the Palestinian cause and its rejection of the proposal to establish diplomatic relations with Israel, Soeharto permitted intelligence cooperation with Israel, particularly in the area of military capacity-building. Following Israel's triumph in the 1967 Six-Day War, the Indonesian military devised the "Alpha Operation," which aimed to repatriate 32 A-4 Skyhawk fighter aircraft from Israel (Poerwoko 2007).

In a statement that was markedly divergent from the prevailing political stance of the majority of Arab countries at that time, Adam Malik, the Indonesian foreign minister, proposed direct negotiations between all Arab countries and Israel to address the prevailing tensions. He further recommended that the United Nations be invited to supervise the situation in Jerusalem and that Arab countries respect Israel's aspirations for territorial and national recognition (Barton and Rubenstein 2005). Soeharto also demonstrated a willingness to engage in diplomatic discourse on occasion, as evidenced by his receipt of a diplomatic visit from Yitzak Rabin in October 1994 in Jakarta. This visit occurred shortly after he received a visit from Yasser Arafat one year earlier. The simultaneous visits of the two heads of state were part of an effort to seek support for the Oslo Declaration agreement reached by Israel and the PLO in 1993 (Suryadinata 1997; Barton and Rubenstein 2005).

Despite Indonesia's continued support for Palestine on the international stage, the Soeharto government's foreign policy underwent significant shifts during his tenure. The pursuit of political pragmatism led the government to pursue a dual strategy. On the one hand, it maintained the continuity of its foreign policy in favor of Palestine and supported the aspirations of the majority of Indonesian citizens. On the other hand, it engaged in to clandestine relations with Israel to modernize its military technology and expand its trade.

During the tenure of President Susilo Bambang Yudhoyono (SBY), Indonesia adopted a more honest foreign policy, which contributed to a notable shift in its international image. SBY's engagement with the Palestinian issue was not limited to expressing support for Palestinian rights in the United Nations and the Organization of Islamic Cooperation. However, during SBY's tenure, there was a notable increase in social activities and philanthropy. Humanitarian approaches and capacity-building became more prominent during this period. The most notable achievement of the SBY administration was the success of the Mer-C philanthropy organization in constructing the As-Syifa Indonesian Hospital in Gaza. This project garnered significant support and a positive image for Indonesia in the eyes of the international community.

In addition to philanthropy, SBY is also committed to capacity-building for the Palestinian people. Since the implementation of KAA in 1955, a crucial missing link has existed, preventing the Asian and African nations from establishing a forum for continued cooperation, including the continuity of their commitment to defend the independence of Palestine. In 2005, the SBY administration initiated the formation of the New Asia Africa Strategic Partnership (NAASP). On April 24, 2005, in Bandung, Indonesia, leaders from 106 countries in Asia and Africa gathered to commemorate the Golden Jubilee of the 1955 Asian-African Conference. The NAASP agreed on a follow-up mechanism for the institutionalization process, which entailed the implementation of a summit every four years, a ministerial meeting every two years, and a business summit (Songbatumis 2021).

In 2008, SBY convened a special NAASP Ministerial Conference on Capacity-Building for Palestine, attended by 218 participants from 56 countries and three international organizations. The Indonesian, South African, and Palestinian Coordinating Unit was instrumental in coordinating the implementation of capacity-building commitments for Palestine. By the end of 2013, SBY had organized 128 capacity-building programs for 1,257 Palestinians (MOFA 2024i; Songbatumis 2021).[3]

Palestine is one of the countries designated as a priority for assistance from the SBY regime through technical cooperation. This is part of the Indonesian government's commitment to provide capacity-building assistance to the Palestinian civil apparatus. This capacity-building program represents a commitment of the Government of Indonesia within the framework of South-South Cooperation, which aims to encourage development cooperation among developing countries (Antara 2017).

17.3 The Jokowi's Reign: Advancing Indonesian Political Approach on Palestine

Jokowi's foreign policy demonstrates a noteworthy commitment to Palestinian independence. The approach taken by Jokowi is markedly distinct from that of Soekarno, who employed a confrontational strategy. Similarly, Jokowi's approach differs from that of Soeharto, who utilized a pragmatic and realistic approach that compelled Indonesia to navigate a complex diplomatic landscape. The foreign policy of the Indonesian government under President Joko Widodo (Jokowi) is more tactical, has a clear positioning, and is more consistent. The most prominent aspect of Jokowi's foreign policy is his use of international forums to initiate, encourage, and mainstream the defense of the Palestinian struggle.

---

[3] NAASP capacity building programs include: *Training Course on Fire Rescue, Training on Project Cycle, Training Course on Gender Mainstreaming for Officers of the Ministry of Women's Affairs, Training Course on Coal and Mineral Resources Management, Apprenticeship Program for Palestine's Small and Medium Enterprises (SMEs) Development, Medical First Responder, dan Collapse Structure Search and Rescue (CSSR).*

17.3.1 *Maximizing International Forum*

In the year following his presidency, Jokowi made his inaugural statement of support for Palestinian independence at the 2015 Asian-African Conference in Jakarta. President Joko Widodo has stated that Indonesia and the international community should not ignore the plight of the Palestinian people, who endure hardship and injustice as a consequence of the prolonged colonial occupation. The AAC summit proved an effective platform for launching the Declaration on Palestine, which called on all AAC state leaders to lend their support to the Palestinian people in their pursuit of an independent state.

Indonesia's active role is evidenced by its consistent appointment of the Permanent Representative of the Republic of Indonesia to the Bureau of the UN Committee for the Exercise of the Inalienable Rights of the Palestinian People (CEIRPP). This body has the objective of advocating for a permanent ceasefire and the delivery of humanitarian aid into Gaza, as well as encouraging full membership for Palestine at the UN. This could facilitate the establishment of a political resolution based on a two-state solution (MOFA 2024g). Moreover, at the conclusion of 2015, Indonesia convened an international conference on the subject of Jerusalem in collaboration with the OIC and CEIRPP. The objective of the conference was twofold. Primarily, it sought to identify academic solutions to the complex issues surrounding Jerusalem and conflict resolution. Secondly, it aimed to raise awareness of Palestinian issues among the international community.

A comparable expression of support was also documented in 2016. Indonesia played a pivotal role in convening the 5th Organization of Islamic Cooperation (OIC) Extraordinary Summit on Palestine and Al-Quds Al-Sharif. The two-day Summit, which took place on March 6–7, 2016, in Jakarta. The Indonesian initiative was instrumental in elevating the Palestinian issue to a prominent position on the forum's agenda. The Indonesia called upon all member states to endorse the "Jakarta Declaration," which outlined recommendations for resolving the Palestinian and al-Quds al-Syarif crises. In his remarks, Jokowi asserted "United for a Just Solution" that the member states of the Organization of Islamic Cooperation must demonstrate unity and solidarity in order to provide effective support to Palestine.

The Summit concluded with remarks from President Jokowi. He stated, "During the course of the Summit, I have been gratified to observe the depth of support and solidarity expressed by the Islamic world towards Palestine. I am heartened by the commitment of the leaders of the Islamic world to unite and reinforce their efforts to provide renewed support for the Palestinian people" (Cabinet Secretariat 2016b).

The OIC Extraordinary Summit has successfully passed two significant documents. The first is a resolution that reaffirms the OIC's principled position and commitment to Palestine and Al-Quds Al-Sharif. This resolution is anticipated to align with the aspirations of the Palestinian people. The second document is the Jakarta Declaration, which was proposed by Indonesia and outlines concrete action plans devised by OIC leaders to address the Palestinian-Israeli conflict and the status of Jerusalem.

In a recent address, President Jokowi outlined five concrete steps that can be taken to provide assistance to Palestine.

(1) It is imperative to reinforce political backing in order to reinvigorate the peace process. The establishment of a new international collective process and the necessity of involving Muslim countries in order to achieve a two-state solution. Indonesia is prepared to participate in this process.
(2) The objective is to exert pressure on Israel, including a boycott of Israeli products produced in the occupied territories.
(3) It is necessary to increase pressure on the UN Security Council to provide international protection for Palestine, and to set a deadline for ending the Israeli occupation.
(4) It is essential to reject restrictions on access to worship et al.-Aqsa Mosque and Israel's actions to change the status quo and demography of Al-Quds Al-Sharif.
(5) Fulfillment of urgent humanitarian assistances (Cabinet Secretariat 2016b).

Despite the lack of tangible outcomes, the Jakarta Declaration and the OIC Extraordinary Summit on Palestine 2016 served to renew and strengthen the commitment of Islamic countries to the Palestinian cause. While they did not result in a change of Israel's attitude and policy in Palestine or the lifting of restrictions on worship activities in Al-Aqsa, they nonetheless played a role in reaffirming the collective stance of Islamic countries on the Palestinian issue.

Following a series of incidents at the Aqsa Mosque in late 2017, Donald Trump made a statement that was met with international condemnation and shock. "*I have determined that it is time to officially recognize Jerusalem as the capital of Israel. While previous presidents have made this a major campaign promise, they failed to deliver. Today, I am delivering.*" According to Trump, the recognition is a realization of the US Congress' 1995 decision to recognize Jerusalem as Israel's capital and move the US Embassy from Tel Aviv to Jerusalem (Ahram 2017).

Trump's policy of unilaterally moving the US embassy to Jerusalem is a step that contradicts the decisions issued by the OIC. Furthermore, it runs counter to Indonesia's political and foreign policy initiatives towards Palestine. This represents a diplomatic setback for the US and narrows the opportunity for a two-state solution, which the US also desires to initiate.

The Indonesian government promptly mobilized support to condemn Trump's decision. In order to facilitate a unified response from the international community, Indonesian Foreign Minister Retno Marsudi lobbied member states of the Organization of Islamic Cooperation (OIC) to convene an Extraordinary Summit. It was ultimately determined that the OIC Extraordinary Summit would be convened in Istanbul, Turkey, on December 13, 2017. At that time, the President of Indonesia, Joko Widodo (Jokowi), unequivocally rejected President Trump's assertion that Jerusalem is the capital of Israel. President Jokowi stated that the recognition was unacceptable and that it must be strongly condemned. He further emphasized the importance of uniting in defense of Palestine, which he said has been "dashed by this unjust decision. He also highlighted that the decision has undermined the prospect of a lasting peace, and thus must be rejected" (Cabinet Secretariat 2017).

In his remarks, President Jokowi articulated six key points pertaining to the proposed stance of OIC member countries. These points are as follows:

(1) The OIC must unequivocally reject the practice of unilateral recognition. The two-state solution is the only viable option, with East Jerusalem designated as the capital of Palestine.
(2) It is recommended that countries with diplomatic missions in Tel Aviv, Israel, refrain from following the United States' lead in relocating their diplomatic missions to Jerusalem.
(3) The OIC countries have the potential to serve as a catalyst for garnering support from countries that have not yet recognized Palestinian independence, with the aim of doing so in the immediate future.
(4) Member countries of the OIC that maintain diplomatic relations with Israel should pursue diplomatic measures. This may entail "reviewing relations with Israel in accordance with various OIC resolutions," as the President indicated.
(5) Member countries of the OIC must undertake collective action to enhance humanitarian assistance, capacity-building, and economic cooperation with Palestine.
(6) The President expressed hope that the OIC will serve as a driving force in various international and multilateral forums to advocate for Palestine, including in the Security Council and the UN General Assembly (Cabinet Secretariat 2017).

In June 2018, Indonesia was selected to serve as a non-permanent member of the United Nations Security Council for the period 2019–2020. It is projected that Indonesia will assume the role of President of the UN Security Council on two occasions within the two-year period of its membership of the UNSC (2019–2020). The initial occurrence is scheduled for May 2019, with the subsequent event anticipated to take place in August or September 2020. Indonesia has consistently identified Palestine as a priority issue, affording it special attention in addition to the four issues that Indonesia has identified as requiring particular focus (MOFA, 2019b; Sitamala 2020).

In its capacity as the current chair of the UNSC, Indonesia advanced the theme "Investing in Peace," which highlighted pivotal considerations essential for attaining peace. These included economic development, community empowerment, the protection of human rights, the promotion of inclusive political processes, and democracy. Foreign Minister Retno Marsudi presided over the Open Session on the Protection of Civilians in Armed Conflict on 23 May 2019. The meeting constituted a tangible manifestation of Indonesia's commitment to support the promotion of human rights and international humanitarian law on the global stage (Ubaedillah et al. 2023).

During its tenure as president of the UNSC, Indonesia reaffirmed its unwavering support for the Palestinian people in their struggle for self-determination. Indonesia persists in its efforts to maintain the Palestinian issue as a priority concern of the UN Security Council, and continues to advocate for the values of humanity and justice in the pursuit of a resolution to the Israeli-Palestinian conflict.

On 9 May 2019, Indonesia held an informal discussion in the Arria Formula format regarding the construction of illegal Israeli settlements in the Palestinian territories. In this informal forum, Indonesia advanced the proposition that the formation of Jewish settlements in Palestine may have contravened the two-states solution, international law, and Palestinian rights. Furthermore, the international community was urged to exert pressure on Israel to bring about an end to its policy. Moreover, on 22 May 2019, the Indonesian Foreign Minister presided over a UN Security Council session in a briefing format on the situation in the Middle East. In this forum, Minister Marsudi underscored the necessity of providing assistance to Palestine, including measures to enhance the protection of civilians and to prompt the international community to provide humanitarian aid to Palestinians without delay (Rezasyah 2019; Ubaedillah et al. 2023).

In the initial phase of the mission, Indonesia advanced a compelling argument at the 19th Summit of the Non-Aligned Movement (NAM) held in Kampala, Uganda, on January 19–20, 2024. The objective was to garner comprehensive support from the 121 NAM member countries for the Palestinian bid for full membership in the United Nations. Such support is likely to have a significant impact, given that the NAM currently represents approximately 55% of the world's population (MOFA 2024b).

The Government of Indonesia has reaffirmed its commitment to continued support for the United Nations Relief and Works Agency for Palestine Refugees in the Middle East (UNRWA). This latest pledge of support to UNRWA includes an increased regular voluntary contribution of US$1.2 million annually from 2024, in addition to a grant of US$2 million. This support is intended to meet the needs outlined in the UNRWA Flash Appeal for the period between April and December 2024. UNRWA highlighted the humanitarian situation in Gaza and stressed the need for an immediate ceasefire and respect for international law. Both stressed UNRWA's important role in the resolution of the conflict through a two-state solution and in maintaining regional stability (MOFA 2024h).

At the 19th Asia Cooperation Dialogue (ACD) Ministerial Meeting in Tehran, Iran, Indonesia once again operationalized the international forum. Indonesia reinforced the ACD with the objective of garnering support for Indonesia's appeal for an end to the humanitarian conflict in Gaza and for Palestinian independence through the two-state solution. Indonesia provided unwavering support in various international forums with the aim of gradually garnering support from all directions, including the ACD as a complementary organization for countries and sub-regional organizations in Asia (MOFA 2024f).

In light of the series of conflicts and wars that have occurred since 7 October, as well as the impact of the numerous humanitarian tragedies that have taken place in this modern century, the Israeli invasion of Gaza can be considered an unacceptable act of genocide that extends beyond the boundaries of conventional warfare. This invasion has deliberately targeted civilians, the elderly, women and children, educational facilities, paramedics and health facilities, and has resulted in significant and inhumane suffering. In February 23, 2024, the International Court of Justice

(ICJ) considered a significant case that resulted in a pivotal decision: that Israel's colonization of Palestine is illegal and violates international law (ICJ 2024).

On that occasion, the Indonesian Foreign Minister delivered an oral statement related to the Palestinian issue before the International Court of Justice in The Hague, Netherlands. The statement encompassed two main aspects. Firstly, in terms of jurisdiction, the Indonesian Foreign Minister emphasized that the International Court of Justice only has the authority to provide advisory opinions. Secondly, in terms of substance, the Foreign Minister emphasized that Israel's policies are contrary to international law and outlined the legal consequences (MOFA 2024c). However, the ICJ's judgment did not include an immediate ceasefire in Gaza, a decision that was met with disappointment by Indonesia and the international community (BBC 2024). Indonesia's advocacy on behalf of Palestine in a multitude of international arenas represents a robust and unwavering dedication to ensuring that Palestinian concerns remain at the forefront, preventing them from being overlooked or forgotten.

Given Indonesia's active defense of Palestine in the international arena, Minister Marsudi rejected the notion that Indonesia's foreign policy under Jokowi is more inward looking and withdrawn from international relations. The Indonesian government, under the presidency of Jokowi, appears to be taking considerable measures not only to enhance its international reputation and engagement but also to maintain its autonomy within the international order. Indonesia's foreign policy places considerable emphasis on its role as a middle power country, with a particular focus on the United Nations (UN) and the Association of Southeast Asian Nations (ASEAN) as key foreign policy priorities (Cabinet Secretariat 2015; Umar 2023). It appears that Indonesian active advocacy on the Palestinian cause is intended to maintain its autonomy in the international order as a middle power state.

17.3.2 Reactive-Responsive Foreign Policy

The term "reactive-responsive foreign policy" is used to describe a foreign policy approach that is characterized by a responsiveness to external stimuli and a capacity to react effectively to changing circumstances. The Indonesian Minister of Foreign Affairs stated that Indonesia's foreign policy is consistently pursued in accordance with the principles of freedom and activity, with a focus on national interests and contributions to global peace, as outlined in the country's constitution. Indonesian diplomacy is characterized by a measured, calculated, action-oriented and result-oriented approach. Concurrently, it upholds unwavering values and principles (Ulya 2024). This is the rationale behind Indonesia's consistent support for and response to all instances of conflict and military action carried out by Israel in Palestine.

In her annual statement, the Minister of Foreign Affairs delineated the accomplishments of three years of Indonesian foreign policy, including a robust affirmation of the Palestinian cause. Retno Marsudi asserted, "Palestine is at the heart of Indonesia's foreign policy, and in every breath of Indonesian diplomacy, there is a struggle for Palestine" (Marsudi 2018). Palestine has been a focal point of Indonesia's foreign policy for almost a decade. Indonesia adheres to the principle of national self-determination and supports the Palestinian people's right to independence. Notably, the Palestinians remain the only participants of the Asian-African Conference in Bandung who have yet to achieve this right (MOFA 2024a). This illustrates the depth of Indonesia's commitment, particularly under the Jokowi administration, to supporting the Palestinian people in their pursuit of self-determination.

In an early 2018 statement, Indonesia's support for Palestine encompasses not only political backing but also economic assistance and technical collaboration. Notably, Indonesia has implemented a "zero tariff" policy for a range of Palestinian products entering the Indonesian market. In 2018, Indonesia plans to enhance its cooperation in the areas of water desalination and health (Marsudi 2018). The program is motivated by the urgent call for capacity-building among the Palestinian community and improving the quality of life of the Palestinian people.

Indonesia highlighted the significant increase in the number of illegal Israeli settlements, which have more than doubled from approximately 110,000 in September 1993 to 620,000 by the end of 2017. This has emerged as a major obstacle to the peace process between Israel and Palestine. In her remarks, Foreign Minister Retno highlighted three key issues pertaining to the illegal settlements.

(1) The construction of illegal settlements in the Palestinian territories, including East Jerusalem, further diminishes the prospect of a two-state solution.
(2) The construction of illegal settlements is a source of various violations of law and human rights against the Palestinian people.
(3) The international community bears a collective responsibility to halt Israel's illegal settlement policy.

In order to achieve this, Foreign Minister Retno emphasized the importance of public pressure from the international community to end the construction of these settlements. One way of maintaining the international community's attention on the issue of settlements would be to establish an international day of solidarity with the victims of illegal settlements (Cabinet Secretariat 2019). In the case of the illicit settlement in Sheikh Jarrah that resulted in a significant intensification of the conflict, Indonesia has consistently condemned Israel's actions (Nurita 2021). Similarly, in response to the recent statements by Israeli officials indicating the country's intention to construct five new settlements in the occupied West Bank, Indonesia has issued a strongly worded condemnation of this provocative move. The establishment of settlements and the continued occupation of Palestinian territory by Israel contravenes international law and relevant UN resolutions, which prohibit the establishment of settlements and the continued occupation of Palestinian territory (Fauziyah 2024).

In the context of the tremendous suffering and human tragedy caused by Israel's invasion of Gaza, Rafah, and the West Bank, Indonesia has issued a condemnation of the war crimes committed by Israel. The aforementioned actions constitute war crimes and genocide, a conclusion that is indisputable in light of the ICJ Judgment of 26 January 2024. This policy was subsequently followed by the International Criminal Court (ICC) Prosecutors in May 21, 2024, who proposed the arrest of two Israeli leaders, Benjamin Netanyahu and Yoav Galant, as well as three Hamas leaders, including Ismail Haniyeh, for war crimes. Consequently, at various forums, including the ministerial-level forum in Brussels, the Indonesian Foreign Minister has emphasized that Indonesia encourages the implementation of substantial measures to halt this crime (CNN 2024; Federman 2024).

In addition to the multifaceted initiatives undertaken by Indonesia in various international, regional, and domestic forums, Minister Retno explained the country's immediate key objectives in extending support to Palestine, *first*, the immediate and sustainable cessation of hostilities. In the absence of a ceasefire, the potential for improvement in the situation will remain unfulfilled. *Secondly*, Indonesia will persist in its support of the operations of UNRWA and will endeavor to persuade other states to maintain their own contributions. Indonesia has set an example by increasing its financial contribution to the UNRWA. *Thirdly*, Indonesia remains committed to the provision of humanitarian assistance. To date, Indonesia has dispatched over 4,500 tons of humanitarian assistance to Palestine. *Fourthly*, Indonesia calls for the implementation of all ICJ orders by Israel. It is Indonesia's hope that the UN Security Council will make a decision that forces Israel to comply with and implement the ICJ decision. *Fifthly*, Indonesia encourages more countries to recognize the State of Palestine, which was one of the objectives of the foreign minister's visit to Europe last week. *Sixthly*, Indonesia continues efforts to finalize the Palestinian membership process at the UN (MOFA 2024e) (Table 17.1).

17.4 Jokowi's Insistence on a Two-State Solution

What are Jokowi's intentions regarding Palestine? This statement, which is open to question, is of great importance in understanding the broader context of Indonesia's foreign policy towards Palestine. In line with the principles of the AAC, Indonesia is strongly supportive of Palestinian independence on the spirit of anti-colonialism. However, following the formal establishment of the State of Israel, Indonesia must consider how to position itself in relation to the achievement of Palestinian independence, taking into account the prevailing contextual conditions.

17.4.1 Leaning on International Institutions

In various international forums, the Indonesian Minister of Foreign Affairs has consistently reiterated that Indonesia's foreign policy is predicated on the two-state proposal leaned with UNSC Resolution 242. Indonesia is firmly committed to facilitating the realization of a two-state solution with the objective of achieving long-term peace between Palestine and Israel in accordance with international parameters and consensus (UN News 2024).

The Director General for Asia-Pacific and Africa at the Ministry of Foreign Affairs (MOFA) indicated that Indonesia is supportive of at least two distinct parameters.

Table 17.1 Comparison of Indonesia's Foreign Policy towards Palestine between Soekarno, Soeharto, SBY, and Jokowi

Aspect	Soekarno	Soeharto	SBY	Jokowi
Approach	Confrontational, active, political approach	Diplomatic pragmatism, humanitarian aid, non-intervention	Second Track diplomacy, Philanthropy, capacity-building	Active diplomatic & political approach, exploring humanitarian issues
Focus	Solidarity with developing countries, against colonialism	Domestic stability, diplomacy	Indonesia image as a friendly state, zero enemy	Middle power approach, Bold Indonesian FP & concern on Palestine, Against occupation and genocide
Tools	High-level conferences, breaking diplomatic relations	International forums, humanitarian aid	International forums, humanitarian aid, civil society engagement	International forums, UN, ICJ, ICC, OIC, UNSC, diplomatic influences
Result	Strengthens developing country solidarity, but could cause tensions	Gaining wider international support, but less confrontational	Indonesia's positive international image, intense humanitarian aid, less impact on political support (internal divisions in Palestine)	Strengthen Indonesia diplomatic position as middle power, leading state on Palestinian issue

The *first parameter*, which is the realization of a sovereign Palestinian state according to the 1967 borders. Indonesia's position on this matter is one of consistency and respect for international law, aligning with the stance of the United Nations as a reflection of global consensus. The delineation of the boundaries between the two countries, as outlined in the 1967 agreement, is founded upon the legal authority of Resolution 242. Consequently, this presents a significant opportunity for Palestine to accept the agreement and be considered by Israel.

Second parameter is the immediate cessation of the development of Israeli settlements deemed illegal by international law. It is acknowledged that, in accordance with UNSC Resolution 242 and the 1967 Borders, any expansion of the occupation by Israel beyond the 1967 Borders is unlawful. The withdrawal of this illegal occupation is contingent upon the robust affirmation of the resolution (Antara 2024).

The clarity of Indonesia's foreign policy stance in favor of a two-state solution has fundamental consequences. Furthermore, it is also cited as a barrier to the future of Indonesia's diplomacy to be more open towards Israel. It must be acknowledged that, despite Indonesia's consistent and unequivocal support for Palestine in its foreign policy, there is a growing pressure on the country to consider the possibility of normalizing relations with Israel. In principle, Indonesia does not sever its diplomatic relations with any country, including Israel. However, every country's foreign policy is shaped by a set of interests. Indonesia's interests in foreign policy are clearly defined. Normalization is a possibility if a two-state solution can be realized, which would entail the establishment of an independent Palestinian state that is recognized by Israel and exists in peaceful coexistence with it (Nugraheny 2024).

At the OIC and European Ministerial Meeting in Brussels, the Indonesian Minister of Foreign Affairs delineated the tangible strategy that Indonesia has devised to advance the two-state solution, which represents the objective of Indonesia's foreign policy towards Israel. The implementation of the two-state solution must take into account the current conditions and developments in the Israeli genocide in Gaza.

(1) It is essential to encourage the immediate and permanent cessation of hostilities. In this regard, it is crucial to urge Israel to comply with the ICJ decision, including the cessation of the Israeli military offensive in Rafah. Without this, the smooth delivery of humanitarian aid will not be possible. Indonesia places particular emphasis on the role of the UN Security Council in this regard.
(2) It encourages the international community to maintain its support for UNRWA, given the crucial role that UNRWA plays in preventing a further deterioration of the humanitarian situation.
(3) It is imperative to acknowledge the state of Palestine and to extend support for Palestinian membership within the United Nations. Indonesia places considerable emphasis on the necessity for all countries to utilize their influence in order to prevent a recurrence of the veto on Palestinian membership in the UN, which occurred in the UN Security Council. Palestinian membership of the UN would facilitate the development of Palestine as a country.
(4) It is often overlooked in efforts to support Palestinian independence that the internal conflict must be ended and Palestinian unity acknowledged. Furthermore, the reforms carried out by the Palestinian Government must be supported. This will contribute to the process of realizing a two-state solution that has the potential to be derailed (MOFA 2024d).

Indonesia is aware that the Palestinian struggle is ongoing and that the efforts of Indonesia and the international community to assist Palestine are similarly ongoing. In order to defend justice and humanity, it is necessary to demonstrate consistency and commitment to the principles of justice and peace. This is the objective of Indonesia's foreign policy (MOFA 2024e).

Furthermore, it is important to acknowledge that the political landscape in Palestine has undergone significant shifts since 1967. The two-state solution is continually undermined by Israel's illegal occupation of Palestinian territories, coupled with the open genocide perpetrated in Gaza and Israel's persistent disregard for international law and ICC-ICJ policies thus far. It is important to note that Israel has made it clear that it views any action by the ICC as an infringement upon its right to self-defense. Israel has stated that it does not anticipate that the ICC will influence its actions and that the ICC's efforts would set a dangerous precedent. In a bid to circumvent the ICC's potential issuance of an arrest warrant against him, Netanyahu reached out to US President Joe Biden, requesting that Washington intervene and prevent such an action. In response, the US spokesperson, Karine Jean-Pierre, articulated the country's definitive stance on the ICC investigation, stating that the US does not endorse it and does not recognize the ICC's authority to make arrests. However, the Israeli government's rejection of these policies represents a significant impediment to the implementation of a two-state solution (Aulia 2024).

17.5 Conclusion

This paper examines why Jokowi's foreign policy towards Palestine is so prominent. What are the desired outcomes of this policy? What differentiates it from the foreign policy of Soekarno, Soeharto, and SBY?

The foreign policy of Indonesia towards Palestine has evolved in a dynamic manner in accordance with the characteristics of the regime in power. Nevertheless, these priorities share a common red-line principle: the overwhelming majority of Indonesian leaders have embraced the foreign policy arrangements that were established during the Soekarno era. Despite the considerable changes in Indonesian leadership over the past 80 years, the country has remained unwavering in its pursuit of a foreign policy objective that is firmly aligned with the goal of Palestinian independence.

In accordance with this red-line, Indonesia's foreign policy is dynamic and parallel to the norms of international order established by the United Nations, including the two-state solution policy. Furthermore, the various iterations of Indonesia's foreign policy are shaped by the characteristics of the ruling regime and the prevailing domestic and international political circumstances. It is evident that Indonesia's foreign policy towards Palestine during the Soekarno era was markedly distinct from that of Suharto. The assertive support for Palestine during Soekarno's tenure must be contextualized within the context of Indonesia as a newly independent nation, and the prevailing currents of the international community that were strongly committed to anti-colonialism.

Furthermore, the foreign policy approach of the Suharto era must be contextualized within the broader framework of Indonesia as a developing country, with a heightened emphasis on domestic stability. Consequently, Indonesia's pragmatic stance in the Palestinian conflict was consistent with the evolving principles of its foreign policy, including the non-interventionist approach that Indonesia championed within the ASEAN framework. SBY's approach must be viewed in the context of a prudent foreign policy strategy and a conciliatory stance. He prioritized non-risky approaches, such as the promotion of humanitarianism and second track diplomacy, which were undoubtedly beneficial for Palestine.

Nevertheless, it is important to acknowledge that there was a notable absence of Indonesian foreign policy engagement with Palestine during the Soeharto and SBY eras. This is a stark contrast to the approach taken during Jokowi's tenure. Indonesia has assumed a prominent role in advocating for and defending the Palestinian cause. While Jokowi's domestic political context has been characterized as anti-Islamic, it is important to acknowledge that he has reinforced Indonesia's consistent and assertive foreign policy stance reminiscent of the Soekarno era, albeit with a more conciliatory approach. This aligns with the influence of Indonesia's foreign policy, which has shaped the country's role as a middle power.

The ongoing efforts of Indonesia to support the Palestinian people in their pursuit of independence maintain the relevance of the two-state solution in the country's foreign policy. As long as the two-state solution remains a central tenet of the United Nations' diplomatic approach, continues to garner substantial international support, and there are no indications of a reversal of this decision, it is unlikely that Indonesia will alter its stance on the two-state solution in its foreign policy interests.

Indonesia's active-reactive policy towards Palestine is to be commended; however, it must be acknowledged that Indonesia has yet to demonstrate a willingness to reformulate the two-state solution policy in a more nuanced manner. It is important to recognize that Indonesia has not yet achieved the level of diplomatic influence that would enable it to exert significant control over the Middle East region. Consequently, Indonesia's diplomatic efforts have not yielded the same level of influence as those of Egypt and Qatar, which have served as direct mediator countries. It is evident that Indonesia's policy strategy pattern towards efforts to support Palestine from the periphery. Secondly, the Indonesian government has not yet fully recognized the Middle East region as a key foreign policy priority. Therefore, despite its specific and differentiated approach to Palestine, Indonesia remains a minor player in the Middle East, particularly in comparison to other major countries such as those in the OIC.

The Indonesian government's policy recommendations should build upon the positive aspects of the policies initiated by President Jokowi. It is important to acknowledge that Indonesia's current influence and strength in the Middle East has increased. Consequently, Indonesia's foreign policy of actively advocating for Palestinian independence must be maintained. Additionally, Indonesia can continue to pursue a two-state solution in alignment with the policies of international bodies such as the United Nations. Third, it would be prudent for Indonesia to consider elevating its foreign policy priorities to the Middle East. As the largest Muslim country and a member of the OIC, Indonesia's position is becoming increasingly influential. Fourth, Indonesia's role as a middle power affords it the potential to become a non-Arab country whose neutrality is highly trusted. This opportunity could encourage Indonesia to expand its role as a peace-making country in the Middle East.

It is evident that there is a minimum of rigorous and comprehensive scholarly research on Indonesia's foreign policy towards Palestine. This study, therefore, assumes great significance in addressing this gap in knowledge. It is hoped that this study will contribute to the ongoing process of reformulating Indonesian foreign policy in a more constructive, evidence-based manner and serve as a valuable reference point for Indonesian policymakers.

References

Ahram. 2017. Full Text of Trump Speech on Jerusalem. Retrieve from October 8, 2024, from http://english.ahram.org.eg/NewsContent/2/0/282895/World/0/Full-text-of-Trumps-speech-on-Jerusalem.aspx.

Alden, Chris. 2017. Critiques of the Rational Actor Model and Foreign Policy Decision Making. In *Oxford Research Encyclopedia of Politics*, eds. William R Thompson and Tereza Capelos. Oxford University Press.

Anderson, B. R. O'G. 2006. *Imagined Communities: Reflections on the Origin and Spread of Nationalism*. Verso.

Antara. 2017. Palestine is Prority Nation to Receive Indonesian Capacity Building Assistance. Retrieved from October 8, 2024, from https://jatim.antaranews.com/berita/200689/palestine-is-prority-nation-to-receive-indonesian-capacity-building-assistance.

Antara. 2024. Indonesia Supports Two-State Solution as Per Global Consensus. Retrieved from October 8, 2024, from https://en.antaranews.com/news/315180/indonesia-supports-two-state-solution-as-per-global-consensus.

Aulia, Luki. 2024. Uji Nyali Mahkamah Kriminal Internasional Menjerat Netanyahu. *Kompas*, April 30. https://www.kompas.id/baca/internasional/2024/04/30/uji-nyali-mahkamah-kriminal-internasional-menjerat-netanyahu.

Azra, Azyumardi. 2006. *Indonesia*. Islam and Democracy: Dynamics in a Global Context. Equinox Publishing.

Barton G., and C. Rubenstein. 2005. Indonesia and Israel: A Relationship in Waiting. *Jewish Political Studies Review* 157–170.

BBC. 2024. Mahkamah Internasional minta Israel cegah genosida di Gaza, namun tak perintahkan gencatan senjata. Retrieved from October 8, 2024, from https://www.bbc.com/indonesia/articles/cpvr5663qgxo.

Cabinet Secretariat of the Republic of Indonesia. 2015, January 8. Bantah Tarik Diri Dari Internasional, Menlu: Indonesia Terus Mainkan Peran Middle Power. https://setkab.go.id/bantah-tarik-dari-dari-internasional-menlu-indonesia-terus-mainkan-peran-midle-power/.

Cabinet Secretariat of the Republic of Indonesia. 2017, December 13. Berbicara di KTT Luar Biasa OKI, Presiden Jokowi Sampaikan Enam Usulan. https://setkab.go.id/berbicara-di-ktt-luar-biasa-oki-presiden-jokowi-sampaikan-enam-usulan/.

Cabinet Secretariat of the Republic of Indonesia. 2019, May 10. Indonesia Calls for End to Illegal Israeli Settlement Expansion. https://setkab.go.id/en/indonesia-calls-for-end-to-illegal-israeli-settlement-expansion/.

Cabinet Secretariat of the Republic of Indonesia. 2016a, March 7. Pidato oleh Presiden Republik Indonesia dalam KTT Luar Biasa dari Organisasi Kerjasama Islam (OKI) tentang Palestina dan Al-Quds Al-Sharif 2016 di Jakarta. https://www.setneg.go.id/baca/index/pidato_oleh_presiden_republik_indonesia.

Cabinet Secretariat of the Republic of Indonesia. 2016b, March 7. Akhiri KTT LB OKI ke-5, Presiden Jokowi Suarakan Pesan Keras Pemimpin Islam terhadap Konflik Palestina. https://setkab.go.id/akhiri-ktt-lb-oki-ke-5-presiden-jokowi-suarakan-pesan-keras-pemimpin-islam-terhadap-konflik-palestina/.

CNN. 2024. RI Kutuk Keras Serangan Israel ke Rafah: Langgar Perintah ICJ. Retrieved from October 28, 2024, from https://www.cnnindonesia.com/internasional/20240528122231-106-1102776/ri-kutuk-keras-serangan-israel-ke-rafah-langgar-perintah-icj.

Fauziyah, Najla Nur. Indonesia 'Strongly Condemns' Israel's Move to Legalize Five Jewish Settlements in Occupied West Bank. *Tempo.co*, 2024, July 1. https://en.tempo.co/read/1886179/indonesia-strongly-condemns-israels-move-to-legalize-five-jewish-settlements-in-occupied-west-bank.

Federman, Josef. 2024, May 21. War crimes prosecutor seeks arrest of Israeli and Hamas leaders, including Netanyahu. AP. https://apnews.com/article/icc-khan-netanyahu-070941d21ccd1f2b9611032b88527575.

Gunn, Geoffrey C. 2022. The Cultural Parameters of Indonesian Foreign Policy since 1945. In *Asian Culture, Diplomacy and Foreign Relations*, ed. C. X. George Wei, volume II. BRILL.

Hasyim, Irsyan. 2023. A Distant Legacy of Rejecting Israel. *Tempo Magazine*, April 3. https://magz.tempo.co/read/cover-story/40466/a-distant-legacy-of-rejectingisrael#:~:text=The%20opening%20of%20the%201962,to%20Digital%20Tempo%2C%20See%20More

ICJ. 2024, January 26. The Convention on the Prevention and Punishment of the Crime of Genocide in the Gaza Strip (South Africa V. Israel). https://www.icj-cij.org/node/203447

MacDonald, Paul K. 2003. Useful Fiction or Miracle Maker: The Competing Epistemological Foundations of Rational Choice Theory. *American Political Science Review* 97 (4): 551–565.

Marsudi, Retno L. P. 2018, January 9. Annual Press Statement of the Minister of Foreign Affairs of the Republic of Indonesia. Issued in Jakarta.

Mintz, Alex, and DeRouen Jr, Karl. 2010. *Rational Actor Model in Understanding Foreign Policy Decision Making*. Cambridge University Press. https://doi.org/10.1017/CBO9780511757761.004.

MOFA of the Republic of Indonesia. 2019a, March 31. Important Milestones in the History of Indonesian Diplomacy. https://kemlu.go.id/portal/en/read/47/tentang_kami/important-milestones-in-the-history-of-indonesian-diplomacy.

MOFA of the Republic of Indonesia. 2019b, April 8. Indonesian Membership on the UN Security Council. https://kemlu.go.id/portal/en/read/147/halaman_list_lainnya/indonesian-membership-on-the-un-security-council.

MOFA of the Republic of Indonesia. 2024a, January 8. Indonesia Consistently Fights for the Rights and Independence of the Palestinian People. https://kemlu.go.id/portal/en/read/5661/view/indonesia-consistently-fights-for-the-rights-and-independence-of-the-palestinian-people.

MOFA of the Republic of Indonesia. (2024b, January 20). 19th NAM Summit: Indonesia Advocates Full Membership for Palestine in the UN. https://kemlu.go.id/portal/en/read/5696/berita/19th-nam-summit-indonesia-advocates-full-membership-for-palestine-in-the-un.

MOFA of the Republic of Indonesia. (2024c, February 23). Indonesia Delivers Oral Statement in Support of Palestine at International Court of Justice. https://kemlu.go.id/portal/en/read/5771/berita/indonesia-delivers-oral-statement-in-support-of-palestine-at-international-court-of-justice.

MOFA of the Republic of Indonesia. 2024d, May 28. Press Briefing Kunjungan Menteri Luar Negeri Ministerial Level Meeting the Follow-Up of the Riyadh Meeting Brussels. https://kemlu.go.id/portal/id/read/5943/siaran_pers/press-briefing-kunjungan-menteri-luar-negeri-ministerial-level-meeting-the-follow-up-of-the-riyadh-meeting-brussels.

MOFA of the Republic of Indonesia. 2024g, July 4. Indonesia has hosted the UN Committee on Palestine Bureau's Delegation Visit. https://kemlu.go.id/portal/en/read/6026/berita/indonesia-has-hosted-the-un-committee-on-palestine-bureaus-delegation-visit.

MOFA of the Republic of Indonesia. 2024h, July 14. Through UNRWA, Indonesia Increases Contribution to Palestinian Refugees. https://kemlu.go.id/portal/en/read/6041/berita/through-unrwa-indonesia-increases-contribution-to-palestinian-refugees.

MOFA of the Republic of Indonesia. 2024e, June 3. Foreign Minister Outlines Indonesia's Diplomacy for Palestine "All Eyes on Rafah" at UGM Yogyakarta. https://kemlu.go.id/portal/en/read/5951/berita/foreign-minister-outlines-indonesias-diplomacy-for-palestine-all-eyes-on-rafah-at-ugm-yogyakarta.

MOFA of the Republic of Indonesia. 2024f, June 24. Indonesia Consistently Supports Palestine at the 19th ACD Ministerial Meeting. https://kemlu.go.id/portal/en/read/5997/berita/indonesia-consistently-supports-palestine-at-the-19th-acd-ministerial-meeting.

MOFA of the Republic of Indonesia. 2024i, July 23. Kemitraan Strategis Baru Asia-Afrika *(NAASP)*. https://manila.kemlu.go.id/portal/id/read/165/halaman_list_lainnya/kemitraan-strategis-baru-asia-afrika-naasp.

UN News. 2024. Rejection of two-State solution by Israeli leadership 'unacceptable', says Guterres. Retrieved from October 10, 2024, from https://news.un.org/en/story/2024/01/1145807

Nugraheny, Dian Erika. 2024. Menlu Retno: Banyak Tekanan agar RI Mulai Normalisasi Hubungan dengan Israel. *Kompas*, April 13. https://nasional.kompas.com/read/2024/04/13/13272441/menlu-retno-banyak-tekanan-agar-ri-mulai-normalisasi-hubungan-dengan-israel.
Nurita, Dewi. 2021. Jokowi: Indonesia Kutuk Keras Serangan Israel yang Renggut Ratusan Nyawa. *Tempo.co*, May 16. https://nasional.tempo.co/read/1462728/jokowi-indonesia-kutuk-keras-serangan-israel-yang-renggut-ratusan-nyawa
Poerwoko, F. Djoko (2007) *Menari di Angkasa*. Jakarta: Kata Hasta Pustaka.
Rachmadita, Amanda. 2023, November 23Soeharto, Yasser Arafat, dan Dukungan untuk Palestina *Historia*. https://historia.id/politik/articles/soeharto-yasser-arafat-dan-dukungan-untuk-palestina-v5JWg/page/2.
Rasyidi, HM. (1989). *Refleksi Pembaharuan Pemikiran Islam: 70 Tahun Harun Nasution*. Lembaga Studi Agama dan Filsafat.
Reid, A. 1974. The Indonesian National Revolution, 1945–1950. *Journal of Southeast Asian Studies* 5 (1): 65–90.
Rezasyah, Teuku. 2019. Peranan Diplomat Indonesia dalam Memperjuangkan Palestina di PBB (Masa Presidency of the UNSC Mei 2019). *Jurnal ICMES* 3(2).
Setiawati, S. M. 2024. The Role of Indonesian Government in Middle East Conflict Resolution: Consistent Diplomacy or Strategic Shifts? *Frontier in Political Science*. 6: 1304108. https://doi.org/10.3389/fpos.2024.1304108.
Sihbudi, R. 1997. *Indonesia dan Timur Tengah: Masalah dan prospek*. Gema Insani Press.
Sitamala, Afandi. 2020. Indonesia as *Non*-Permanent Member of United Nations Security Council, Guarding the Peace and Stability in Asean. *Lajil Journal* 2 (2): 121–128.
Songbatumis, A. M. (2021). The Role of Islam in Indonesian Foreign Policy: A Case of Susilo Bambang Yudhoyono. Polish Political Science Yearbook, 50. https://doi.org/10.15804/ppsy202119
Sukma, Rizal. 1995. The Evolution of Indonesia's Foreign Policy: An Indonesian View. *Asian Survey* 35 (3): 304–315. https://doi.org/10.2307/2645547.
Suryadinata, Leo. 1995. Islam and Suharto's Foreign Policy: Indonesia, the Middle East, and Bosnia. *Asian Survey* 35 (3): 291–303. https://doi.org/10.2307/2645546.
Suryadinata, L. 1997. *Indonesia's Foreign Policy under Suharto: Aspiring to International Leadership*. Times Academic Press.
Suryadinata, Leo. 2023. Indonesia's Foreign Policy under Suharto: *Aspiring to International Leadership*. ISEAS Yusof Ishak Institute.
Tempo. 1993. Menyambut Arafat dengan Demo dan Qunut Nazilah. *Tempo Magazine*, October 2, 1993. Retrieved from October 15, 2024, from https://www.datatempo.co/MajalahTeks/detail/ARM201806124541/menyambut-arafat-dengan-demo-dan-qunut-nazilah.
Ubaedillah, Achmad., Ali, Muhamad, and Arfino, Bimo. 2023. Articulation of Islam: President Joko Widodo's Foreign Policy 2014–2019. *Jurnal Hubungan Internasional* 11(2).
Ulya, Fika Nurul. 2024. Menlu Retno: Politik Luar Negeri Indonesia Bukan Transaksional. *Kompas*, January 8. https://nasional.kompas.com/read/2024/01/08/19304181/menlu-retno-politik-luar-negeri-indonesia-bukan-transaksional.
Umar, Ahmad Rizky Mardhatillah. 2023. The rise of the Asian middle powers: Indonesia's conceptions of international order. *International Affairs* 99(4):1459–1476. https://doi.org/10.1093/ia/iiad167.
Yani, Yanyan Mochamad. 2009. Change and Continuity in Indonesian Foreign Policy. *Sosiohumaniora* 11(1):1–16.

Nostalgiawan Wahyudhi is Researcher at Research Centre for Politics, the National Research and Innovation Agency, Jakarta. He also serves as a lecturer at International Relations Department, Syarif Hidayatullah State Islamic University, Jakarta. He works on several issues relates to political Islam, Middle East, transnational Islamic movements, Southeast Asian Islam, Islam in Indonesia. He writes several books on Political Islam in Indonesia, Algeria, Sudan, Saudi Arabia,

Qatar, and Gulf State. His latest publication are The New Faces of Political Islam in the Middle East: A Survey of 15 Countries After the Arab Spring (Springer 2025), and Beyond Radicalism: Islamist Attitudes and Democratic Support in Indonesia (Sage 2025).

Hamdan Basyar is a senior researcher at Research Centre for Politics, the National Research and Innovation Agency, Jakarta. He also serves as a lecturer at International Relations Department (Islam and the Middle East) University of Indonesia, Jakarta. He also works at Department of Global and Strategic Studies at the same university.

Dhurorudin Mashad is a senior researcher at Research Centre for Politics, the National Research and Innovation Agency, Jakarta. He is prospected and promoted as a professor from the same institution on religious and political studies. He works on several book concerning about religion and politics in the Middle East, South Asia, Southeast Asia, and Indonesia.

Fakhry Ghafur is a researcher at Research Centre for Politics, the National Research and Innovation Agency, Jakarta. He is a PhD candidate at Syarif Hidayatullah State Islamic University, Jakarta. He was educated in Libya for Islamic Studies. His works are related to Muslim Brotherhood in Middle East.

Defbry Margiansyah is a researcher at Research Centre for Politics, the National Research and Innovation Agency, Jakarta. He is honored as a master from Global Studies Program, Institut Für Soziologie, Albert Ludwig-Universitaet Freiburg, Germany.

Epilogue

This book compiles a range of academic essays exploring different dimensions of Indonesia's foreign policy and diplomacy under President Joko Widodo. It serves as a scholarly overview of the diverse challenges and opportunities encountered in shaping and executing Indonesia's foreign policy during Jokowi's presidency, as well as the legacy of his tenure. President Joko Widodo has a unique characteristic that sets him apart from his predecessors. His focus on prioritizing national interests is balanced by his skill in managing relationships with foreign counterparts. This approach is evident in the seventeen detailed examinations discussed in each chapter of this book.

Since no issue is insignificant, it was a deliberate decision for the volume on President Joko Widodo's Foreign Policy and Diplomatic Legacy to address two groundbreaking aspects: his Down-to-Earth Diplomacy, which became the defining characteristic of his foreign policy style, and the appointment of the first female foreign minister, Retno Marsudi, who made a lasting impact by advancing a gender mainstreaming agenda during her ten-year tenure. The chapters further delve into strategic issues across five key segments: Jokowi and diplomacy, economic diplomacy, diplomacy related to citizen protection, sovereignty, and Indonesia's role at regional and global levels.

Each chapter of this volume effectively addresses both praise and criticism on the topics discussed. Several highlights are worth noting. First, President Joko Widodo's focus on delivering tangible benefits to Indonesians is evident in policies addressing issues that significantly impact society. Second, when dealing with longstanding challenges inherited from previous administrations, President Joko Widodo and his government have approached them seriously. However, given the complexity of these issues, the political will to resolve them peacefully must be sustained by future administrations. Third, while President Joko Widodo is known for initiating and supporting new programs, these initiatives tend to be deprioritized when political will is insufficient. Lastly, in enhancing Indonesia's global image, President Joko Widodo has effectively utilized diplomatic resources to maintain the country's reputation in international fora.

I. Kartini and Kanisa (eds.), *Joko Widodo's Foreign Policy and Diplomatic Legacy*,
https://doi.org/10.1007/978-981-95-2085-5

Reflecting on the four previous highlights, President Joko Widodo's tenure has left a significant mark on Indonesia's foreign policy. Indonesia has gained international attention through several key diplomatic achievements, including the maritime-focused foreign policy during his first term, the promotion of investment through economic diplomacy, swift coordination in addressing vaccine issues during the pandemic, and the execution of back-to-back roles as a multilateral leader. However, when considering his legacy, one must ask: what do these successes mean for the future of Indonesia's foreign policy?

To answer this question, the editors aspire for this volume to offer insights into how the discussed themes and research questions might unfold, particularly if certain recommendations are overlooked under the new administration. The authors of several chapters in the first key segments have recommended some issues that need to be addressed by Jokowi's successor.

Concerning down-to-earth diplomacy, the author concludes that President Jokowi successfully implemented this approach to achieve the Nawa Cita vision. This strategy has effectively met the needs of the people while maintaining Indonesia's leadership in regional and international spheres. The authors suggest that this principle of Indonesian foreign policy should remain relevant in the future to guide the country through regional and global challenges.

Moreover, on the issue of gender-mainstreaming in Indonesian foreign policy, the authors have highlighted challenges in advancing and promoting the gender equality agenda, such as limited meaningful representation, an inadequate support system for female diplomats, and sociocultural obstacles. The authors have warned that neglecting these challenges may affect the successful implementation of domestic gender mainstreaming and our rank in the Gender Gap Index. Specifically, women's representation in foreign policy is part of political empowerment, in which Indonesia ranks the lowest among other indicators in the Global Gender Gap Report 2024. The authors go on to recommend that ensuring the sustainability of the gender-mainstreaming agenda despite the changing administration is an important task. In this sense, sustainability is not only on internal MoFA but also on maintaining Indonesia's role in the global forum by utilizing gender issues as a new arena for greater global engagement.

On the issue of maritime diplomacy, particularly through the Global Maritime Fulcrum (GMF), the authors conclude that the GMF is fundamentally rooted in reviving Indonesia's identity as a maritime nation. As a national doctrine, it embodies both domestic and international ambitions. The authors then recommend that future studies consider the essential question of how has Indonesia achieved notable international success in maritime diplomacy amid significant domestic challenges. Furthermore, regarding the development of a middle power identity, the author suggests that if expectations of a more proactive and assertive foreign policy under Prabowo prove accurate, Indonesia is poised to make significant strides in establishing and solidifying its status as a middle power in the next decade.

In the second key segments that discusses economic diplomacy towards Africa, the authors warned that changing the perception of Africa from a negative to a more positive and realistic view is a challenging task, given Indonesia's limited knowledge of the continent. Although the Bandung Conference of 1955 laid the groundwork for Afro-Asian solidarity, it predominantly highlighted Asian leaders, with limited focus on African figures. Consequently, there is a lack of comprehensive understanding and awareness about Africa among Indonesians. The authors conclude that Indonesia's official economic diplomacy towards Africa has not adequately acknowledged the role of the African traders in Indonesia. This omission might stem from a lack of awareness or a deliberate choice to focus solely on governmental initiatives rather than recognizing existing informal trade networks.

Regarding gastrodiplomacy which has much potential for cultural influence and economic growth, the authors recommend that a new strategy is needed to improve Indonesian gastrodiplomacy. Rather than establishing fine-dining Indonesian restaurants, the implementation of food trucks and small food stalls introducing Indonesian cuisine as a grassroots movement of nation branding may serve as an effective initial approach to nation branding and achieve broader visibility across all societal strata overseas. Not to mention, with the rise of healthy lifestyles in certain countries abroad, Indonesian gastrodiplomacy needs to think of adding one of Indonesian plant-based dishes to its "national food" list.

Moving on to the third key segments that discusses protecting citizen diplomacy, particularly on handling Trafficking in Persons (TIPs) bilaterally, regionally, and multilaterally, the authors recommend that Indonesian foreign policy under the Prabowo administration needs to continue to prioritize human rights diplomacy as a tool and goal. The authors conclude that entrepreneurial diplomacy is important to be implemented by strengthening collaboration with non-state actors to achieve the goal and to be in line with Indonesia's domestic policy. National interests are important to balance the interests of security politics and the interests of mainstreaming human rights for the victims.

On the issue of health diplomacy, the authors recommend that the new administration under President Prabowo take health diplomacy seriously, especially since Indonesia was one of the countries that pushed the norm of integrating health into the scope of foreign policy and since Indonesia still needs a qualified health system. The Indonesian government should adopt a more unison pandemic preparedness policy to anticipate the next pandemic. The authors conclude that the COVID-19 experience in Indonesia has shown the need for more preparedness and trust in the scientific community in recommending the pandemic policy.

Regarding the digital diplomacy issue, the author recommends that the Indonesian Ministry of Foreign Affairs (MoFA) needs to update its strategy and also regulations to be more accommodating towards new trends. In this sense, the Indonesian MoFA is acting more as a follower instead of a leader. In terms of regulation, the Indonesian MoFA needs to formulate a grand strategy for Indonesian digital diplomacy, in which it needs to include the aspect of consular diplomacy. The author concludes that Indonesian digital diplomacy focuses more on the conduct of digital diplomacy for public diplomacy. Thus, the current regulation in digital media is not addressing specifically about citizen protection.

In the fourth key segments that discusses sovereignty diplomacy, the authors focus on border diplomacy and the issue of Papua. On the issue of border diplomacy, especially on the land border negotiations between Indonesia-Timor Leste and Indonesia-Malaysia, the authors conclude that Jokowi's administration has successfully finalized the conflictual segment between Indonesia-Malaysia, in Sebatik (West pillar to East pillar) and under MoU preparation. However, the conflictual segment between Indonesia-Timor Leste has not been finalized due to a deadlock between the two countries. Therefore, the authors recommend that under the Prabowo administration, these land boundary issues need to be addressed seriously and resolved. Tackling these issues is difficult, especially due to the technical challenges present on the ground, which are closely linked to possible future social impacts.

Regarding the issue of Papua, the authors conclude that Jokowi has made several significant accomplishments in the areas of infrastructure, social and economic development, the division of Papua into six provinces, the amendment of the Special Autonomy Law 2001, and his outreach and engagement through regular trips to Papua. However, criticism from both state and non-state actors has persisted due to the inability of Indonesia to resolve human rights abuses and protracted conflict. In this regard, the authors remind us that under the Prabowo administration, Jokowi's successor will still have to navigate the unfinished and unresolved issues. The authors then recommend some important points that Prabowo should adopt a constructive policy approach that engages not only with the Papuan community but also with multilateral institutions and intergovernmental organizations. Practically, Prabowo needs to move beyond a centralized top-down approach towards a consultative approach which places greater emphasis on strategic partnership with the Papuan community. Infrastructure development in Papua and humanitarian approaches will help the Prabowo government focus on resolving past human rights violations and initiate the reconciliation process to foster peace in Papua. In addition, the relevant stakeholders, including Papuan customary leaders, religious leaders, and indigenous communities, should also be more involved along the way with the government to build a better future for Papua under the Prabowo government.

Moving on to the last part of the volume that examines Indonesia's role on regional and global levels, especially on the issue of the South China Sea (SCS), the author recommends that politically, Indonesia must be able to become a mediator between the US and China on SCS disputes. Indonesia's diplomacy could strengthen ASEAN's solidity to become a joint force in preventing China's and/or the US' intention to engage in open war. Relations with China must be maintained despite Indonesia's desire to demilitarize the SCS. Moreover, although Indonesia and the US have more commonality in terms of interests and opportunities for cooperation in SCS, especially in matters of safeguarding sovereignty and freedom of navigation, the Indonesian government must be able to demonstrate a neutral foreign policy as much as possible. Indonesia can also strengthen the function and capability of the Indonesia Coast Guard (ICG) as a legal instrument in defending its sovereignty and territorial claims in accordance with international legal rules.

Regarding the issue in the Indo-Pacific region, the authors recommend that as far as Indonesia is concerned, its preoccupation with domestic priorities should not lead to neglect of regional and international dynamics, given the myriad geopolitical, geo-economic, and geo-strategic challenges that exist, particularly in the Indo-Pacific region.

Furthermore, on the issue of minilateralism, the author concludes that Indonesia's involvement in minilateralism remains a preliminary attempt in Jokowi's diplomatic endeavors that are focused on economic growth, social welfare and political independence. For Indonesia, which has vast and long experience in ASEAN, minilateralism continues to be attractive, including in the cross-regional framework because of its specific objectives, especially in bringing minilateralism as an alternative and creative alignment.

In accordance with minilateralism, Indonesia's involvement in multilateralism has become Jokowi's administration concerned. The authors on this issue conclude that President Joko Widodo has not abandoned the potential of multilateralism. However, the authors recommend that multilateralism engagement must be maintained not only when Indonesia is the chair. It is a long-running diplomatic effort that needs strong commitment. Indonesia's continuous role in ASEAN is a good example of how it works with the organization and makes the organization as the cornerstone of Indonesia's foreign policy. Indonesia's role in the G20 might not be as substantial as in ASEAN, but Indonesia can play in leading the developing countries' voice in the organization.

Finally, on the issue of Palestine, the authors conclude that Indonesia's current influence and strength in the Middle East has increased. Consequently, Indonesia's foreign policy of actively advocating for Palestinian independence must be maintained. The authors then recommend that Indonesia continue to pursue a two-state solution in alignment with the policies of international bodies such as the United Nations. Furthermore, the authors suggest that it would be prudent for Indonesia to consider elevating its foreign policy priorities to the Middle East. As the largest Muslim country and a member of the OIC, Indonesia's position is becoming increasingly influential. Lastly, Indonesia's role as a middle power affords it the potential to become a non-Arab country whose neutrality is highly trusted. This opportunity could encourage Indonesia to expand its role as a peace-making country in the Middle East.

We hope this edited volume serves as a valuable resource for anyone interested in Indonesian foreign policy and diplomacy. The findings and recommendations presented by the contributing authors represent some of the first published scholarly reflections on this significant and timely topic, coinciding with the transition from Jokowi's presidency to the new administration under President Prabowo.

MIX
Papier aus verantwortungsvollen Quellen
Paper from responsible sources
FSC® C105338

If you have any concerns about our products,
you can contact us on
ProductSafety@springernature.com

In case Publisher is established outside the EU,
the EU authorized representative is:
Springer Nature Customer Service Center GmbH
Europaplatz 3, 69115 Heidelberg, Germany

Printed by Libri Plureos GmbH
in Hamburg, Germany